Computational Handbook of Statistics

Third Edition

James L. Bruning
Ohio University

B. L. Kintz
Western Washington University

D0149084

HarperCollins*Publishers*
Glenview, Illinois
London

Library of Congress Cataloging-in-Publication Data

Bruning, James L.
  Computational handbook of statistics.

  Bibliography: p.
  Includes index.
  1. Social sciences—Statistical methods.
2. Statistics. I. Kintz, B. L. II. Title.
HA29.B835 1987   519.5   86-31593
ISBN 0-673-18407-2

    8 —MPC—94

950704

Preface

Most statistics textbooks concentrate on theoretical discussions and mathematical proofs of the various concepts presented. The result of this approach is that students often have little understanding of how actually to apply statistical tests to their experimental findings. The intent of the first and second editions of *Computational Handbook of Statistics* was to reverse this approach and to present statistical concepts and tests *as they are applied.* This emphasis on application will be continued in the third edition, the major changes being the addition of multivariate analyses, the latest computer applications to statistics, expanded computer programs, and greater treatment of how to write them.

Our experience has indicated that carefully worked and clearly described examples can be easily generalized to a wide variety of situations. Consequently, the method of presentation employed throughout this book is primarily that of teaching by example. The mathematical proofs and theoretical discussions that appear in most textbooks are, for the most part, excluded. What has been substituted is a step-by-step analysis of the computational procedures needed to test the significance of experimental findings. Each test or analysis is illustrated by a clearly worked example, and all computational steps and procedures are expressed verbally rather than in mathematical symbols. Thus, while the analyses range from the simplest *t*-test or correlation to the most complex multivariate analysis or multiple correlation, almost no previous mathematical knowledge is required to use this book effectively.

Although the theoretical discussion presented in this book is minimal, the assumptions and basic computational formulas for the various analyses are presented, either at the beginning of each section or in the example used to demonstrate the computations. An alternative approach is taken, however, for the sections on analysis of variance, multiple comparisons, and covariance. In place of the assumptions and basic formulas, the Textbook Reference Chart, following the appendixes, allows the reader to cross-check the sections in the most widely used advanced statistics texts to find the theoretical discussions, assumptions, and formulas. The reasons for this approach are twofold. First, the design names and the terms used to describe the analyses vary widely from textbook to textbook, and, second, there is no consistent system of mathematical notation and formula presentation, so that translation from the system used by one author to that of another is often quite difficult. Thus, the presentation of all the necessary formulas and terms would have consumed more time and space than was deemed appropriate in a text of this type.

While most of its examples deal with the analysis of experimental findings in the behavioral and social sciences, this text is equally appropriate for courses in education, business, medical research, or any other area in which statistical analysis of data is required. In addition, because of its broad range of coverage, it is appropriate for courses at both the undergraduate and the graduate levels.

Since this text requires little mathematical or formal statistical background, it should prove invaluable to instructors in beginning laboratory courses in which students have completed only the most basic statistics course. Use of this text will permit the design and analysis of more advanced experiments in which additional variables are manipulated or repeated measures taken. We have found that the student need only be told which analysis should be applied to the data. From that point, nearly all students can follow the step-by-step presentation to completion of the entire analysis.

We have also found that assistants with limited statistical backgrounds can use this text to carry out the analysis of an experiment. With the aid of this text, even the most inexperienced research assistant can compute highly complex analyses with little or no additional supervision.

One other important use of this text has been noted by those interested in computer programming. Since the computational steps of each analysis are indicated separately, they represent a detailed flow chart of the instructions that must be included in a computer program. To illustrate this application, sample programs, written from the computational steps of selected analyses, are presented in Appendix P. These programs represent the most complex problems that might be encountered in the programming of analyses of variance. In all instances, the basic approach was simply to translate the verbal steps outlined in this text into the BASIC language system.

At the back of the third edition of this text is an order form that the reader can use to order a computer disk directly from the authors. This formatted disk permits the use of an Apple PC, IBM PC (or any IBM-compatible machine), or Atari 800 (XL) to undertake all the analyses. The programs on the disk are designed specifically for checking the hand calculations assigned for pedagogical reasons and for use in the analysis of collected experimental data.

It would be impossible to thank everyone who has made either a direct or an indirect contribution to this book. There are, however, several persons whose influence warrants special recognition. We again want to thank Dr. Dee W. Norton, whose methods of teaching design and analysis of experiments led to our interest in this area. Four persons at Ohio University—Dr. Caryn Christensen, Dr. Charles Harrington, Dr. Frank Bellezza, and Dr. Larry Waters—deserve special thanks. Dr. Christensen offered assistance in the development of the multivariate analyses. Dr. Harrington gave advice and assistance in the development of the computer disk available as an adjunct to this book. Drs. Bellezza and Waters provided source material and advice for the new sections in this edition.

Our publisher-author relationship has been all that could be desired, and we would like to extend thanks to Mr. Chris Jennison for his interest, help, and encouragement. To our wives, Marlene and Maria, go a special statement of appreciation for general encouragement in every phase of the manuscript revision and development.

We are grateful to the Literary Executor of the late Sir Ronald A. Fisher, F.R.S., to Dr. Frank Yates, F.R.S., and to Longman Group Ltd., London, for permission to reprint Tables III, V, and VI from their book *Statistical Tables for Biological, Agricultural and Medical Research* (6th edition, 1974).

JAMES L. BRUNING
Ohio University

B. L. KINTZ
Western Washington University

Contents

PART 1

Organizing Data and Some Simple Computations   *1*

1.1 Blocking and Tabling of Data   *2*

1.2 Range and Standard Deviation   *4*

1.3 Standard Error of the Mean   *6*

1.4 Sample Size and the Power of Hypothesis Tests   *6*

1.5 The *t*-Test for a Difference Between a Sample Mean and the Population Mean   *7*

1.6 The *t*-Test for a Difference Between Two Independent Means   *9*

1.7 The *t*-Test for Related Measures   *12*

1.8 Sandler's *A* Test   *16*

PART 2

Analysis of Variance   *18*

2.1 Completely Randomized Design   *24*

2.2 Factorial Design: Two Factors   *27*

2.3 Factorial Design: Three Factors   *32*

2.4 Treatments-by-Levels Design   *39*

2.5 Treatments-by-Subjects, or Repeated-Measures, Design   *44*

2.6 Treatments-by-Treatments-by-Subjects, or Repeated-Measures: Two-Factors, Design   *48*

2.7 Two-Factor Mixed Design: Repeated Measures on One Factor   *55*

2.8 Three-Factor Mixed Design: Repeated Measures on One Factor   *62*

2.9 Three-Factor Mixed Design: Repeated Measures on Two Factors *73*

2.10 Latin Square Design: Simple *85*

2.11 Latin Square Design: Complex *93*

2.12 Analysis of Variance for Dichotomous Data *107*

2.13 Analysis of Variance for Rank-Order Data *108*

PART 3
Supplemental Computations for Analysis of Variance *110*

3.1 Test for Difference Between Variances of Two Independent Samples (Test for Homogeneity of Independent Variances) *112*

3.2 Test for Difference Between Variances of Two Related Samples (Test for Homogeneity of Related Variances) *113*

3.3 Test for Differences Among Several Independent Variances (F-Maximum Test for Homogeneity of Variances) *115*

3.4 The *t*-Test for Differences Among Several Means *116*

3.5 Duncan's Multiple-Range Test *119*

3.6 The Newman-Keuls' Multiple-Range Test *122*

3.7 The Tukey Test *124*

3.8 Scheffé's Test *127*

3.9 Dunnett's Test *130*

3.10 *F*-Tests for Simple Effects *132*

3.11 Use of Orthogonal Components in Tests for Trend *145*

PART 4
Correlation and Related Topics *174*

4.1 Pearson Product-Moment Correlation *176*

4.2 Spearman Rank-Order Correlation *(rho)* *180*

4.3 Kendall Rank-Order Correlation *(tau)* *183*

4.4 Point-Biserial Correlation *187*

4.5 The Correlation Ratio *(eta)* *191*

4.6 Partial Correlation: Three Variables *192*

4.7 Partial Rank-Order Correlation (Using Kendall's *tau*) *194*

4.8   Multiple Correlation: Three Variables   *195*

4.9   Simple Regression: *X* Variable to Predict *Y*   *197*

4.10   Multiple Regression: Two *X* Variables   *200*

4.11   Simple Analysis of Covariance: One Treatment Variable   *204*

4.12   Factorial Analysis of Covariance: Two Treatment Variables   *210*

4.13   Reliability of Measurement: The Test As a Whole (Test-Retest, Parallel Forms, and Split Halves)   *222*

4.14   Reliability of Measurement: The Individual Items (Kuder-Richardson and Hoyt)   *223*

4.15   Test for Difference Between Independent Correlations   *226*

4.16   Test for Difference Between Dependent Correlations   *228*

PART 5

Multivariate Analyses   *230*

5.1   Difference Between Sample Centroid and Population Values   *231*

5.2   Single-Sample Matched Pairs   *236*

5.3   Difference Between Two Independent Groups   *239*

5.4   Difference in the Change Scores of Two Groups   *245*

5.5   More Than Two Groups   *250*

5.6   Two-Factor Comparisons   *256*

PART 6

Nonparametric Tests, Miscellaneous Tests of Significance, and Indexes of Relationship   *269*

6.1   Test for Significance of a Proportion   *271*

6.2   Test for Significance of Difference Between Two Proportions   *272*

6.3   The Mann-Whitney *U*-Test for Differences Between Independent Samples   *275*

6.4   A Sign Test (Wilcoxon) for Differences Between Related Samples   *278*

6.5   Simple Chi-Square and the *Phi* Coefficient   *280*

6.6 Complex Chi-Square and the Contingency Coefficient (C)   *283*

6.7 Tests for Trends, Runs, and Randomness   *288*

APPENDIXES   *292*

A   Normal-Curve Areas   *292*

B   Critical Values of "Student's" *t* Statistic   *294*

C   Critical Values for Sandler's *A* Statistic   *296*

D   Centile Values of the Chi-Square Statistic   *298*

E   Per Cent Points in the *F* Distribution   *300*

F   Fisher's *z* Transformation Function for Pearson's *r* Correlation Coefficient   *306*

G   Critical Values of Pearson's *r* Correlation Coefficient for Five Alpha Significance Levels   *308*

H   Critical Values of the *U* Statistic of the Mann-Whitney Test   *310*

I   Critical Values for Hartley's Maximum *F*-Ratio Significance Test for Homogeneity of Variances   *313*

J   Significant Studentized Ranges for Duncan's New Multiple-Range Test   *315*

K   Significant Studentized Ranges for the Newman-Keuls' and Tukey Multiple-Comparison Tests   *320*

L   Dunnett's Test: Comparison of Treatment Means with a Control   *324*

M   Critical Values of Wilcoxon's *T* Statistics for the Matched-Pairs Signed-Ranks Test   *326*

N   Coefficients for Orthogonal Polynomials   *327*

O   Probability for Total Number of Runs Up or Down   *328*

P   Sample Computer Programs   *331*
    P–1   Sums and sums of squares
    P–2   Chi-squared for goodness of fit
    P–3   Mean and standard deviation (Section 1.2)
    P–4   Chi-squared for independence (Section 5.6)
    P–5   Grouped frequency distribution, used with large distributions
    P–6   Rank-ordered numbers, from small to large
    P–7   Mann-Whitney U-test (Section 5.3)
    P–8   t-test for independent groups (Section 1.6)
    P–9   t-test for matched groups (Section 1.7)
    P–10   The correlation coefficient (Section 4.1)
    P–11   Intercorrelations—many variables
    P–12   Percentile ranks and z-scores
    P–13   Completely randomized design (Section 2.1)

P–14   *Factorial design: Two factors (Section 2.2)*
P–15   *Treatments-by-treatments-by-subjects design (Section 2.6)*
P–16   *Factorial design: Three factors (Section 2.3)*
P–17   *Type 1: Two-factor mixed design and TXS (Sections 2.5 and 2.7)*
P–18   *Type 3: Three-factor mixed design—two betweens (Section 2.8)*
P–19   *Type 6: Three-factor mixed design—two withins (Section 2.9)*
P–20   *Covariance analyses (Sections 4.10 and 4.11)*
P–21   *Multiple regression: Two predictors (Section 4.9)*
P–22   *MANOVA for sample and population (Section 5.1)*
P–23   *MANOVA for one group: Matched pairs (Section 5.2)*
P–24   *MANOVA for two independent groups (Section 5.3)*
P–25   *MANOVA for two groups: Change scores (Section 5.4)*
P–26   *MANOVA: More than two groups (Section 5.5)*

TEXTBOOK REFERENCE CHART   *361*

INDEX   *368*

To the User

As explained in the Preface, this book approaches statistics from the applied, rather than the theoretical, point of view. We hope that the extensive use of examples and the step-by-step presentation of the computational procedures will not only provide models to be followed in computing analyses but also help the user to gain insights into how each segment or step fits into the final analysis. We also hope that by presenting nearly all the material in verbal rather than mathematical form, this book will help dispel the fears that many have of statistics. At the same time, we have included as aids to the user the basic assumptions and computational formulas for each analysis. These are either presented at the beginning of each section or incorporated into the computational example, except in the sections dealing with analysis of variance, multiple comparisons, and covariance. For these sections, textbook sources for the basic formulas and assumptions are presented in the Textbook Reference Chart in the back of the book. This chart permits the user to identify the chapters in current advanced statistics textbooks where a complete discussion of the designs can be found. The following example demonstrates the methods of approach and presentation that are used throughout this book.

Example

Assume that a physical education teacher has measured the heights of five students in one class and now wishes to compute the average, or mean, height of this group.

The mathematical formula for computation of the mean is

$$\overline{X} = \frac{\Sigma X}{N}$$

where $\overline{X}$ = mean of the scores

ΣX = sum of the scores

N = number of scores

Step 1. Table the scores as follows. No particular order is necessary.

Subject	Height (inches)
S_1	60
S_2	53
S_3	57
S_4	52
S_5	58

Step 2. Add all the scores to get the sum for the group.

$$60 + 53 + 57 + 52 + 58 = \mathbf{280}$$

Step 3. Computation of the mean: Divide the value of Step 2 by the number of scores (heights) which were added. (In the present example, there were 5 scores.)

$$\overline{X} = \frac{280}{5} = \mathbf{56\ inches}$$

While it is sometimes possible to use short cuts and combine steps (as in the previous example), we recommend that, to avoid error and to more fully understand the meaning of each step, the user closely follow the steps we have outlined. During initial study or computation, the user should also label each step since this greatly facilitates the correction of errors and review of points that were not clear.

Part 1

Organizing Data and Some Simple Computations

Research involves the making of systematic observations, organizing these observations in some meaningful fashion, and then, usually on the basis of statistical tests, drawing conclusions. Raw data—observations in their original form—are usually a mass of numbers or symbols that have little or no meaning. To allow even the most gross kinds of interpretations, the data must be organized in some fashion. Probably the simplest organization is the classification of data into categories. In this instance, a count is usually made of the observations that fall into one class as opposed to another. A good example is an observation of a crowd that is later classified according to the number of males versus females present. Statistical procedures to aid in the interpretation of data collected and organized in this fashion are *usually* (but not always) of the nonparametric type. These statistical tests are discussed in Part 6 of this book.

The form of data organization that is of interest in the first five parts of this book involves the use of measures that are recorded for each of several observations. These measures almost always imply some notion of varying magnitude or amount. This type of classification differs from the first in that it implies varying amounts of an observation rather than simply indicating whether or not a particular observation was made. In the following sections, organization and computation of measures will be considered in relation to various simple statistical tests that experimenters use as aids in drawing conclusions regarding the outcome of their research.

Organizing Data (Sections 1.1, 1.2, 1.3, and 1.4)

These first four sections discuss some of the basic procedures for making decisions regarding sample size, organizing the data, and then putting it in an understandable form. While it is recognized that there are

1

many other procedures for organizing data (such as frequency distributions, etc.), the ones presented in these chapters are those used most often.

Some Simple Computations (Sections 1.5, 1.6, 1.7, and 1.8)

The first three of these sections are concerned primarily with demonstrating the basic uses of the t-test. These include the t for testing the difference between a sample and a population mean, the t for testing the difference between two sample means, and the t for related measures. The last section also tests for related measures and is an alternative to the t-test.

SECTION 1.1
Blocking and Tabling of Data

In most experiments involving the collection and analysis of data, the data are grouped, or blocked, to reduce variability and to make the scores more manageable for statistical analysis. The most commonly used methods are to take the mean, median, or mode of several scores. The value obtained is then treated as a single score.

For example, if the number of correct responses on each of fifteen trials was the measure recorded, the raw data for Subject 1 might appear as follows:

Trial	1	2	3	4	5	6	7	8	9	10	11	12	13	14	15
Correct responses for S_1	1	4	2	6	1	3	5	9	7	9	4	9	14	10	14

Since there are fifteen trials, the most reasonable grouping would probably be to combine each consecutive set of three or five trials. If only one score were desired, the entire fifteen trials would be combined. The size of the group, or block, of data is entirely arbitrary.

The most common technique is to take the mean of each successive set of trials. In the above example, assume that the experimenter decides to block by taking the mean of each successive five scores. Thus, the fifteen scores of S_1 would be reduced to three scores, as follows:

Trial Block	1	2	3
Mean of 5 trials for S_1	2.8	6.6	10.2

If an overall mean were desired, the experimenter would simply add all the correct responses and divide by the number of trials given. For S_1 this would be 98/15, which equals 6.53.

Less frequently, the median or mode is used to block raw data. Use of these measures usually occurs only if the data within each set of scores are severely skewed. In the present example, the median would equal the middle score of each block of trials after the scores within each block had been rank ordered. Using the data presented above, the median of trials one through five would be 2, since there are two scores smaller than 2 (1, 1) and two larger (4, 6). The same type of computation would be made for trials six through ten and eleven through fifteen.

Trial Block	1	2	3
Median of 5 trials for S_1	2	7	10

The overall median in this example would be 6.

The mode is simply the score that occurs most often. When applied to blocks of five trials, the scores for S_1 become:

Trial Block	1	2	3
Mode of 5 trials for S_1	1	9	14

The overall mode would be 9, since that score occurs three times during the fifteen trials.

After the data have been blocked, they must be tabled so that statistical analyses can be undertaken. While the specific method of tabling depends upon the statistical analyses to be used, the most common technique involves simply identifying each subject and then entering the subject's score or scores in a row to the right. For example, using the mean scores in the above example:

Subject	Trial Block		
	1	2	3
S_1	2.8	6.6	10.2
S_2	1.4	1.8	4.3
S_3	2.1	6.7	6.4

While it is apparent that different methods of blocking scores can result in different measures being recorded for each subject, the major point to note is that whenever raw data are blocked in some fashion, the statistical analyses may be based entirely on the blocked scores. In the above example, all analyses may be computed as though only three trials had been given or, if an overall score were used, as if only one score had been recorded. It is also important to note that for purposes of determining degrees of freedom (df), numbers of scores in a table, group means, etc., all computations and numerical counts are based on the blocked scores and not on the raw data.

SECTION 1.2
Range and Standard Deviation

By grouping and arranging scores, it is possible to get a fairly good understanding of the data that have been collected. In addition to grouping, however, it is usually desirable to obtain some idea of the dispersion of the scores. The simplest such measure is the *range*. This value is computed very simply since it is equal to the difference between the largest and smallest scores in the distribution plus 1. For example, in the sample data presented below, the tallest child is 72 inches and the shortest is 43 inches; the range, then, is 29 plus 1, or 30 inches.

Although the range does give some information regarding the dispersion of the scores, its usefulness is limited. By far the most common and useful measure of dispersion is the *standard deviation* (s.d.). There are two ways to compute the standard deviation: directly, using difference scores; and indirectly, using the computational formula. As is shown in nearly all introductory textbooks, these two formulas produce the same numerical answer. Since the method that employs the computational formula is more generally applicable, it will be presented here.

Example

This example is drawn from the area of developmental psychology. Assume that an experimenter is interested in obtaining information about the heights of twelve-year-olds. To do this, the experimenter collects a sample of data (the heights of twenty randomly selected twelve-year-olds) and proceeds to compute the standard deviation.

The mathematical formula for the standard deviation is

$$\text{s.d.} = \sqrt{\frac{\Sigma X^2 - \frac{(\Sigma X)^2}{N}}{N - 1}}$$

where ΣX^2 = the sum of the squared score values

$(\Sigma X)^2$ = the square of the sum of all the scores

N = the total number of scores used in the computation

Step 1. Table the data as follows. No particular order is necessary.

Subject	Height (inches)
S_1	64
S_2	48
S_3	55
S_4	68
S_5	72
S_6	59
S_7	57
S_8	61
S_9	63
S_{10}	60
S_{11}	60
S_{12}	43
S_{13}	67
S_{14}	70
S_{15}	65
S_{16}	55
S_{17}	56
S_{18}	64
S_{19}	61
S_{20}	60

Step 2. Add all the scores. (*Note:* If you are using a calculator, you can do Step 2 and Step 3 at the same time.)

$$64 + 48 + \cdots + 60 = 1208$$

Step 3. Square all the scores, and add the squared values.

$$64^2 + 48^2 + \cdots + 60^2 = 73{,}894$$

Step 4. Square the sum obtained in Step 2, and divide this value by the number of scores that were added to obtain the sum. The resultant value is called the *correction term.*

$$\frac{1208^2}{20} = \frac{1{,}459{,}264}{20} = 72{,}963$$

Step 5. Subtract the value obtained in Step 4 from the sum in Step 3.

$$73{,}894 - 72{,}963 = 931$$

Step 6. Divide the value obtained in Step 5 by $N - 1$.* (In this example, it would be $20 - 1 = 19$.) The resultant value is called the *variance.*

$$\frac{931}{19} = 49$$

---

*This term is almost universally used since the division by $N - 1$ gives the unbiased *estimate* of the population variance rather than the *actual* variance of the sample. If the actual sample variance is desired, divide by N rather than $N - 1$. Step 7 is the same, regardless.

Step 7. Take the square root of the value obtained in Step 6. This value is the *standard deviation.*

$$\sqrt{49} = 7 \text{ in.}$$

SECTION 1.3
Standard Error of the Mean

The standard error of the mean is extremely simple to compute once the steps outlined in Section 1.2 have been completed. All that must be done is to follow Steps 1 through 7 of Section 1.2 and then divide the value obtained in Step 7 by the square root of the size of the sample.

In the example used in Section 1.2, the value obtained in Step 7 was 7 inches. Since there were 20 subjects in the sample, simply divide 7 by the square root of 20 to obtain the standard error of the mean.

$$\frac{7}{\sqrt{20}} = \frac{7}{4.47} = 1.57$$

SECTION 1.4
Sample Size and the Power of Hypothesis Tests

The power of any hypothesis test to reject a false hypothesis depends upon four contributing factors:

1. the value of alpha (the smaller the alpha, the lower the power);
2. the variance of the measures (the greater the variance, the lower the power);
3. the size the treatment effect must be in order to be judged meaningful or useful (the greater the required effect, the lower the power);
4. the size of the sample (the smaller the sample, the lower the power).

In practice, factors 1, 2, and 3 above will be constant for any particular experiment. Only factor 4 can be changed, and in most experiments the experimenter can select larger or smaller samples without too much difficulty.

A great many experienced researchers use a rule-of-thumb sample size of approximately twenty. Smaller samples often result in low-power values, while larger samples often result in a waste of time and money. Very small samples (smaller than five per group) usually produce power so low that false hypotheses are often accepted. On the other hand, very large samples usually produce power so great that very tiny treatment effects are detected. While such differences achieve statistical significance, they may be of very little practical importance. If the reader is interested in a more complete discussion of power, almost any of the advanced statistics textbooks by authors such as Keppel, Kirk, Myers, or Winer are good sources. Complete citations precede the Textbook Reference Chart at the end of the book.

SECTION 1.5
The *t*-Test for a Difference Between a Sample Mean and the Population Mean

One of the most commonly used tests of significance is the *t*-test. One use of this test assumes that the mean for some population or theoretical value, such as "poverty level of income," is known. Knowing the population or theoretical value, the experimenter can then determine whether the mean of the sample chosen is significantly different from the population mean.

Example

This example uses the data presented in Section 1.2. Assume that after collecting measures of the heights of twenty randomly selected twelve-year-olds, the experimenter notices that while the standard deviation is large, the average height of the children in this sample is above the population average for twelve-year-olds. By use of the *t*-test, it can be determined whether this difference is significant.

The basic formula for testing the significance of a difference between a sample mean and population mean is

$$t = \frac{\overline{X} - \mu}{\text{standard error of the mean}} \quad \text{(Sections 1.2 and 1.3)}$$

or, in simplified form,

$$t = \frac{\overline{X} - \mu}{\sqrt{\dfrac{\Sigma X^2 - \dfrac{(\Sigma X)^2}{N}}{N(N-1)}}}$$

where $\overline{X}$ = the mean of the sample

μ = the mean of the population

ΣX^2 = the sum of the squared score values

$(\Sigma X)^2$ = the square of the sum of all the scores

N = the number of scores used in the analysis

Step 1. Table the data as follows. No particular order is necessary.

Subject	Height (inches)
S_1	64
S_2	48
S_3	55
S_4	68
S_5	72
S_6	59
S_7	57
S_8	61
S_9	63
S_{10}	60
S_{11}	60
S_{12}	43
S_{13}	67
S_{14}	70
S_{15}	65
S_{16}	55
S_{17}	56
S_{18}	64
S_{19}	61
S_{20}	60

Step 2. Add all the scores. (*Note:* If you are using a calculator, you can do Step 2 and Step 3 at the same time.)

$$64 + 48 + \cdots + 60 = \mathbf{1208}$$

Step 3. Square all the scores, and add these squared values.

$$64^2 + 48^2 + \cdots + 60^2 = \mathbf{73{,}894}$$

Step 4. Square the sum obtained in Step 2. Then divide this value by the number of scores that were added to get the sum. (In the present example, 20 scores were added.) The resultant value is called the *correction term*.

$$\frac{1208^2}{20} = \frac{1{,}459{,}264}{20} = \mathbf{72{,}963}$$

Step 5. Subtract the result of Step 4 from the sum in Step 3.

$$73{,}894 - 72{,}963 = \mathbf{931}$$

Step 6. Divide the value obtained in Step 5 by N times $N - 1$. In the present example, it would be $20 \times (20 - 1) = 380$.

$$\frac{931}{380} = 2.45$$

Step 7. Take the square root of the value obtained in Step 6.

$$\sqrt{2.45} = 1.57$$

Step 8. Subtract the sample mean (sum in Step 2 divided by N) from the population mean (known from some other source).

$$\frac{1208}{20} = 60.4 = \text{sample mean}$$

$$58.0 = \text{population mean found in growth-curve tables}$$

$$58.0 - 60.4 = 2.4$$

(*Note:* For computational purposes, only the absolute difference between the means is important.)

Step 9. Divide the value obtained in Step 8 by the value obtained in Step 7. This yields the *t* value.

$$t = \frac{2.4}{1.57} = 1.53$$

Step 10. To determine whether the *t* value is significant, the degrees of freedom must be obtained. For a test of significance of a sample mean and a population mean, the *df* are equal to $N - 1$. In the present example, this would be $20 - 1 = 19$. The *t* for significance at the .05 level (see Appendix B) is 2.093. Since the *t* value in the present example is smaller than 2.093, it is concluded that there is no significant difference between this sample mean and the population mean.

SECTION 1.6

The *t*-Test for a Difference Between Two Independent Means

Probably the most common use of the *t*-test is to determine whether the performance difference between two groups of subjects is significant. In most experimental situations, the subjects are randomly assigned to the two groups; one of the groups is manipulated experimentally, and the effects of this manipulation are analyzed by comparing the performance of the two groups. However, there are many instances where the groups are already constituted (e.g., males versus females) and the experimenter wishes to determine whether they differ with respect to some other variable (e.g., height, weight, etc.).

Example

Assume that a principal wishes to determine whether there is a significant difference between the IQ scores of boys and girls in a particular grade. The score on the Wechsler Intelligence Scale for Children (WISC) is recorded for each student.

The basic computational formula for the t-test of a difference between two independent means is

$$t = \frac{\overline{X}_1 - \overline{X}_2}{\sqrt{\left[\frac{\Sigma X_1^2 - \frac{(\Sigma X_1)^2}{N_1} + \Sigma X_2^2 - \frac{(\Sigma X_2)^2}{N_2}}{(N_1 + N_2) - 2}\right] \cdot \left[\frac{1}{N_1} + \frac{1}{N_2}\right]}}$$

where $\overline{X}_1$ = the mean of the first group of scores

$\overline{X}_2$ = the mean of the second group of scores

ΣX_1^2 = the sum of the squared score values of the first group

ΣX_2^2 = the sum of the squared score values of the second group

$(\Sigma X_1)^2$ = the square of the sum of the scores in the first group

$(\Sigma X_2)^2$ = the square of the sum of the scores in the second group

N_1 = the number of scores in the first group

N_2 = the number of scores in the second group

Step 1. Table the data as follows. No particular order within groups is necessary.

Group 1 (Boys)		Group 2 (Girls)	
Subject	Score	Subject	Score
S_1	107	S_{13}	109
S_2	96	S_{14}	94
S_3	88	S_{15}	127
S_4	131	S_{16}	76
S_5	109	S_{17}	115
S_6	84	S_{18}	121
S_7	79	S_{19}	87
S_8	105	S_{20}	92
S_9	108	S_{21}	91
S_{10}	92	S_{22}	98
S_{11}	96	S_{23}	104
S_{12}	101	S_{24}	96
		S_{25}	110
		S_{26}	108

Step 2. Add the scores in Group 1. (*Note:* If you are using a calculator, Step 2 and Step 3 can be done at the same time.)

$$107 + 96 + \cdots + 101 = \mathbf{1196}$$

Step 3. Square every score in Group 1, and add these squared values.

$$107^2 + 96^2 + \cdots + 101^2 = \mathbf{121{,}318}$$

Step 4. Square the value obtained in Step 2. Then divide the squared value by the number of scores on which the sum was based. (In this example, 12 scores were added.)

$$\frac{1196^2}{12} = \frac{1{,}430{,}416}{12} = \mathbf{119{,}201}$$

Step 5. Subtract the value obtained in Step 4 from the sum in Step 3.

$$121{,}318 - 119{,}201 = \mathbf{2117}$$

Step 6. Add the scores in Group 2.

$$109 + 94 + \cdots + 108 = \mathbf{1428}$$

Step 7. Square every score in Group 2, and add these squared values.

$$109^2 + 94^2 + \cdots + 108^2 = \mathbf{148{,}202}$$

Step 8. Square the value obtained in Step 6. Then divide the squared value by the number of scores on which the sum was based. (In the present example, 14 scores were added.)

$$\frac{1428^2}{14} = \frac{2{,}039{,}184}{14} = \mathbf{145{,}656}$$

Step 9. Subtract the value obtained in Step 8 from the sum in Step 7.

$$148{,}202 - 145{,}656 = \mathbf{2546}$$

Step 10. Add the values obtained in Step 5 and Step 9.

$$2117 + 2546 = \mathbf{4663}$$

Step 11. Divide the value obtained in Step 10 by $(n_1 + n_2) - 2$. [In the present example, $(12 + 14) - 2 = 24$.]

$$\frac{4663}{24} = \mathbf{194}$$

Step 12. Multiply the value obtained in Step 11 by $1/n_1 + 1/n_2$. (In the present example, this would equal $1/12 + 1/14$.)

$$194 \left(\frac{1}{12} + \frac{1}{14} \right) = 194 \times \frac{13}{84} = 30.0$$

Step 13. Take the square root of the value obtained in Step 12.

$$\sqrt{30.0} = 5.48$$

Step 14. Obtain the mean of Group 1 (sum in Step 2 divided by n_1) and the mean of Group 2 (sum in Step 6 divided by n_2).

$$\frac{1196}{12} = 99.67 = \text{mean of Group 1}$$

$$\frac{1428}{14} = 102.00 = \text{mean of Group 2}$$

Step 15. Subtract the mean of Group 1 from the mean of Group 2.

$$102.00 - 99.67 = 2.33$$

(*Note:* For computational purposes, only the absolute difference between the means is important.)

Step 16. Divide the value obtained in Step 15 by the value obtained in Step 13. This yields the *t* value.

$$t = \frac{2.33}{5.48} = .43$$

Step 17. To determine whether the *t* value is significant, the degrees of freedom (*df*) must be computed. For a test of significance between two means, the *df* are equal to $(n_1 + n_2) - 2$. In the present example, $(12 + 14) - 2 = 24$. The *t* value which is significant at the .05 level for 24 *df* (see Appendix B) is equal to 2.064. Since the *t* value in the present example is less than 2.064, it is concluded that there is no significant difference between the IQ scores of boys and girls in this particular class.

SECTION 1.7
The *t*-Test for Related Measures

A somewhat rarer use of the *t*-test is to determine the significance of a difference between two correlated means. It is most commonly used in this way when two scores are recorded for the same individuals. For instance, IQ scores might be taken at the beginning and end of a special training program to determine if there has been any improvement in test scores. The second use of this test is in the instance where pairs of subjects in two different groups are "matched" on the basis of some variable—like age, sex, race, years of education, etc.—to ensure that the pairs of subjects in each group are the "same" before experimental manipulations are begun.

Points to Consider When Using the *t* for Related Measures

1. Both experimental groups will have the same number of measures since they represent two measures on the same subjects or matched pairs of subjects.
2. If two measures are taken on the same subjects, the treatments-by-subjects analysis (see Section 2.5) is usually more appropriate.

Example

This example is taken from the field of education. Assume that the experimenter is interested in determining the effects of a special education program on the intelligence test scores of underprivileged children. The first step is to match several pairs of students on the basis of their Wechsler Intelligence Test scores. One student from each pair is randomly assigned to either the special training group or the control group that receives no special treatment; the remaining student in each pair is assigned to the other group. After six weeks of training, an alternate form of the Wechsler test is given to the students in both groups to determine the effects of the program. The score on this second test is recorded for each student. In the present example, the pairs of matched subjects will be represented by S and S'; the test score of each S by X; and the test score of each S' by Y.

There are two formulas that are used for the *t* for related measures. The first involves computation of a correlation and is given by

$$t = \frac{\overline{X} - \overline{Y}}{\sqrt{\dfrac{S_X^2 + S_Y^2 - 2rS_XS_Y}{N}}}$$

where S_X^2 = variance of the X scores

S_Y^2 = variance of the Y scores

r = correlation between X and Y

N = number of pairs of scores

The second formula, which gives *exactly* the same results, is much simpler and will be used to demonstrate the computational steps.

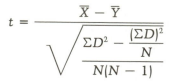

$$t = \frac{\overline{X} - \overline{Y}}{\sqrt{\dfrac{\Sigma D^2 - \dfrac{(\Sigma D)^2}{N}}{N(N - 1)}}}$$

where D = difference score between each X and Y pair

N = number of pairs of scores

Step 1. Table the data as follows. No particular order is necessary, except that each pair $(S$ and $S')$ of scores must be in the same relative position.

Group 1 (No Special Treatment)		Group 2 (Special Treatment)	
Subject	Score	Subject	Score
S_1	89	S'_1	94
S_2	86	S'_2	94
S_3	96	S'_3	101
S_4	100	S'_4	105
S_5	94	S'_5	100
S_6	86	S'_6	84
S_7	81	S'_7	81
S_8	93	S'_8	96
S_9	87	S'_9	90
S_{10}	89	S'_{10}	88
S_{11}	110	S'_{11}	115
S_{12}	95	S'_{12}	100
S_{13}	107	S'_{13}	110
S_{14}	96	S'_{14}	102

Step 2. Obtain the difference between each pair of scores.

$$89 - 94 = -5$$
$$86 - 94 = -8$$
$$96 - 101 = -5$$
$$100 - 105 = -5$$
$$94 - 100 = -6$$
$$86 - 84 = +2$$
$$81 - 81 = \quad 0$$
$$93 - 96 = -3$$
$$87 - 90 = -3$$
$$89 - 88 = +1$$
$$110 - 115 = -5$$
$$95 - 100 = -5$$

$$107 - 110 = -3$$
$$96 - 102 = -6$$

Step 3. Square all the difference scores recorded in Step 2, and add these squared values.

$$(-5)^2 + (-8)^2 + \cdots + (-6)^2 = 293$$

Step 4. Obtain the *algebraic* sum of the difference scores obtained in Step 2. Square this value, and divide by the number of difference scores recorded.

$$(-5) + (-8) + \cdots + (-6) = -51$$

$$\frac{-51^2}{14} = \frac{2601}{14} = 186$$

Step 5. Subtract the value obtained in Step 4 from the sum in Step 3.

$$293 - 186 = 107$$

Step 6. Divide the value obtained in Step 5 by $N - 1$. (In the present example, this is $14 - 1 = 13$, since N refers to pairs of scores.)

$$\frac{107}{13} = 8.23$$

Step 7. Take the square root of the value obtained in Step 6.

$$\sqrt{8.23} = 2.87$$

Step 8. Divide the value of Step 7 by $\sqrt{N}$. (In the present example, $\sqrt{14} = 3.74$.)

$$\frac{2.87}{3.74} = .767$$

Step 9. Obtain the mean score of each of the two groups—that is, add all the scores in each group and divide each sum by the number of scores added to obtain it.

$$89 + 86 + \cdots + 96 = 1309 = \text{sum for Group 1}$$

$$94 + 94 + \cdots + 102 = 1360 = \text{sum for Group 2}$$

$$\frac{1309}{14} = 93.50 = \text{mean for Group 1}$$

$$\frac{1360}{14} = 97.14 = \text{mean for Group 2}$$

Step 10. Subtract the mean for Group 2 from the mean for Group 1.

$$93.50 - 97.14 = \textbf{3.64}$$

(*Note:* For computational purposes, only the absolute difference between the means is important.)

Step 11. Divide the value obtained in Step 10 by the value obtained in Step 8. This yields the *t* value.

$$t = \frac{3.64}{.767} = \textbf{4.74}$$

Step 12. To determine whether the *t* value is significant, the degrees of freedom (*df*) must first be obtained. For the *t* for related measures, the $df = N - 1$ where *N* is the number of *pairs* of scores. In the present example, $N - 1 = 13$. From the *t* tables (see Appendix B), we find that the *t* value that is significant at the .05 level for 13 *df* is 2.16. Since the obtained *t* is larger than 2.16, it is concluded that the special training program improved the IQ test scores.

SECTION 1.8
Sandler's *A* Test

An alternate and somewhat simpler technique for determining whether the difference between correlated samples is significant is Sandler's *A* test, which is derived directly from the *t*-test. Thus, the results of an analysis using the *A* test will be the same as those of the *t*-test for related measures described in Section 1.7.

The formula for *A* is

$$A = \frac{\text{Sum of the squared difference scores}}{\text{Sum of the difference scores squared}}, \text{ or } \frac{\Sigma D^2}{(\Sigma D)^2}$$

To demonstrate the computation of the *A* statistic, the data presented in Section 1.7 will be used. Steps 1, 2, and 3 for the computation of *A* are the same as the computation of *t* for related measures.

Step 1. Table the data as in Section 1.7.

Step 2. Compute the difference scores as in 1.7.

Step 3. Square and sum all of the difference scores. In the example from 1.7, this sum was equal to 293.

Step 4. Obtain the algebraic sum of the difference scores computed in Step 2, and square that value. In the example used in Section 1.7, this sum computed in Step 2 was equal to -51.

$$-51^2 = \mathbf{2601}$$

(*Note:* You do *not* divide by the number of difference scores as in Step 4 of Section 1.7.)

Step 5. Divide the value obtained in Step 3 by the value obtained in Step 4.

$$A = \frac{293}{2601} = \mathbf{.113}$$

Step 6. Turn to Appendix C and read into the table for 13 *df* (since there are 14 pairs of scores, $N - 1 = 13$). From this table we find that for an *A* value (two-tailed) to be significant at the .05 level, it must be equal to, *or smaller than*, .270. Since the obtained *A* is smaller (.113) than the tabled value, it is concluded that the difference is significant.

Part 2

Analysis of Variance

The statistical tests presented in the previous sections are designed to determine the significance of a difference between two groups. In actual practice, however, most experiments involve more than two groups. While it would be possible to test the difference between, say, four experimental groups by using t-tests of all possible combinations of two group means, the number of tests necessary to make these comparisons would be large. (The actual number of tests is equal to $N(N - 1)/2$. In the case of an experiment involving four experimental groups, six separate tests would be needed.)

In addition to posing computational problems, many multigroup experiments require that several variables be manipulated at the same time. For example, four groups of subjects might be presented with a learning task and their performance recorded over a series of several trials. The experimenter would very likely be interested not only in overall performance but also in whether the rate of learning of the several groups was the same. To determine this, statistical procedures are needed that are far more complex than the simple, two-group t statistic.

In the following sections, several statistical analyses are introduced that permit evaluation of much more complicated experimental designs. To aid the user in determining which analysis is appropriate for a particular experimental design, a descriptive list of all the analyses is presented below. In addition to indicating the type of problem for which each analysis is appropriate, the list also indicates some of the major strengths and weaknesses of each.

Completely Randomized Design (Section 2.1)

This design is basically an extension of the t-test to experiments involving three or more groups. This design would typically be used, for example, if an experimenter were interested in determining reactions to several different drugs. In this case, subjects would be randomly assigned to the several groups and a different drug administered to each group. After the performance of subjects in all groups had been measured, statistical analyses of the data would be undertaken to determine the differential effects of the drugs.

Factorial Design: Two Factors (Section 2.2)

This extension of the completely randomized design permits investigation of one set of variables in combination with some other set. For example, instead of being interested only in the effects of drugs, an experimenter might be interested in determining the effects of drugs in combination with varying amounts of sleep loss. In the simplest case, two drugs, A and B, would be paired with no sleep loss versus twenty-four-hour loss. This would result in the formation of four groups: Group 1—Drug A and no sleep loss; Group 2—Drug A and twenty-four-hour loss; Group 3—Drug B and no sleep loss; Group 4—Drug B and twenty-four-hour loss. Thus, it is possible to determine not only the effects of drugs and sleep loss but also whether the effects of Drugs A and B were enhanced or suppressed by sleep loss. Stated another way, it is possible to determine whether the variables interact.

The experimental procedure for the two-factor design is basically the same as for the completely randomized design. The subjects would be randomly assigned to the four (in this example) groups, sleep loss would be manipulated, and then the drugs would be administered. Again, after measuring performance, statistical analyses of the data would be undertaken.

Factorial Design: Three Factors (Section 2.3)

This is a commonly used extension of the two-factor design. The only difference is that a third factor, or dimension, is added. For example, an experimenter working with Drugs A and B and sleep versus sleep loss might also be interested in determining the effects of these factors in combination with a third factor, alcohol concentration in the blood. If only two concentrations are used, 0.0 and 0.2 per cent, the experimental groups would be constituted as follows: Group 1—Drug A, no sleep loss, no alcohol; Group 2—Drug A, no sleep loss, alcohol; Group 3—Drug A, sleep loss, no alcohol; Group 4—Drug A, sleep loss, alcohol. Groups 5, 6, 7, and 8 would receive the above combinations of sleep loss and alcohol in combination with Drug B. The assignment of subjects, measurement of performance, and analysis of the results are basically the same as in the two-factor design.

Treatments-by-Levels Design (Section 2.4)

This represents a slight variation on the two-factor design discussed above. All the computations needed to analyze the treatments-by-levels design are the same as for the two-factor design; the only difference is in the intent of the design and interpretation of the results. In the treatments-by-levels, one of the factors usually represents some variable already known to affect performance. For example, if it were known that sleep loss generally heightens the effects of drugs, it would probably be wise for the experimenter to include different levels of sleep loss as a variable in the drug experiments. By so doing, variability due to the specific effects of different amounts of sleep loss could be statistically separated from the effects of the drugs.

Treatments-by-Subjects, or Repeated-Measures, Design (Section 2.5)

The treatments-by-subjects design is a variation of the completely randomized design. Instead of using several different groups of subjects with each group receiving a single drug, only one group of subjects would be used and each subject would receive all the drug treatments. If the treatments-by-subjects design were to be used for the drug experiment, the single group of subjects would probably receive one drug, perform the criterion task, and wait until the effects wore off. Then a second drug would be administered, the criterion task would be performed, and another waiting period would ensue. This same procedure would be repeated until every subject in the group had received every drug. Analysis of the findings would be based on the subjects' performance while under the influence of the several drugs. The major advantage of this design over the completely randomized design is that fewer numbers of subjects are required.* The major disadvantage is that there might be carry-over effects from one treatment to the next. In the present example, it is possible that exposure to one drug would heighten or lessen sensitivity to other drugs administered later. Also, the subjects might become progressively more proficient at performing the criterion task and show an improvement in performance that would be due to learning and not to the effects of the drugs.

---

*Very often the advantage of increased statistical power is also gained by using a repeated-measures design. The reason for this is that the random variability of a single subject from one measuring period to the next is usually much less than the variability introduced by measuring and comparing different subjects.

Treatments-by-Treatments-by-Subjects Design (Section 2.6)

This two-factor design bears the same relationship to the treatments-by-subjects design as the factorial bears to the completely randomized design. The treatments-by-treatments-by-subjects design is basically the same as the two-factor factorial design, with one notable exception. Instead of different subjects being assigned to the experimental groups, one group of subjects performs under all experimental conditions. Using the example from the two-factor factorial design, an experiment using the treatments-by-treatments-by-subjects design would probably be conducted as follows. First, all subjects would receive Drug A with no prior sleep loss, perform the criterion task, and then wait for the effects to wear off. After a suitable period, the same subjects would again receive Drug A, but this time the drug would follow a twenty-four-hour sleep loss. After performance measures had been recorded and a sufficient period allowed for the effects of the drug to wear off, the same procedures would be repeated with Drug B. The major advantage of this design over the factorial is that far fewer subjects are required. As in the treatments-by-subjects design, the major disadvantage is that there may be carry-over effects from one treatment to the next.

Two-Factor Mixed Design: Repeated Measures on One Factor (Section 2.7)

This design is essentially a combination of the completely randomized and treatments-by-subjects designs. It involves the assignment of subjects to several experimental groups. However, instead of taking only one measure, as in the completely randomized design, the performance of the subjects would be measured repeatedly to determine the effects of the experimental variables over a prolonged period of time. This recording of repeated measures is, of course, identical with the procedure used in the treatments-by-subjects design.

To further clarify, let us continue with the example used to illustrate the completely randomized design. In the two-factor mixed design, subjects would again be assigned to the several experimental groups, and each group would be administered a different drug. Then, the subjects' performance on the criterion task would be measured several times.

Since measures are recorded over several successive test periods, this design permits (1) comparison of the overall performance of the experimental groups (as in the completely randomized design), (2) evaluation of performance changes from one measuring period to the next (as in the treatments-by-subjects design), and (3) evaluation of the drug effects in relation to the passage of time between measuring periods. This particular design is very widely used in research in the behavioral sciences.

Three-Factor Mixed Design: Repeated Measures on One Factor (Section 2.8)

This extension of the two-factor mixed design is essentially a combination of the factorial and repeated-measures designs. In the three-factor mixed design, the experimental groups are formed and subjects are assigned in exactly the same manner as in the factorial design. The basic difference (using the example of the factorial design) is that the effects of the drugs in combination with sleep loss would be measured several times rather than only once. Consequently, this design permits not only the evaluation of the overall experimental effects (as in the factorial design) but also the evaluation of general changes and interactions of the variables over the passage of time (as in the repeated-measures designs).

This particular design is probably the one used most often in behavioral science research. There are several slight variations of the format presented here, the most common of which involves presenting subjects with a task to be learned during a predetermined set of practice trials. In this situation, the repeated measures represent the learning rates of the subjects in the several experimental groups.

Three-Factor Mixed Design: Repeated Measures on Two Factors (Section 2.9)

This design is essentially a combination of the completely randomized and the treatments-by-treatments-by-subjects designs. As in the completely randomized design, subjects are randomly assigned to two or more experimental groups. For example, one group might be subjected to a twenty-four-hour sleep loss just prior to the experiment and the other group to no sleep loss. Each of these different groups would then be treated in the same fashion as in the treatments-by-treatments-by-subjects design. That is, two additional factors (say, Drug A versus Drug B and 0.0 versus 0.2 per cent alcohol) would be combined and presented to each subject in both of the experimental groups. Consequently, all subjects in the sleep-loss and no-sleep-loss groups would receive Drug A with *no* alcohol, perform the criterion task, wait for the effects to dissipate, then receive Drug A *with* alcohol, etc., until all variable combinations had been experienced. The major advantage of the three-factor mixed design over the regular three-factor design discussed above is that many fewer subjects are required to conduct the experiment. Carry-over effects probably represent the most serious drawback.

Latin Square Design: Simple (Section 2.10)

This design is a variation of the treatments-by-subjects, or repeated-measures, design. The basic difference in the Latin square design is that the presentation of the treatments is deliberately counterbalanced. Thus, instead of giving subjects the various drugs in the same order (Drug A, then B, then C, etc.) or randomly assigning the order of presentation of the drugs for each subject, a systematic order of presentations is set up so that order and carry-over effects can be ascertained. Assume that three drugs are to be investigated. If the simple Latin square design were to be used, one third of the subjects would receive Drug A first, then B, then C; a second third would receive Drug C first, then A, then B; and the final third would receive Drug B first, then C, then A. As indicated above, this design permits the analysis not only of the effects of the drugs but also of the order in which they were given.

Latin Square Design: Complex (Section 2.11)

This design can be viewed as a variation of the two-factor mixed design or as an extension of the simple Latin square. As in the two-factor mixed design, subjects would be randomly assigned to different experimental groups (like sleep loss versus no sleep loss). The drugs would then be administered to all subjects in both groups in the same manner as in the simple Latin square. Thus, one third of the subjects in the no-sleep-loss group would receive Drug A, then B, then C; a second third would receive Drug C, then A, then B; and a final third would receive Drug B, then C, then A. The same procedure would be used with the subjects in the sleep-loss group. Thus, this design permits not only the evaluation of the effects of the drugs and of sleep loss but also the evaluation of the effects of the order of drug administration in relation to the other two factors.

For most of Parts 2 and 3 of this book, only the verbal instructions for computing the analyses and the computed examples are presented. In place of symbols and formulas, the textbook sources of the several analyses are listed in the Textbook Reference Chart at the end of this book. The reasons for this are twofold. First, several of the more complex analyses require up to twenty separate computational formulas, plus additional ones for retabling the data. Second, and more important, there is no standard set of terms and symbols to indicate analysis-of-variance operations. These tend to be specific to each text, and translation from one text to another is difficult. To include all the symbols and terms currently in use or to develop a set of our own is beyond the scope of this book.

In Part 3, specific supplemental computations and the uses of multiple comparisons employed in conjunction with analyses of variance are discussed. Because of the relative specificity of these computations, the reader is encouraged to read the introductory comments at the beginning of Part 3 to determine whether a particular test is appropriate to use in a statistical analysis of specific experimental findings.

Dichotomous Data (Section 2.12)

This does not represent a special design but, rather, a situation in which the data recorded are usually "1" or "0." The actual computational steps are the same as those presented in earlier sections.

Rank-Order Data (Section 2.13)

In this situation, the data represent rank-ordered preferences. Once again, the computational steps are the same as those presented earlier.

SECTION 2.1
Completely Randomized Design

One of the more simple experimental designs that permits comparison of several groups is the *simple randomized*, or *completely randomized*, *design*.

Points to Consider When Using the Completely Randomized Design

1. For each subject in the experimental groups, there can be only *one* score. If many measures are taken for each subject, these measures must be combined in some fashion (added, etc.) so that there is only one score for each subject.
2. Although it is not necessary, it is usually best to have an equal

number of subjects in each of the experimental groups. In most experiments, ten to fifteen subjects per group are used.
3. The number of experimental groups that may be compared is arbitrary. However, it is rare that more than four or five groups are compared in one experiment.

Example

The example used to demonstrate the computational procedures is drawn from psychology. Assume that the experimenter is interested in determining the effect of shock on the time required to solve a set of difficult problems. Subjects are randomly assigned to four experimental conditions. Subjects in Group 1 receive no shock; Group 2, very-low-intensity shocks; Group 3, medium shocks; and Group 4, high-intensity shocks. The total time required to solve all the problems is the measure recorded for each subject.

Step 1. After the experiment has been conducted and the data collected, table the data as follows.

Group 1 (No Shock)		Group 2 (Low Shock)		Group 3 (Medium Shock)		Group 4 (High Shock)	
Subject	Time (Min.)	Subject	Time (Min.)	Subject	Time (Min.)	Subject	Time (Min.)
S_1	10	S_{13}	3	S_{25}	19	S_{37}	23
S_2	7	S_{14}	8	S_{26}	12	S_{38}	14
S_3	9	S_{15}	7	S_{27}	16	S_{39}	16
S_4	8	S_{16}	5	S_{28}	14	S_{40}	18
S_5	15	S_{17}	6	S_{29}	7	S_{41}	12
S_6	3	S_{18}	10	S_{30}	8	S_{42}	13
S_7	8	S_{19}	12	S_{31}	13	S_{43}	16
S_8	9	S_{20}	4	S_{32}	10	S_{44}	17
S_9	11	S_{21}	7	S_{33}	19	S_{45}	19
S_{10}	9	S_{22}	6	S_{34}	9	S_{46}	14
S_{11}	5	S_{23}	5	S_{35}	15	S_{47}	16
S_{12}	17	S_{24}	15	S_{36}	14	S_{48}	17

Step 2. Add the scores in each group to get the sum of *each group.* (*Note:* If you are using a calculator, Steps 2 and 3 can be done at the same time.)

	Group 1	Group 2	Group 3	Group 4
	10	3	19	23
	7	8	12	14
	⋮	⋮	⋮	⋮
Sums:	111	88	156	195

Step 3. Square each number in the table, and add these squared values together.

$$10^2 + 7^2 + \cdots + 17^2 + 3^2 + 8^2 + \cdots + 15^2 + 19^2 + 12^2$$
$$+ \cdots + 14^2 + 23^2 + 14^2 + \cdots + 17^2 = \mathbf{7434}$$

Step 4. Add all the group sums (Step 2) together to obtain a grand total for the entire table.

$$111 + 88 + 156 + 195 = \mathbf{550}$$

Step 5. Square the grand total (Step 4), and divide the squared value by the total number of measures recorded (see Step 1). (In the present example, the latter number is 48, since there are 4 groups of 12 subjects each.) The value obtained is the correction term.

$$\frac{550^2}{48} = \frac{302{,}500}{48} = \mathbf{6302}$$

Step 6. Subtract the correction term (Step 5) from the sum of the squared values obtained in Step 3. The resultant value is the *sum of squares total*, or SS_t.

$$7434 - 6302 = \mathbf{1132}$$

Step 7. Square the sum of each of the groups (Step 2), divide by the number of measures in each group, and then add these values.

$$\frac{111^2}{12} + \frac{88^2}{12} + \frac{156^2}{12} + \frac{195^2}{12} = \frac{82{,}426}{12} = \mathbf{6869}$$

Step 8. Subtract the correction term (Step 5) from the value obtained in Step 7. The resultant value is the *sum of squares between groups*, or SS_b.

$$6869 - 6302 = \mathbf{567}$$

Step 9. Subtract the sum of squares between groups (Step 8) from the sum of squares total (Step 6). This yields the *sum of squares within groups*, or SS_w.

$$1132 - 567 = \mathbf{565}$$

Step 10. All computations based directly on the data have now been completed. However, since the test of significance (*F*-test) is a ratio of the mean squares, these values must be computed. To compute the needed mean squares, the degrees of freedom (*df*) for each of the components (SS_t, SS_b, and SS_w) must be determined, as follows.

df for SS_t = the total number of measures (see Step 1) minus 1

$$48 - 1 = \mathbf{47}$$

df for SS_b = the number of experimental groups minus 1

$$4 - 1 = \mathbf{3}$$

df for SS_w = the total df minus the between df.

$$47 - 3 = \mathbf{44}$$

Step 11. The mean squares are then computed as SS/df.

$$ms_t = \frac{SS_t}{df} \quad \text{(This value is not needed for the analysis.)}$$

$$ms_b = \frac{SS_b}{df} = \frac{(\text{Step 8})}{3} = \frac{567}{3} = \mathbf{189}$$

$$ms_w = \frac{SS_w}{df} = \frac{(\text{Step 9})}{44} = \frac{565}{44} = \mathbf{12.84}$$

Step 12. The test of significance (F) is equal to ms_b/ms_w.

$$F = \frac{189}{12.84} = \mathbf{14.71}$$

Step 13. Table the final analysis as follows.

Source	SS	df	ms	F	p
Total	1132	47	—	—	—
Between groups	567	3	189	14.71	<.001
Within groups	565	44	12.84	—	—

Since the F value (see Appendix E) of 14.71, with df of 3 and 44, would occur by chance less than once in one thousand times, it is concluded that level of shock intensity does affect the time required to solve these problems.

SECTION 2.2

Factorial Design: Two Factors

Many times an experimenter wishes to determine the effects of two variables in combination with each other. For example, instead of investigating the effects of differing list difficulty on performance (memorizing the lists) in one experiment and then doing another experiment to determine the effects of varying room noise on performance, the experimenter may well wish to investigate these variables in combination with each other. The two-factor design permits such comparisons.

Points to Consider When Using the Two-Factor Factorial Design

1. For each subject in the experimental groups, there can be only *one* score. As in the simple randomized design, if many measures are taken for each subject, these must be combined (added, etc.) so that there is only one score for each subject.
2. It is usually best to have an equal number of subjects in each of the experimental groups. If equality cannot be obtained, the number of subjects in each of the groups should be proportional. In most experiments, ten to fifteen subjects per group are used.
3. The number of treatment groups within each factor that may be compared is arbitrary. However, it is rare that more than four or five groups are included in either of the two factors.

Example

The example used to demonstrate the computational procedures is drawn from psychology. Assume that an experimenter is interested in determining the effects of high- versus low-intensity shock on the memorization of a hard versus an easy list of nonsense syllables. Subjects are randomly assigned to four experimental conditions. Subjects in Group 1 receive periodic low-intensity shocks and must memorize an easy list; Group 2, high shock and an easy list; Group 3, low shock and a hard list; and Group 4, high shock and a hard list. The total number of errors made by each subject is the measure recorded.

Step 1. After the experiment has been conducted, table the data as follows. (*Note:* If more than a 2×2 factorial—e.g., a 2×3, 3×3, etc.—is to be analyzed, simply table the additional treatment groups in their appropriate row or column. All other steps are the same.)

Group 1 (Low Shock–Easy List)		Group 2 (High Shock–Easy List)	
Subject	No. of Errors	Subject	No. of Errors
S_1	9	S_7	15
S_2	16	S_8	13
S_3	14	S_9	9
S_4	11	S_{10}	9
S_5	12	S_{11}	8
S_6	8	S_{12}	11

	Group 3 (Low Shock–Hard List)		Group 4 (High Shock–Hard List)	
Subject	No. of Errors		Subject	No. of Errors
S_{13}	10		S_{19}	19
S_{14}	12		S_{20}	16
S_{15}	18		S_{21}	18
S_{16}	16		S_{22}	23
S_{17}	17		S_{23}	14
S_{18}	15		S_{24}	15

Step 2. Add the scores in each group. (*Note:* If you are using a calculator, steps 2 and 3 can be done at the same time.)

	Group 1	Group 2	Group 3	Group 4
	9	15	10	19
	16	13	12	16
	⋮	⋮	⋮	⋮
Sums:	**70**	**65**	**88**	**105**

Step 3. Square each number in the entire table (Step 1), and add these squared values.

$$9^2 + 16^2 + \cdots + 15^2 + 13^2 + \cdots + 10^2 + 12^2$$
$$+ \cdots + 19^2 + 16^2 + \cdots + 15^2 = \mathbf{4832}$$

Step 4. Add the sums of all the groups (Step 2) to get the grand sum of the entire table.

$$70 + 65 + 88 + 105 = \mathbf{328}$$

Then, square this value, and divide by the total number of measures recorded in the table (see Step 1). This yields the correction term.

$$\frac{328^2}{24} = \frac{107,584}{24} = \mathbf{4483}$$

Step 5. Computation of the total sum of squares (SS_t): Simply subtract the correction term (Step 4) from the sum of the squared measures (Step 3).

$$4832 - 4483 = \mathbf{349} = SS_t$$

Step 6. Computation of the effects of the first factor (the overall effects of low shock versus high shock): First, add the sums of the two shock groups (see Step 2), disregarding the difficulty-of-list dimension.

$$70 + 88 = \mathbf{158} = \text{sum of the low-shock groups}$$
$$65 + 105 = \mathbf{170} = \text{sum of the high-shock groups}$$

Then, square the above sums, divide by the number of measures on which each of the sums was based, and add the quotients.

$$\frac{158^2}{12} + \frac{170^2}{12} = \frac{53,864}{12} = \mathbf{4489}$$

Then, subtract the correction term from the above value.

$$4489 - 4483 = \mathbf{6} = SS_{shock}$$

Step 7. Computation of the effects of the second factor (the overall effects of hard versus easy list): First, add the sums of the two same-list-difficulty groups (see Step 2), disregarding the shock dimension.

$$70 + 65 = \mathbf{135} = \text{sum of easy-list groups}$$
$$88 + 105 = \mathbf{193} = \text{sum of hard-list groups}$$

Then, square the above sums, divide by the number of measures on which each of the sums was based, and add the quotients.

$$\frac{135^2}{12} + \frac{193^2}{12} = \frac{55,474}{12} = \mathbf{4623}$$

Then, subtract the correction term from the above value.

$$4623 - 4483 = \mathbf{140} = SS_{list}$$

Step 8. Computation of the interactive effects of shock and list difficulty: First, square the sums of each of the experimental groups (Step 2), divide by the number of measures on which each sum was based, and then add the quotients.

$$\frac{70^2}{6} + \frac{65^2}{6} + \frac{88^2}{6} + \frac{105^2}{6} = \frac{27,894}{6} = \mathbf{4649}$$

Then, from this value, subtract the correction term (Step 4), SS_{shock} (Step 6), and SS_{list} (Step 7).

$$4649 - 4483 - 6 - 140 = \mathbf{20} = SS_{shock \times list}$$

Step 9. Computation of the error-term sum of squares (SS_{error}): Simply subtract SS_{shock} (Step 6), SS_{list} (Step 7), and $SS_{shock \times list}$ (Step 8) from SS_{total} (Step 5).

$$349 - 6 - 140 - 20 = \mathbf{183} = SS_{error}$$

Step 10. All computations based directly on the data have now been completed. However, since the F ratios are ratios of mean squares, the degrees of freedom (df) for each of the components must be computed.

df for SS_t = the total number of measures (see Step 1) minus 1.

$$24 - 1 = \textbf{23}$$

df for SS_{shock} = the number of shock conditions minus 1.

$$2 - 1 = \textbf{1}$$

df for SS_{list} = the number of list conditions minus 1.

$$2 - 1 = \textbf{1}$$

df for $SS_{shock \times list}$ = the df for SS_{shock} times the df for SS_{list}.

$$1 \times 1 = \textbf{1}$$

df for SS_{error} = the df for SS_t minus the df for SS_{shock}, SS_{list}, and $SS_{shock \times list}$.

$$23 - 1 - 1 - 1 = \textbf{20}$$

Step 11. The mean squares are then computed as SS/df.

     ms_t (This value is not needed for the analysis.)

$$ms_{shock} = \frac{(\text{Step 6})}{1} = \frac{6}{1} = \textbf{6}$$

$$ms_{list} = \frac{(\text{Step 7})}{1} = \frac{140}{1} = \textbf{140}$$

$$ms_{shock \times list} = \frac{(\text{Step 8})}{1} = \frac{20}{1} = \textbf{20}$$

$$ms_{error} = \frac{(\text{Step 9})}{20} = \frac{183}{20} = \textbf{9.15}$$

Step 12. The several F ratios are then computed as

$$\frac{ms_{shock}}{ms_{error}} \times \frac{ms_{list}}{ms_{error}} \times \frac{ms_{shock \times list}}{ms_{error}}$$

Step 13. Table the final analysis as follows.

Source	SS	df	ms	F	p
Total	349	23	—	—	—
Shock	6	1	6	.66	n.s.
List	140	1	140	15.30	<.001
Shock × List	20	1	20	2.19	n.s.
Error	183	20	9.15	—	—

Therefore, we conclude (see Appendix E) that shock had no significant effect on performance and that the interaction of shock by list difficulty was also nonsignificant. However, the effect of list difficulty was significant and would be expected to occur by chance less than once in one thousand times.

SECTION 2.3
Factorial Design: Three Factors

This design is a commonly used extension of the two-factor design discussed in Section 2.2. The one basic difference is that instead of determining the effects of two variables in combination with each other, this design permits evaluation of three variables.

Points to Consider When Using the Three-Factor Factorial Design

1. For each subject in the experimental groups, there can be only *one* score. As in the two-factor design, if many measures are taken for each subject, these must be combined (added, etc.) so that there is only one score for each subject.
2. It is usually best to have an equal number of subjects in each experimental group. In most experiments, ten to fifteen subjects per group are used.
3. The number of treatment groups within each factor that may be compared is arbitrary. However, it is rare that more than four or five groups are included in any of the factors.

Example

The example used to demonstrate the computational procedures is drawn from the field of psychology and is essentially an extension of the one used in Section 2.2. Assume that the experimenter, in addition to determining the effects of high- versus low-intensity shock on the memorization of easy versus difficult lists, is also concerned with determining the effects of rate of list presentation on learning in the above conditions. Obviously, the experimenter could conduct one two-factor experiment at a fast rate of presentation and another at a slow rate. As an alternative, rate of presentation could be included as an additional dimension and the entire experiment conducted as one three-factor design.

In the present example, then, the variables are high versus low shock, hard versus easy list, and fast versus slow rate of presentation. Subjects are randomly assigned to one of eight experimental conditions. Subjects in Group 1 receive periodic low-intensity shock and must memorize an easy list presented at a slow rate; Group 2—low-intensity shock, easy list, fast rate; Group 3—low-intensity shock, hard list, slow rate; Group 4—low-intensity shock, hard list, fast rate. Groups 5, 6, 7, and 8 receive the same treatment as Groups 1, 2, 3, and 4, except that high-intensity shocks are administered: Group 5—high-intensity shock, easy list, slow rate; Group 6—high-intensity shock, easy list, fast rate; Group 7—high-intensity shock, hard list, slow rate; Group 8—high-intensity shock, hard list, fast rate.

The total number of errors made by each subject is the measure recorded.

Step 1. After the experiment has been completed, table the data as follows. (*Note:* If other than a $2 \times 2 \times 2$ factorial—e.g., $2 \times 2 \times 3$, $3 \times 2 \times 3$, etc.—is to be analyzed, simply table the additional treatment groups in their appropriate row or column. All other steps are the same.)

Group 1 (Low, Easy, Slow)		Group 2 (Low, Easy, Fast)		Group 3 (Low, Hard, Slow)		Group 4 (Low, Hard, Fast)	
Subject	No. of Errors	Subject	No. of Errors	Subject	No. of Errors	Subject	No. of Errors
S_1	15	S_7	23	S_{13}	11	S_{19}	25
S_2	8	S_8	16	S_{14}	16	S_{20}	27
S_3	9	S_9	17	S_{15}	8	S_{21}	29
S_4	7	S_{10}	17	S_{16}	14	S_{22}	34
S_5	8	S_{11}	14	S_{17}	20	S_{23}	38
S_6	4	S_{12}	11	S_{18}	16	S_{24}	26

Group 5 (High, Easy, Slow)		Group 6 (High, Easy, Fast)		Group 7 (High, Hard, Slow)		Group 8 (High, Hard, Fast)	
S_{25}	14	S_{31}	24	S_{37}	11	S_{43}	39
S_{26}	6	S_{32}	16	S_{38}	17	S_{44}	32
S_{27}	3	S_{33}	10	S_{39}	12	S_{45}	38
S_{28}	5	S_{34}	18	S_{40}	14	S_{46}	39
S_{29}	6	S_{35}	13	S_{41}	21	S_{47}	33
S_{30}	7	S_{36}	17	S_{42}	18	S_{48}	31

Step 2. Add the scores in each group. (*Note:* If you are using a calculator, Step 2 and Step 3 can be done at the same time.)

	Group 1	Group 2	Group 3	Group 4
	15	23	11	25
	8	16	16	27
	⋮	⋮	⋮	⋮
Sums:	51	98	85	179
	Group 5	Group 6	Group 7	Group 8
	14	24	11	39
	6	16	17	32
	⋮	⋮	⋮	⋮
Sums:	41	98	93	212

Step 3. Square each number in the entire table (Step 1), and add these squared values.

$$15^2 + 8^2 + \cdots + 31^2 = \textbf{20,083}$$

Step 4. Add all the group sums (Step 2) to get the grand sum of the entire table.

$$51 + 98 + 85 + 179 + 41 + 98 + 93 + 212 = \textbf{857}$$

Then, square the above value and divide by the number of scores added (recorded in Step 1) to obtain the grand sum. (In the present example, 48 scores were added.) This yields the correction term.

$$\frac{857^2}{48} = \frac{734,449}{48} = \textbf{15,301}$$

Step 5. Computation of the total sum of squares (SS_t): Simply subtract the correction term (Step 4) from the sum of the squared measures (Step 3).

$$20,083 - 15,301 = \textbf{4782} = SS_t$$

Step 6. Computation of the effects of the first factor (the overall effects of high versus low shock): First, add the scores of the two same-intensity-shock groups (see Step 2), disregarding the difficulty-of-list and rate-of-presentation dimensions.

$$51 + 98 + 85 + 179 = \textbf{413} = \text{sum of the low-shock groups}$$
$$41 + 98 + 93 + 212 = \textbf{444} = \text{sum of the high-shock groups}$$

Then, square the above sums, divide by the number of measures on which each of the sums was based, and add the quotients.

$$\frac{413^2}{24} + \frac{444^2}{24} = \frac{367,705}{24} = \textbf{15,321}$$

Then, subtract the correction term (Step 4) from the above value.

$$15{,}321 - 15{,}301 = 20 = \text{SS}_{\text{shock}}$$

Step 7. Computation of the effects of the second factor (the overall effects of hard versus easy list): First, sum the scores of the two same-list-difficulty groups (Step 2), disregarding the shock-intensity and rate-of-presentation dimensions.

$$51 + 98 + 41 + 98 = \textbf{288} = \text{sum of the easy-list groups}$$
$$85 + 179 + 93 + 212 = \textbf{569} = \text{sum of the hard-list groups}$$

Then, square the above sums, divide by the number of measures on which each was based, and add the quotients.

$$\frac{288^2}{24} + \frac{569^2}{24} = \frac{406{,}705}{24} = \textbf{16{,}946}$$

Then, subtract the correction term (Step 4) from the above value.

$$16{,}946 - 15{,}301 = \textbf{1645} = \text{SS}_{\text{list}}$$

Step 8. Computation of the effects of the third factor (the overall effects of fast versus slow rate of presentation): First, sum the scores of the two same-rate-of-presentation groups (Step 2), disregarding the shock-intensity and list-difficulty dimensions.

$$51 + 85 + 41 + 93 = \textbf{270} = \text{sum of the slow-rate groups}$$
$$98 + 179 + 98 + 212 = \textbf{587} = \text{sum of the fast-rate groups}$$

Then, square the above sums, divide by the number of measures on which each was based, and add the quotients.

$$\frac{270^2}{24} + \frac{587^2}{24} + \frac{417{,}469}{24} = \textbf{17{,}394}$$

Then, subtract the correction term from the above value.

$$17{,}394 - 15{,}301 = \textbf{2093} = \text{SS}_{\text{rate}}$$

Step 9. Computation of the interactive effects of the first and second factors (shock × list): First, sum the scores of the groups which have the same pairings of shock intensity and list difficulty, disregarding the rate-of-presentation dimension.

$$51 + 98 = \textbf{149} = \text{sum of the low-shock, easy-list groups}$$
$$85 + 179 = \textbf{264} = \text{sum of the low-shock, hard-list groups}$$
$$41 + 98 = \textbf{139} = \text{sum of the high-shock, easy-list groups}$$
$$93 + 212 = \textbf{305} = \text{sum of the high-shock, hard-list groups}$$

Then, square the above sums, divide by the number of scores on which each sum was based, and add the quotients.

$$\frac{149^2}{12} + \frac{264^2}{12} + \frac{139^2}{12} + \frac{305^2}{12} = \frac{204,243}{12} = \textbf{17,020}$$

Then, from this value, subtract the correction term (Step 4), the SS_{shock} (Step 6), and the SS_{list} (Step 7).

$$17,020 - 15,301 - 20 - 1645 = \textbf{54} = SS_{shock \times list}$$

Step 10. Computation of the interactive effects of the first and third factors (shock $\times$ rate of presentation): First, sum the scores of the groups that have the same pairings of shock intensity and rate of presentation. Disregard the difficulty-of-list dimension.

$$51 + 85 = \textbf{136} = \text{sum of the low-shock, slow-rate groups}$$
$$98 + 179 = \textbf{277} = \text{sum of the low-shock, fast-rate groups}$$
$$41 + 93 = \textbf{134} = \text{sum of the high-shock, slow-rate groups}$$
$$98 + 212 = \textbf{310} = \text{sum of the high-shock, fast-rate groups}$$

Then, square the above sums, divide by the number of scores on which each is based, and add the quotients.

$$\frac{136^2}{12} + \frac{277^2}{12} + \frac{134^2}{12} + \frac{310^2}{12} = \frac{209,281}{12} = \textbf{17,440}$$

Then, from the above value, subtract the correction term (Step 4), the SS_{shock} (Step 6), and the SS_{rate} (Step 8).

$$17,440 - 15,301 - 20 - 2093 = \textbf{26} = SS_{shock \times rate}$$

Step 11. Computation of the interactive effects of the second and third factors (difficulty $\times$ rate of presentation): First, sum the scores of the groups that have the same pairings of difficulty of material and rate of presentation. Disregard the shock dimension.

$$51 + 41 = \textbf{92} = \text{sum of the easy-list, slow-rate groups}$$
$$98 + 98 = \textbf{196} = \text{sum of the easy-list, fast-rate groups}$$
$$85 + 93 = \textbf{178} = \text{sum of the hard-list, slow-rate groups}$$
$$179 + 212 = \textbf{391} = \text{sum of the hard-list, fast-rate groups}$$

Then, square the above sums, divide by the number of scores on which each is based, and add the quotients.

$$\frac{92^2}{12} + \frac{196^2}{12} + \frac{178^2}{12} + \frac{391^2}{12} = \frac{231,445}{12} = \textbf{19,287}$$

Then, from the above value, subtract the correction term (Step 4), the SS_{list} (Step 7), and the SS_{rate} (Step 8).

$$19{,}287 - 15{,}301 - 1645 - 2093 = \mathbf{248} = SS_{list \times rate}$$

Step 12. Computation of the interactive effects of the first, second, and third factors (shock × list difficulty × rate of presentation): First, square the sums of each of the experimental groups (Step 2), divide by the number of measures on which each sum was based, and then add the quotients.

$$\frac{51^2}{6} + \frac{98^2}{6} + \frac{85^2}{6} + \frac{179^2}{6} + \frac{41^2}{6} + \frac{98^2}{6} + \frac{93^2}{6} + \frac{212^2}{6}$$

$$= \frac{116{,}349}{6} = \mathbf{19{,}391}$$

Then, from this value, subtract the correction term (Step 4), the SS_{shock} (Step 6), the SS_{list} (Step 7), the SS_{rate} (Step 8), the $SS_{shock \times list}$ (Step 9), $SS_{shock \times rate}$ (Step 10), and the $SS_{list \times rate}$ (Step 11).

$$19{,}391 - 15{,}301 - 20 - 1645 - 2093$$
$$- 54 - 26 - 248 = \mathbf{4} = SS_{shock \times list \times rate}$$

Step 13. Computation of the error-term sum of squares (SS_{error}): Simply subtract SS_{shock} (Step 6), SS_{list} (Step 7), SS_{rate} (Step 8), $SS_{shock \times list}$ (Step 9), $SS_{shock \times rate}$ (Step 10), $SS_{list \times rate}$ (Step 11), and $SS_{shock \times list \times rate}$ (Step 12) from the SS_{total} (Step 5).

$$4782 - 20 - 1645 - 2093 - 54 - 26 - 248 - 4 = \mathbf{692} = SS_{error}$$

Step 14. All of the computations based directly on the data have been completed. However, since the F ratios are ratios of mean squares, the degrees of freedom (df) for each of the components must be computed.

df for SS_t = the total number of measures (see Step 1) minus 1.

        $48 - 1 = \mathbf{47}$

df for SS_{shock} = the number of shock conditions minus 1.

        $2 - 1 = \mathbf{1}$

df for SS_{list} = the number of list difficulties minus 1.

        $2 - 1 = \mathbf{1}$

df for SS_{rate} = the number of rates of presentation minus 1.

        $2 - 1 = \mathbf{1}$

df for $SS_{shock \times list}$ $\quad$ = the df for shock times the df for list.

$$1 \times 1 = 1$$

df for $SS_{shock \times rate}$ $\quad$ = the df for shock times the df for rate.

$$1 \times 1 = 1$$

df for $SS_{list \times rate}$ $\quad$ = the df for list times the df for rate.

$$1 \times 1 = 1$$

df for $SS_{shock \times list \times rate}$ = the df for shock times the df for list times the df for rate.

$$1 \times 1 \times 1 = 1$$

df for SS_{error} $\quad$ = the df for SS_t minus the dfs for SS_{shock}, SS_{list}, SS_{rate}, $SS_{shock \times list}$, $SS_{list \times rate}$, $SS_{shock \times rate}$, and $SS_{shock \times list \times rate}$.

Step 15. The mean squares are then computed as SS/df.

ms_t $\qquad\qquad$ (This value is not needed for the analysis.)

$$ms_{shock} \quad = \frac{(\text{Step 6})}{1} = \frac{20}{1} = 20$$

$$ms_{list} \quad = \frac{(\text{Step 7})}{1} = \frac{1645}{1} = 1645$$

$$ms_{rate} \quad = \frac{(\text{Step 8})}{1} = \frac{2093}{1} = 2093$$

$$ms_{shock \times list} \quad = \frac{(\text{Step 9})}{1} = \frac{54}{1} = 54$$

$$ms_{shock \times rate} \quad = \frac{(\text{Step 10})}{1} = \frac{26}{1} = 26$$

$$ms_{list \times rate} \quad = \frac{(\text{Step 11})}{1} = \frac{248}{1} = 248$$

$$ms_{shock \times list \times rate} = \frac{(\text{Step 12})}{1} = \frac{4}{1} = 4$$

$$ms_{error} \quad = \frac{(\text{Step 13})}{40} = \frac{692}{40} = 17.30$$

Step 16. The several F ratios are then computed as

$$\frac{ms_{shock}}{ms_{error}} \qquad \frac{ms_{shock \times list}}{ms_{error}}$$

$$\frac{ms_{list}}{ms_{error}} \qquad \frac{ms_{shock \times rate}}{ms_{error}}$$

$$\frac{ms_{rate}}{ms_{error}} \qquad \frac{ms_{list \times rate}}{ms_{error}}$$

$$\frac{ms_{shock \times list \times rate}}{ms_{error}}$$

Step 17. Table the final analysis as follows.

Source	SS	df	ms	F	p
Total	4782	47	—	—	—
Shock	20	1	20	1.16	n.s.
List	1645	1	1645	95.09	<.001
Rate	2093	1	2093	120.98	<.001
Shock × list	54	1	54	3.12	n.s.
Shock × rate	26	1	26	1.50	n.s.
List × rate	248	1	248	14.34	<.001
Shock × list × rate	4	1	4	.23	n.s.
Error	692	40	17.30	—	—

Therefore, it is concluded (see Appendix E) that list difficulty and presentation rate affect error making in a word-memorizing situation and also that the effects of list difficulty and presentation rate are interactive.

SECTION 2.4
Treatments-by-Levels Design

Occasionally a researcher is faced with a group of subjects who either come from distinctly different populations or are tested at different times and places. In this case, it might be reasonable to expect the subjects in the various samples to differ in some systematic fashion. To control for this source of bias and confusion, a variation of the factorial design, the treatments-by-levels design, is appropriate.

Points to Consider When Using the Treatments-by-Levels Design

1. For each subject in the experimental groups, there can be only *one* score. As in the factorial designs, if many measures are taken for

each subject, these must be combined so that there is only one score for each subject.

2. It is best to have an equal number of subjects in each experimental group and within each level or replication. In most experiments, ten to fifteen subjects per group are used.

3. The number of treatment groups is arbitrary. The same is true of the number of levels or replications employed. However, it is rare that more than four or five groups or levels are employed.

Example

One of the most common uses of the treatments-by-levels design occurs in the field of education when children are drawn from various schools that differ markedly in quality and location. For this example, assume that an experimenter wishes to investigate the effectiveness of three different methods of teaching arithmetic. Because there is reason to believe that the success of the teaching methods depends on the overall quality of the school and the type of students enrolled, one school is selected from an upper-class neighborhood, another from a middle-class area, and a third from a slum area. In each school, one class is taught by the usual teacher-only method, a second class by a teacher plus teaching machines, and the third by television plus teaching machines. An achievement test is given to each student at the beginning of the term and again at the end. The score recorded for each student is the amount of improvement (in test points) shown.

Step 1. After the experiment has been completed, table the data as follows.

Level 1 (Upper-Class School)					
Group 1 (Teacher Only)		Group 2 (Teacher + Machines)		Group 3 (Television + Machines)	
Subject	Points	Subject	Points	Subject	Points
S_1	74	S_7	94	S_{13}	87
S_2	79	S_8	96	S_{14}	76
S_3	94	S_9	92	S_{15}	94
S_4	90	S_{10}	84	S_{16}	92
S_5	88	S_{11}	85	S_{17}	98
S_6	87	S_{12}	89	S_{18}	71
Level 2 (Middle-Class School)					
Group 4		Group 5		Group 6	
S_{19}	70	S_{25}	76	S_{31}	70
S_{20}	78	S_{26}	91	S_{32}	86
S_{21}	96	S_{27}	94	S_{33}	91

	Level 2 (Middle-Class School)				
Group 4		**Group 5**		**Group 6**	
Subject	Points	Subject	Points	Subject	Points
S_{22}	86	S_{28}	86	S_{34}	74
S_{23}	80	S_{29}	87	S_{35}	88
S_{24}	92	S_{30}	91	S_{36}	77

	Level 3 (Lower-Class School)				
Group 7		**Group 8**		**Group 9**	
S_{37}	92	S_{43}	91	S_{49}	70
S_{38}	86	S_{44}	84	S_{50}	86
S_{39}	84	S_{45}	80	S_{51}	81
S_{40}	71	S_{46}	73	S_{52}	71
S_{41}	74	S_{47}	69	S_{53}	65
S_{42}	78	S_{48}	71	S_{54}	63

Step 2. Add the scores in each group. (*Note:* If you are using a calculator, Step 2 and Step 3 can be done at the same time.)

	Level 1		
	Group 1	Group 2	Group 3
	74	94	87
	79	96	76
	⋮	⋮	⋮
Sums:	**512**	**540**	**518**
	Level 2		
	Group 4	Group 5	Group 6
	70	76	70
	78	91	86
	⋮	⋮	⋮
Sums:	**502**	**525**	**486**
	Level 3		
	Group 7	Group 8	Group 9
	92	91	70
	86	84	86
	⋮	⋮	⋮
Sums:	**485**	**468**	**436**

Step 3. Square each number in the entire table (Step 1), and add these squared values.

$$74^2 + 79^2 + \cdots + 63^2 = \mathbf{374{,}776}$$

Step 4. Add the sums of each group (Step 2) to get the grand sum of the entire table.

$$512 + 540 + 518 + 502 + 525 + 486 + 485 + 468 + 436 = \mathbf{4472}$$

Then, square this value, and divide by the total number of measures recorded in the table (see Step 1). This yields the correction term.

$$\frac{4472^2}{54} = \frac{19{,}998{,}784}{54} = \mathbf{370{,}347}$$

Step 5. Computation of the total sum of squares (SS_t): Simply subtract the correction term (Step 4) from the sum of the squared measures (Step 3).

$$374{,}776 - 370{,}347 = \mathbf{4429} = SS_t$$

Step 6. Computation of the effects of levels (the overall performance of students in the upper- versus middle- versus lower-class schools): First, add the sums of the three levels, disregarding the other dimension, method of teaching (see Step 2).

$$512 + 540 + 518 = \mathbf{1570} = \text{sum for the upper-class school}$$
$$502 + 525 + 486 = \mathbf{1513} = \text{sum for the middle-class school}$$
$$485 + 468 + 436 = \mathbf{1389} = \text{sum for the lower-class school}$$

Then, square the above sums, divide by the number of measures on which each of the sums was based, and add the quotients.

$$\frac{1570^2}{18} + \frac{1513^2}{18} + \frac{1389^2}{18} = \frac{6{,}683{,}390}{18} = \mathbf{371{,}299}$$

Then, subtract the correction term from the above value.

$$371{,}299 - 370{,}347 = \mathbf{952} = SS_{\text{levels}}$$

Step 7. Computation of the treatment effects (the overall effects of teacher only versus teacher plus machines versus television plus machines): First, add the sums of the three treatment conditions, disregarding the levels dimension (see Step 2).

$$512 + 502 + 485 = \mathbf{1499} = \text{sum for teacher-only groups}$$
$$540 + 525 + 468 = \mathbf{1533} = \text{sum for teacher-plus-machines groups}$$
$$518 + 486 + 436 = \mathbf{1440} = \text{sum for television-plus-machines groups}$$

Then, square the above sums, divide by the number of measures on which each was based, and add the quotients.

$$\frac{1499^2}{18} + \frac{1533^2}{18} + \frac{1440^2}{18} = \frac{6{,}670{,}690}{18} = \mathbf{370{,}594}$$

Then, subtract the correction term from the above value.

$$370,594 - 370,347 = \textbf{247} = SS_{treatments}$$

Step 8. Computation of the interactive effects of treatments and levels: First, square the sums of each of the experimental groups (Step 2), divide by the number of measures on which each of the sums was based, and then add the quotients.

$$\frac{512^2}{6} + \frac{540^2}{6} + \frac{518^2}{6} + \frac{502^2}{6} + \frac{525^2}{6} + \frac{486^2}{6} + \frac{485^2}{6} + \frac{468^2}{6} + \frac{436^2}{6}$$

$$= \frac{2,230,238}{6} = \textbf{371,706}$$

Then, subtract the correction term (Step 4), the SS_{levels} (Step 6), and the $SS_{treatments}$ (Step 7) from the above value.

$$371,706 - 370,347 - 952 - 247 = \textbf{160} = SS_{treatments \times levels}$$

Step 9. Computation of the error-term sum of squares (SS_{error}): Simply subtract SS_{levels} (Step 6), $SS_{treatments}$ (Step 7), and $SS_{treatments \times levels}$ (Step 8) from SS_t (Step 5).

$$4429 - 952 - 247 - 160 = \textbf{3070} = SS_{error}$$

Step 10. All computations based directly on the data have now been completed. Since the *F* ratios are ratios of mean squares, the degrees of freedom (*df*) must be computed for each of the components.

df for SS_t = the total number of measures (see Step 1) minus 1.

$$54 - 1 = \textbf{53}$$

df for SS_{levels} = the number of levels used in the experiment minus 1.

$$3 - 1 = \textbf{2}$$

df for $SS_{treatments}$ = the number of treatment conditions minus 1.

$$3 - 1 = \textbf{2}$$

df for $SS_{treatments \times levels}$ = *df* for treatments times the *df* for levels.

$$2 \times 2 = \textbf{4}$$

df for SS_{error} = the *df* for SS_t minus the *dfs* for SS_{levels}, $SS_{treatments}$, and $SS_{treatments \times levels}$.

$$53 - 2 - 2 - 4 = \textbf{45}$$

Step 11. The mean squares are then computed as SS/df.

ms_t (This value is not needed for the analysis.)

$$ms_{levels} = \frac{(Step\ 6)}{2} = \frac{952}{2} = 476$$

$$ms_{treatments} = \frac{(Step\ 7)}{2} = \frac{247}{2} = 123.5$$

$$ms_{treatments \times levels} = \frac{(Step\ 8)}{4} = \frac{160}{4} = 40$$

$$ms_{error} = \frac{(Step\ 9)}{45} = \frac{3070}{45} = 68.22$$

Step 12. The several F ratios are then computed as

$$\frac{ms_{levels}}{ms_{error}} \qquad \frac{ms_{treatments}}{ms_{error}} \qquad \frac{ms_{treatments \times levels}}{ms_{error}}$$

Step 13. Table the final analysis as follows.

Source	SS	df	ms	F	p
Total	4429	53	—	—	—
Levels	952	2	476	6.98	<.005
Treatments	247	2	123.5	1.81	n.s.
Treatments × levels	160	4	40	.58	n.s.
Error	3070	45	68.22	—	—

Therefore, it is concluded (see Appendix E) that while the expected differences in the overall arithmetic achievements of the students in the three schools were present, there were no differences due to the three methods of instruction and that the methods of instruction did not interact with the achievement level of the students.

SECTION 2.5
Treatments-by-Subjects, or Repeated-Measures, Design

Whenever two or more different treatments are given to the same subjects, a very powerful statistical design can be used. This design, which permits analysis of repeated measures on the same individuals, is known by various names and is treated in considerable detail in most textbooks.

Points to Consider When Using the Treatments-by-Subjects Design

1. Each subject is tested under, and a score is entered for, each treatment condition.
2. The number of subjects employed and the number of treatments given each subject is arbitrary. However, most experiments use ten to fifteen subjects, and it is rare that the effects of more than four or five treatments are compared.

Example

In addition to analyzing the effects of successive experimental manipulations, this design is often used to indicate and evaluate the stability of measures over a prolonged period. This use will be illustrated in the following example. Assume that an experimenter wishes to know whether the mean scores on an intelligence test remain stable from year to year or whether they differ to a degree larger than could be expected by chance alone. To answer this question, the experimenter chooses twenty subjects, all twelve years old, and obtains an IQ score for each subject. Then, each year after that, for three years, another IQ test is given to the same twenty students. The IQ scores on each of these four tests are the data recorded.

Step 1. At the end of four years, when the data have been collected, they are tabled as follows.

Subject	IQ—Age 12	IQ—Age 13	IQ—Age 14	IQ—Age 15
S_1	98	102	113	108
S_2	104	100	105	111
S_3	126	131	128	136
S_4	74	79	84	81
S_5	86	85	90	88
S_6	92	92	88	95
S_7	98	96	101	103
S_8	110	115	107	118
S_9	118	122	129	120
S_{10}	97	96	104	100
S_{11}	83	85	80	88
S_{12}	124	128	135	138
S_{13}	133	132	130	138
S_{14}	101	112	115	114
S_{15}	112	108	116	114
S_{16}	96	91	95	98
S_{17}	89	96	90	97
S_{18}	95	101	99	104
S_{19}	87	95	94	97
S_{20}	93	99	92	102

Step 2. Add the scores in each treatment (age level in the present example) to get the sum for each treatment. (*Note:* If you are using a calculator, Step 2 and Step 3 can be done at the same time.)

Subject	Age 12	Age 13	Age 14	Age 15
S_1	98	102	113	108
S_2	104	100	105	111
⋮	⋮	⋮	⋮	⋮
Sums:	**2016**	**2065**	**2095**	**2150**

Step 3. Square each number in the table (Step 1), and add these squared values together.

$$98^2 + 104^2 + \cdots + 102^2 = \textbf{886,376}$$

Step 4. Add the sums for each treatment (Step 2) to obtain a grand total for the entire table.

$$2016 + 2065 + 2095 + 2150 = \textbf{8326}$$

Step 5. Square the grand total (Step 4), and divide by the total number of measures recorded (see Step 1). This yields the correction term.

$$\frac{8326^2}{80} = \textbf{866,528}$$

Step 6. Computation of the total sum of squares (SS_t): Subtract the correction term (Step 5) from the sum of the squared values obtained in Step 3.

$$886,376 - 866,528 = \textbf{19,848} = SS_t$$

Step 7. Add the four scores recorded for each subject, and write the twenty sums in a column.

Subject	Age 12		Age 13		Age 14		Age 15	Sums
S_1	98	+	102	+	113	+	108	**421**
S_2	104	+	100	+	105	+	111	**420**
⋮	⋮		⋮		⋮		⋮	⋮
S_{20}	93	+	99	+	92	+	102	**386**

Step 8. Square all the sums obtained in Step 7, and add the squares.

$$421^2 + 420^2 + \cdots + 386^2 = \textbf{3,540,914}$$

Step 9. Divide the value obtained in Step 8 by the number of scores that were added to get each of the sums in Step 7. (In the present example, 4 numbers were added to get each of the sums.)

$$\frac{3,540,914}{4} = \textbf{885,228}$$

Step 10. Computation of the sum of squares for subjects ($SS_{subjects}$): Subtract the correction term (Step 5) from the value obtained in Step 9.

$$885,228 - 866,528 = \textbf{18,700} = SS_{subjects}$$

Step 11. Square each of the values obtained in Step 2, and add the squares.

$$2016^2 + 2065^2 + 2095^2 + 2150^2 = \textbf{17,340,006}$$

Then, divide by the number of scores on which each sum in Step 2 was based. (In the present example, this would be 20.)

$$\frac{17,340,006}{20} = \textbf{867,000}$$

Step 12. Computation of the sum of squares for treatments ($SS_{treatments}$): Subtract the correction term (Step 5) from the value of Step 11.

$$867,000 - 866,528 = \textbf{472} = SS_{treatments}$$

Step 13. Computation of the error-term sum of squares (SS_{error}): Subtract the $SS_{subjects}$ (Step 10) and the $SS_{treatments}$ (Step 12) from the SS_t (Step 6).

$$19,848 - 18,700 - 472 = \textbf{676} = SS_{error}$$

Step 14. All computations based directly on the data have been completed. However, since the test of significance (F ratio) is a ratio of the mean squares, these values must be computed. To do this, the degrees of freedom (df) must be determined for each of the components, as follows.

df for SS_t = the total number of measures (see Step 1) minus 1.

$$80 - 1 = \textbf{79}$$

df for $SS_{subjects}$ = the total number of subjects used in the experiment minus 1.

$$20 - 1 = \textbf{19}$$

df for $SS_{treatments}$ = the total number of treatments given each subject minus 1.

$$4 - 1 = 3$$

df for SS_{error} = the total df minus the df for subjects and the df for treatments.

$$79 - 19 - 3 = 57$$

Step 15. The mean squares are then computed as SS/df.

ms_t (This value is not needed for analysis.)

$ms_{subjects}$ (This is also not needed.)

$$ms_{treatments} = \frac{(Step\ 12)}{3} = \frac{472}{3} = 157.33$$

$$ms_{error} \quad \frac{(Step\ 13)}{57} = \frac{676}{57} = 11.86$$

Step 16. Table the analysis as follows. The test of significance (F) is equal to $ms_{treatment}/ms_{error}$.

Source	SS	df	ms	F	p
Total	19,848	79	—	—	—
Subjects	18,700	19	—	—	—
Treatments	472	3	157.33	13.27	<.001
Error	676	57	11.86	—	—

Since the F value of 13.27, with df of 3 and 57, would occur by chance less than once in one thousand times (see Appendix E), it is concluded that the IQ scores were not stable over the four-year period.

SECTION 2.6

Treatments-by-Treatments-by-Subjects, or Repeated-Measures: Two Factors, Design

The treatments-by-treatments-by-subjects design is a very efficient analysis that can be used when all treatments in a two-factor experiment are administered to each subject.

Points to Consider When Using the Treatments-by-Treatments-by-Subjects Design

1. Each subject is tested under, and a score is entered for, each treatment condition within both of the factors.
2. The number of subjects employed and the number of treatments given within each factor are arbitrary. However, most experiments use ten to fifteen subjects, and it is rare that the effects of more than four or five treatments within either or both of the factors are compared.

Example

Consider a memorization experiment where the experimenter wants to study the relationship between short-term and long-term retention on the one hand and easy and difficult items on the other. The experimenter has every subject learn both easy and difficult items, measuring the retention of these items over short and long periods of time. In this experiment, the experimenter is not interested in analyzing performance over several trials. Therefore, the experimenter will consider only the total number of correct responses. (*Note:* If he were also to consider trials, this would add another dimension and make this a treatments-by-treatments-by-treatments-by-subjects design.) The scores recorded for each subject are the subject's total number of correct responses on the easy and difficult items when the subject is tested after a short and after a long period of time.

Step 1. Table the data as follows. Remember, each subject performs under each treatment condition within both factors.

| Subject | Short-Term Retention A_1 | | Long-Term Retention A_2 | |
	Easy Items B_1	Difficult Items B_2	Easy Items B_1	Difficult Items B_2
S_1	83	100	75	40
S_2	100	83	71	41
S_3	100	100	50	43
S_4	89	92	74	47
S_5	100	90	83	62
S_6	100	100	100	70
S_7	92	92	72	44
S_8	100	89	83	62
S_9	50	54	33	38
S_{10}	100	100	72	50

Step 2. Add the scores in each treatment for both factors. (*Note:* If

you are using a calculator, Step 2 and Step 3 can be done at the same time.)

Subject	STR–Easy	STR–Difficult	LTR–Easy	LTR–Difficult
S_1	83	100	75	40
S_2	100	83	71	41
⋮	⋮	⋮	⋮	⋮
Sums:	914	900	713	497

Step 3. Square each number in the table (Step 1), and add the squares.

$$83^2 + 100^2 + \cdots + 50^2 = \textbf{248,212}$$

Step 4. Add the sums for each treatment for both factors (Step 2) to obtain a grand total for the entire table.

$$914 + 900 + 713 + 497 = \textbf{3024}$$

Step 5. Square the grand total of Step 4, and divide by the total number of measures recorded in Step 1 (40 in this example). This yields the correction term.

$$\frac{3024^2}{40} = \frac{9,144,576}{40} = \textbf{228,614.4}$$

Step 6. Subtract the correction term (Step 5) from the sum of the squares obtained in Step 3.

$$248,212 - 228,614.4 = \textbf{19,597.6} = SS_t$$

Step 7. Add all the scores recorded for each subject in Step 1, and write these sums in a column.

	Short-Term Retention				Long-Term Retention				
Subject	Easy Items		Difficult Items		Easy Items		Difficult Items		Sums
S_1	83	+	100	+	75	+	40		298
S_2	100	+	83	+	71	+	41		295
⋮	⋮		⋮		⋮		⋮		⋮
S_{10}	100	+	100	+	72	+	50		322

Step 8. Square all the sums obtained in Step 7, and add the squares.

$$298^2 + 295^2 + \cdots + 322^2 = \textbf{937,872}$$

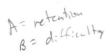
A = retention
B = difficulty

Step 9. Divide the value obtained in Step 8 by the number of scores that were added to get each of the sums in Step 7 (4 scores were added in this example).

$$\frac{937,872}{4} = \textbf{234,468}$$

Step 10. Subtract the correction term of Step 5 from the value of Step 9.

$$234,468 - 228,614.4 = \textbf{5853.6} = \text{SS}_{\text{subjects}}$$

Step 11. Computation of the sum of squares for the first factor (the differential effects of STR versus LTR—$\text{SS}_{\text{retention}}$): First, add the sums of the STR and the LTR treatments (Step 2), disregarding the difficulty-of-items dimension.

$$914 + 900 = \textbf{1814} = \text{sum of the STR treatment}$$
$$713 + 497 = \textbf{1210} = \text{sum of the LTR treatment}$$

Then, square the above sums, divide by the number of measures on which each sum was based (20 in this example), and add the quotients.

$$\frac{1814^2}{20} + \frac{1210^2}{20} = \frac{4,754,696}{20} = \textbf{237,734.8}$$

Then, subtract the correction term (Step 5) from the above value.

$$237,734.8 - 228,614.4 = \textbf{9120.4} = \text{SS}_{\text{retention}}$$

Step 12. Computation of the effects of the second factor (the effects of easy versus difficult items—$\text{SS}_{\text{difficulty}}$): First, sum the scores on the easy items and on the difficult items (Step 2), disregarding the retention dimension.

$$914 + 713 = \textbf{1627} = \text{sum of the easy items}$$
$$900 + 497 = \textbf{1397} = \text{sum of the difficult items}$$

Then, square the above sums, divide by the number of measures on which each sum was based (20 in this example), and add the quotients.

$$\frac{1627^2}{20} + \frac{1397^2}{20} = \frac{4,598,738}{20} = \textbf{229,936.9}$$

Then, subtract the correction term (Step 5) from the above value.

$$229,936.9 - 228,614.4 = \textbf{1322.5} = \text{SS}_{\text{difficulty}}$$

Step 13. Computation of the interactive effects of the first and second factors ($\text{SS}_{\text{retention} \times \text{difficulty}}$): First, square the sums of each of the experimental treatments (Step 2), divide by the number of measures on which each sum was based (10 in this example), and add the quotients.

$$\frac{914^2}{10} + \frac{900^2}{10} + \frac{713^2}{10} + \frac{497^2}{10} = \frac{2,400,774}{10} = \textbf{240,077.4}$$

Then, subtract the correction term (Step 5), the sum of squares for retention (Step 11), and the sum of squares for difficulty (Step 12) from the above value.

$$240,077.4 - 228,614.4 - 9120.4 - 1322.5$$
$$= \textbf{1020.1} = SS_{retention \times difficulty}$$

Step 14. Computation of the error term for retention ($SS_{error\ for\ retention}$): This requires that the data be retabled as follows. For each subject, add the number of correct responses made to the tests of STR and those of LTR (disregarding the difficulty-of-items dimension).

Subject	STR	LTR
S_1	183 (83 + 100)	115 (75 + 40)
S_2	183 (100 + 83)	112 (71 + 41)
S_3	200	93
S_4	181	121
S_5	190	145
S_6	200	170
S_7	184	116
S_8	189	145
S_9	104	71
S_{10}	200	122

Then, square all the sums in the above table, and add the squares.

$$183^2 + 183^2 + \cdots + 122^2 = \textbf{489,622}$$

Then, divide the above value by the number of scores added to get each of the sums entered in the table in Step 14 (2 in this example).

$$\frac{489,622}{2} = \textbf{244,811}$$

Then, from this value, subtract the correction term (Step 5), the sum of squares for subjects (Step 10), and the sum of squares for retention (Step 11).

$$244,811 - 228,614.4 - 5853.6 - 9120.4 = \textbf{1222.6} = SS_{error\ for\ retention}$$

Step 15. Computation of the error term for item difficulty ($SS_{error\ for\ difficulty}$): Again, the data must first be retabled. For each subject, add the number of correct responses made on the easy and on the difficult items (disregarding the retention conditions).

Subject	Easy	Difficult
S_1	158 (83 + 75)	140 (100 + 40)
S_2	171 (100 + 71)	124 (83 + 41)
S_3	150	143
S_4	163	139
S_5	183	152
S_6	200	170
S_7	164	136
S_8	183	151
S_9	83	92
S_{10}	172	150

Then, square all the sums in the above table, and add the squares.

$$158^2 + 171^2 + \cdots + 150^2 = \mathbf{472{,}632}$$

Then divide this value by the number of scores added to get each of the sums in the above table (2 in this example).

$$\frac{472{,}632}{2} = \mathbf{236{,}316}$$

Then, from this value, subtract the correction term (Step 5), the sum of squares for subjects (Step 10), and the sum of squares for difficulty (Step 12).

$$236{,}316 - 228{,}614.4 - 5853.6 - 1322.5 = \mathbf{525.5} = SS_{\text{error for difficulty}}$$

Step 16. Computation of the error term for retention-by-difficulty interaction ($SS_{\text{error for retention} \times \text{difficulty}}$): From the total sum of squares (Step 6), subtract the sum of squares for subjects (Step 10), sum of squares for retention (Step 11), sum of squares for difficulty (Step 12), sum of squares for difficulty by retention (Step 13), error for retention (Step 14), and error for difficulty (Step 15).

$$19{,}597.6 - 5853.6 - 9120.4 - 1322.5 - 1020.1 - 1222.6 - 525.5$$
$$= \mathbf{532.9} = SS_{\text{error for retention} \times \text{difficulty}}$$

Step 17. All computations based directly on the data have been completed. However, since the test of significance (F ratio) is a ratio of mean squares, the degrees of freedom must be determined for each of the components.

df for SS_t = the total number of measures recorded in Step 1 minus 1.

$$40 - 1 = \mathbf{39}$$

df for $SS_{subjects}$ = the number of subjects used in the experiment minus 1.

$$10 - 1 = 9$$

df for $SS_{retention}$ = the number of retention conditions minus 1.

$$2 - 1 = 1$$

df for $SS_{error\ for\ retention}$ = the df for $SS_{subjects}$ time the df for $SS_{retention}$.

$$9 \times 1 = 9$$

df for $SS_{difficulty}$ = the number of difficulty conditions minus 1.

$$2 - 1 = 1$$

df for $SS_{error\ for\ difficulty}$ = the df for $SS_{subjects}$ times the df for $SS_{difficulty}$.

$$9 \times 1 = 9$$

df for $SS_{retention \times difficulty}$ = the df for $SS_{retention}$ times the df for $SS_{difficulty}$.

$$1 \times 1 = 1$$

df for $SS_{error\ for\ retention \times difficulty}$ = the df for $SS_{subjects}$ times the df for $SS_{retention}$ times the df for $SS_{difficulty}$.

$$9 \times 1 \times 1 = 9$$

Step 18. The desired mean squares are then computed as SS/df.

$$ms_{retention} = \frac{(Step\ 11)}{1} = \frac{9120.4}{1} = 9120.4$$

$$ms_{difficulty} = \frac{(Step\ 12)}{1} = \frac{1322.5}{1} = 1322.5$$

$$ms_{difficulty \times retention} = \frac{(Step\ 13)}{1} = \frac{1020.1}{1} = 1020.1$$

$$ms_{retention\ error} = \frac{(Step\ 14)}{9} = \frac{1222.6}{9} = 135.844$$

$$ms_{difficulty\ error} = \frac{(Step\ 15)}{9} = \frac{525.5}{9} = 58.389$$

$$ms_{difficulty \times retention\ error} = \frac{(Step\ 16)}{9} = \frac{532.9}{9} = 59.211$$

Step 19. Table the analysis as follows. The tests of significance (F ratios) are computed as

$$\frac{ms_{retention}}{ms_{error\ for\ retention}} \quad \frac{ms_{difficulty}}{ms_{error\ for\ difficulty}}$$

$$\frac{ms_{difficulty \times retention}}{ms_{error\ for\ difficulty \times retention}}$$

Source	SS	df	ms	F	p
Total	19,597.6	39	—	—	—
Subjects	5853.6	9	—	—	—
Retention A	9120.4	1	9120.4	67.14	<.001
Difficulty B	1322.5	1	1322.5	22.65	<.005
Difficulty × retention A × B	1020.1	1	1020.1	17.23	<.005
Error retention	1222.6	9	135.84	—	—
Error difficulty	525.5	9	58.39	—	—
Error retention × difficulty	532.9	9	59.21	—	—

Therefore, it is concluded (see Appendix E) that:

1. The length of the retention interval had a significant effect on the amount retained.
2. The difficulty of the learned material significantly affected the amount retained.
3. The effects of difficulty of material and retention interval interacted to a significant degree.

SECTION 2.7
Two-Factor Mixed Design: Repeated Measures on One Factor

This two-factor mixed design is basically a combination of the completely randomized design and the treatments-by-subjects design. Not only does this design permit comparison of the differences in the overall performance of the subjects in the several experimental groups, but it also permits evaluation of the changes in performance shown by the subjects during the experimental session.

Points to Consider When Using the Two-Factor Mixed Design

1. If trials or observations are blocked, each block is treated as one score.

2. Although it is not necessary, it is usually best to have an equal number of subjects in each experimental group.
3. The number of experimental groups compared, the number of subjects per group, and the number of trials or observations (scores) recorded for each subject are arbitrary. However, it is rare that more than four or five experimental groups are compared in one experiment and that more than five or six scores are entered and analyzed for each subject. In most experiments, ten to fifteen subjects are assigned to each experimental group.

Example

Assume that an experimenter wishes to determine the effects of meaningfulness of material on rate of learning. The subjects are randomly assigned to three groups, one of which must learn a low-meaningful list of nonsense syllables, the second a medium-meaningful list, and the third a high-meaningful list. All subjects are given fifteen trials, and the number of correct responses per trial is recorded. To simplify the analysis, the experimenter blocks the number of correct responses into three blocks of five trials each. Thus, the scores used in the analysis represent the number of correct responses for each successive block of five trials.

Step 1. After the experiment has been conducted, table the data as follows.

Subject	Trial Block 1	Trial Block 2	Trial Block 3
Group 1 (Low-Meaningful Material)			
S_1	3	3	4
S_2	3	4	4
S_3	1	3	4
S_4	1	2	3
S_5	2	3	5
S_6	4	5	6
S_7	4	5	6
S_8	1	4	5
S_9	1	4	5
S_{10}	2	3	3
S_{11}	2	2	4
Group 2 (Medium-Meaningful Material)			
S_{12}	2	3	6
S_{13}	1	4	5
S_{14}	3	6	7
S_{15}	2	4	6

Subject	Group 2 (Medium-Meaningful Material)		
	Trial Block 1	Trial Block 2	Trial Block 3
S_{16}	1	2	3
S_{17}	4	5	5
S_{18}	1	3	4
S_{19}	1	2	4
S_{20}	2	3	4
S_{21}	3	5	6
S_{22}	4	6	7

Subject	Group 3 (High-Meaningful Material)		
S_{23}	2	4	8
S_{24}	3	6	3
S_{25}	4	7	7
S_{26}	1	7	4
S_{27}	3	7	12
S_{28}	1	4	2
S_{29}	1	5	3
S_{30}	2	5	6
S_{31}	3	6	6
S_{32}	1	5	12
S_{33}	4	9	7

Step 2. Add the scores in each group for each trial block (hereafter, referred to simply as "trials").

Subject	Group 1		
	Trial 1	Trial 2	Trial 3
S_1	3	3	4
S_2	3	4	4
⋮	⋮	⋮	⋮
Sums:	24	38	49

Subject	Group 2		
S_{12}	2	3	6
S_{13}	1	4	5
⋮	⋮	⋮	⋮
Sums:	24	43	57

Subject	Group 3		
S_{23}	2	4	8
S_{24}	3	6	3
⋮	⋮	⋮	⋮
Sums:	25	65	70

Step 3. Obtain the sum for each group by adding the sums of the individual trials (Step 2).

$$24 + 38 + 49 = \textbf{111} = \text{sum of Group 1}$$
$$24 + 43 + 57 = \textbf{124} = \text{sum of Group 2}$$
$$25 + 65 + 70 = \textbf{160} = \text{sum of Group 2}$$

Step 4. Add the scores for *each* subject in each group. (*Note:* If you are using a calculator, Step 5 can be done at the same time as Step 4.)

Subject	Trial 1		Trial 2		Trial 3	Sums
Group 1						
S_1	3	+	3	+	4	10
S_2	3	+	4	+	4	11
⋮	⋮		⋮		⋮	⋮
S_{11}	2	+	2	+	4	8
Group 2						
S_{12}	2	+	3	+	6	11
S_{13}	1	+	4	+	5	10
⋮	⋮		⋮		⋮	⋮
S_{22}	4	+	6	+	7	17
Group 3						
S_{23}	2	+	4	+	8	14
S_{24}	3	+	6	+	3	12
⋮	⋮		⋮		⋮	⋮
S_{33}	4	+	9	+	7	20

Step 5. Square each score in the entire table (Step 1), and add these squared values to get a grand sum of the squared values.

$$3^2 + 3^2 + \cdots + 7^2 = \textbf{2043}$$

Step 6. Add the group totals (Step 3) to get the grand sum of the entire table.

$$111 + 124 + 160 = \textbf{395}$$

Step 7. Square the grand sum (Step 6), and divide it by the total number of measures in the entire table—i.e., the number of subjects times the number of measures per subject (see Step 1). (In the present example, this would be $33 \times 3 = 99$.) The resultant value is the correction term.

$$\frac{395^2}{99} = \frac{156,025}{99} = \textbf{1576}$$

Step 8. Computation of the total sum of squares (SS_t): Simply subtract the correction term (Step 7) from the sum of the squared scores (Step 5).

$$2043 - 1576 = \mathbf{467} = SS_t$$

Between-Subjects Effects (Steps 9–11)

Step 9. Computation of the between-subjects sum of squares (SS_b): First, square the sum of each subject's scores (Step 4), and add these squared values.

$$10^2 + 11^2 + \cdots + 20^2 = \mathbf{5271}$$

Then, divide this value by the number of trials given each subject.

$$\frac{5271}{3} = \mathbf{1757}$$

Then, subtract the correction term from the above value.

$$1757 - 1576 = \mathbf{181} = SS_b$$

Step 10. Computation of the effects of the experimental conditions on overall performance (SS_c): (In the present example, these are the meaningfulness effects.) First, square the sum for each experimental group (Step 3), and divide these values by the number of measures in each experimental group. (In the present example, this would be 11 subjects $\times$ 3 trials per subject, or 33.) Then, add these quotients.

$$\frac{111^2}{33} + \frac{124^2}{33} + \frac{160^2}{33} = \frac{53{,}297}{33} = \mathbf{1615}$$

Then, subtract the correction term from the above value.

$$1615 - 1576 = \mathbf{39} = SS_c$$

Step 11. Computation of the between-subjects error term $(error_b)$: Simply subtract the SS_c (Step 10) from the SS_b (Step 9).

$$181 - 39 = \mathbf{142} = error_b$$

Within-Subjects Effects (Steps 12–15)

Step 12. Computation of the within-subjects sum of squares (SS_w): Simply subtract the SS_b (Step 9) from the SS_t (Step 8).

$$467 - 181 = \mathbf{286} = SS_w$$

Step 13. Computation of the sum of squares for trials (SS_{tr}): First, add the sums of Trial 1 (Step 2) for all experimental groups.

$$24 + 24 + 25 = \mathbf{73} = \text{sum for Trial 1}$$

Do the same for Trial 2 and Trial 3.

$$38 + 43 + 65 = \mathbf{146} = \text{sum for Trial 2}$$
$$49 + 57 + 70 = \mathbf{176} = \text{sum for Trial 3}$$

Then, square each of these sums, and divide the products by the number of measures on which each sum was based.

$$\frac{73^2}{33} + \frac{146^2}{33} + \frac{176^2}{33} = \frac{57,621}{33} = \mathbf{1746}$$

Then, subtract the correction term from the above value.

$$1746 - 1576 = \mathbf{170} = SS_{tr}$$

Step 14. Computation of the sum of squares for the trials-by-conditions interaction ($SS_{tr \times c}$): First, square the sums of each trial in each of the experimental groups (Step 2), divide by the number of measures on which each sum was based, and add these quotients.

$$\frac{24^2}{11} + \frac{38^2}{11} + \frac{49^2}{11} + \frac{24^2}{11} + \frac{43^2}{11} + \frac{57^2}{11} + \frac{25^2}{11}$$

$$+ \frac{65^2}{11} + \frac{70^2}{11} = \frac{19,845}{11} = \mathbf{1804}$$

Then, from this value, subtract the correction term (Step 7), SS_c (Step 10), and SS_{tr} (Step 13).

$$1804 - 1576 - 39 - 170 = \mathbf{19} = SS_{tr \times c}$$

Step 15. Computation of the within-subjects error term (error_w): Simply subtract SS_{tr} (Step 13) and $SS_{tr \times c}$ (Step 14) from the SS_w (Step 12).

$$286 - 170 - 19 = \mathbf{97} = \text{error}_w$$

Step 16. All computations based directly on the data have been completed:

$$
\begin{array}{lll}
SS_t & = & 467 \ (\text{Step 8}) \\
SS_b & = & 181 \ (\text{Step 9}) \\
SS_c & = & 39 \ (\text{Step 10}) \\
SS_{error_b} & = & 142 \ (\text{Step 11}) \\
SS_w & = & 286 \ (\text{Step 12}) \\
SS_{tr} & = & 170 \ (\text{Step 13})
\end{array}
$$

$$SS_{tr \times c} \; = \; 19 \; (\text{Step } 14)$$
$$SS_{error_w} \; = \; 97 \; (\text{Step } 15)$$

However, since the tests of significance (F-tests) are ratios of *mean squares*, we must compute these values. To do this, the degrees of freedom (*df*) for each of the components must be determined, as follows.

df for SS_t = the total number of measures recorded minus 1.

$$99 - 1 = \textbf{98}$$

df for SS_b = the total number of subjects minus 1.

$$33 - 1 = \textbf{32}$$

df for SS_c = the total number of experimental groups minus 1.

$$3 - 1 = \textbf{2}$$

df for SS_{error_b} = the *df* for SS_b minus the *df* for SS_c.

$$32 - 2 = \textbf{30}$$

df for SS_w = the *df* for SS_t minus the *df* for SS_b.

$$98 - 32 = \textbf{66}$$

df for SS_{tr} = the number of trials given each subject minus 1.

$$3 - 1 = \textbf{2}$$

df for $SS_{tr \times c}$ = the *df* for SS_{tr} times the *df* for SS_c.

$$2 \times 2 = \textbf{4}$$

df for SS_{error_w} = the *df* for SS_w minus the *df* for SS_{tr} minus the *df* for $SS_{tr \times c}$.

$$66 - 2 - 4 = \textbf{60}$$

Step 17. The mean squares are then computed as SS/df.

ms_t (This value is not needed.)

ms_b (This value is not needed.)

$$ms_c \;\;\; = \; \frac{(\text{Step } 10)}{2} = \frac{39}{2} = \textbf{19.5}$$

$$ms_{error_b} \; = \; \frac{(\text{Step } 11)}{30} = \frac{142}{30} = \textbf{4.73}$$

ms_w (This value is not needed.)

$$ms_{tr} \;\;\; = \; \frac{(\text{Step } 13)}{2} = \frac{170}{2} = \textbf{85}$$

$$ms_{tr \times c} = \frac{(\text{Step 14})}{4} = \frac{19}{4} = 4.75$$

$$ms_{error_w} = \frac{(\text{Step 15})}{60} = \frac{97}{60} = 1.62$$

Step 18. Table the final analysis as follows. The tests of significance (*F*-tests) are equal to

$$\frac{ms_c}{ms_{error_b}} \times \frac{ms_{tr}}{ms_{error_w}} \times \frac{ms_{tr \times c}}{ms_{error_w}}$$

Source	SS	df	ms	F	p
Total	467	98	—	—	—
Between subjects	181	32	—	—	—
Conditions	39	2	19.50	4.12	<.05
Error$_b$	142	30	4.73	—	—
Within subjects	286	66	—	—	—
Trials	170	2	85.00	52.46	<.001
Trials × conditions	19	4	4.75	2.93	<.05
Error$_w$	97	60	1.62	—	—

Hence, it is concluded (see Appendix E) that:

1. Meaningfulness of material significantly affected the overall amount learned.
2. Subjects learned as a function of practice (trials).
3. The groups of subjects learned at different rates.

SECTION 2.8

Three-Factor Mixed Design: Repeated Measures on One Factor

The three-factor mixed design is basically a combination of the factorial design and the treatments-by-subjects design. Not only does this design permit examination of the effects of two factors in combination with each other, but it also permits examination of performance variations shown by the subjects during the experimental session.

Points to Consider When Using the Three-Factor Mixed Design

1. As in the other designs, if trials are blocked, each block is treated as one score.

2. It is usually best to have an equal number of subjects in each of the experimental groups. If equality cannot be obtained, the number of subjects in each group should be proportional. In most experiments, ten to fifteen subjects are assigned per group.
3. The number of experimental groups within each factor is arbitrary. However, it is rare that more than four or five groups are included in either of the two factors. Similarly, it is rare that more than five or six scores are entered and analyzed for each of the subjects.

Example

Assume that an experimenter wishes to determine the effects of high versus low performance expectancy in combination with failure versus success feedback on the learning of verbal material. Subjects are randomly assigned to one of four groups. The subjects in Group 1 (high expectancy/failure) are initially told that they are likely to perform the task very well. During the learning period, however, they are periodically told that they are actually performing quite poorly and must try harder. Group 2 subjects (high expectancy/success) are also initially told that they are likely to perform well. During the learning period, they are told that they are indeed doing well and should continue their excellent performance. Group 3 subjects (low expectancy/failure) are initially told that their performance is likely to be very low. During the practice period, they are told that they are doing poorly, as expected, but if possible, they should try harder. Group 4 subjects (low expectancy/success) are also told that their performance is likely to be very low. During the practice period, however, they are told that they are actually doing very well and should continue their excellent performance.

The subjects in all four groups are given a total of twenty trials to learn a list of low-meaningful nonsense syllables. The number of syllables recited correctly on each trial is recorded. For analysis purposes, the experimenter blocks the scores on each of the twenty trials into four blocks of five trials each.

The number of correct responses within each of these blocks of five trials is the score recorded.

Step 1. After the experiment has been conducted, table the data as follows.

Group 1 (High Expectancy/Failure)			
Subject	Trial Block 1	Trial Block 2	Trial Block 3	Trial Block 4
S_1	4	6	7	9
S_2	6	7	9	11
S_3	5	8	10	13

	Group 1 (High Expectancy/Failure)			
Subject	Trial Block 1	Trial Block 2	Trial Block 3	Trial Block 4
S_4	3	4	6	9
S_5	4	6	9	10
S_6	5	5	6	8
S_7	7	8	9	12

	Group 2 (High Expectancy/Success)			
S_8	5	8	9	12
S_9	3	5	10	14
S_{10}	7	10	12	13
S_{11}	5	8	11	15
S_{12}	4	9	13	15
S_{13}	3	5	9	12
S_{14}	6	7	8	11

	Group 3 (Low Expectancy/Failure)			
S_{15}	3	4	4	5
S_{16}	5	6	8	9
S_{17}	5	6	6	8
S_{18}	5	6	7	8
S_{19}	3	4	4	5
S_{20}	7	7	8	7
S_{21}	6	8	8	8

	Group 4 (Low Expectancy/Success)			
S_{22}	4	8	10	14
S_{23}	5	10	14	18
S_{24}	3	7	12	18
S_{25}	5	8	11	17
S_{26}	6	11	15	19
S_{27}	6	8	10	15
S_{28}	5	11	17	20

Step 2. Add the scores in each group for each trial block (hereafter referred to only as "trials").

	Group 1			
Subject	Trial 1	Trial 2	Trial 3	Trial 4
S_1	4	6	7	9
S_2	6	7	9	11
⋮	⋮	⋮	⋮	⋮
Sums:	**34**	**44**	**56**	**72**

	Group 2			
Subject	Trial 1	Trial 2	Trial 3	Trial 4
S_8	5	8	9	12
S_9	3	5	10	14
⋮	⋮	⋮	⋮	⋮
Sums:	33	52	72	92

	Group 3			
S_{15}	3	4	4	5
S_{16}	5	6	8	9
⋮	⋮	⋮	⋮	⋮
Sums:	34	41	45	50

	Group 4			
S_{22}	4	8	10	14
S_{23}	5	10	14	18
⋮	⋮	⋮	⋮	⋮
Sums:	34	63	89	121

Step 3. Obtain the sum for each group by adding the sums of the individual trials.

$$34 + 44 + 56 + 72 = \textbf{206} = \text{sum for Group 1}$$
$$33 + 52 + 72 + 92 = \textbf{249} = \text{sum for Group 2}$$
$$34 + 41 + 45 + 50 = \textbf{170} = \text{sum for Group 3}$$
$$34 + 63 + 89 + 121 = \textbf{307} = \text{sum for Group 4}$$

Step 4. Add the scores for *each* subject in each group. (*Note:* If you are using a calculator, Step 5 can be done at the same time as Step 4.)

		Group 1							
Subject	Trial 1		Trial 2		Trial 3		Trial 4	Sums
S_1	4	+	6	+	7	+	9	26
S_2	6	+	7	+	9	+	11	33
⋮	⋮		⋮		⋮		⋮	⋮
S_7	7	+	8	+	9	+	12	36

		Group 2							
S_8	5	+	8	+	9	+	12	34
S_9	3	+	5	+	10	+	14	32
⋮	⋮		⋮		⋮		⋮	⋮
S_{14}	6	+	7	+	8	+	11	32

			Group 3					
Subject	Trial 1		Trial 2		Trial 3		Trial 4	Sums
S_{15}	3	+	4	+	4	+	5	16
S_{16}	5	+	6	+	8	+	9	28
$\vdots$	$\vdots$		$\vdots$		$\vdots$		$\vdots$	$\vdots$
S_{21}	6	+	8	+	8	+	8	30
			Group 4					
S_{22}	4	+	8	+	10	+	14	36
S_{23}	5	+	10	+	14	+	18	47
$\vdots$	$\vdots$		$\vdots$		$\vdots$		$\vdots$	$\vdots$
S_{28}	5	+	11	+	17	+	20	53

Step 5. Square each score in the entire table (Step 1), and add these squared values to get the grand sum of the squared numbers.

$$4^2 + 6^2 + \cdots + 20^2 = \textbf{9412}$$

Step 6. Add the group totals (Step 3) to get the grand sum of the entire table.

$$206 + 249 + 170 + 307 = \textbf{932}$$

Step 7. Square the grand sum, and divide it by the total number of measures in the entire table—ie., the number of subjects times the number of measures per subject (see Step 1). The value obtained is the correction term.

$$\frac{932^2}{112} = 7755$$

Step 8. Computation of the total sum of squares (SS_t): Simply subtract the correction term (Step 7) from the sum of squares (Step 5).

$$9412 - 7755 = \textbf{1657} = SS_t$$

Between-Subjects Effects (Steps 9–13)

Step 9. Computation of the between-subjects sum of squares (SS_b): First, square the sum of each subject's scores (Step 4), and add these squared values.

$$26^2 + 33^2 + \cdots + 53^2 = \textbf{33,324}$$

Then, divide this value by the number of trials given to each subject.

$$\frac{33,324}{4} = \textbf{8331}$$

Then, subtract the correction term (Step 7) from the above value.

$$8331 - 7755 = \textbf{576} = SS_b$$

Step 10. Computation of the effects of the first experimental condition on overall performance (the overall effects of high versus low expectancy—$SS_{expectancy}$): First, sum the scores of the two expectancy groups (Step 3), disregarding the success or failure feedback.

$$\begin{array}{rl} 206 = & \text{sum for Group 1 (high expectancy/failure)} \\ + \ 249 = & \text{sum for Group 2 (high expectancy/success)} \\ \hline \textbf{455} = & \text{sum for all high expectancy subjects} \end{array}$$

$$\begin{array}{rl} 170 = & \text{sum for Group 3 (low expectancy/failure)} \\ + \ 307 = & \text{sum for Group 4 (low expectancy/success)} \\ \hline 477 = & \text{sum for all low expectancy subjects} \end{array}$$

Then, square the sums of the two experimental conditions obtained above, and divide each product by the number of measures added to get each of the sums. (In the present example, the divisor is 56 in both cases: 7 subjects per group $\times$ 2 groups $\times$ 4 trials per subject.) Then, add the quotients.

$$\frac{455^2}{56} + \frac{477^2}{56} = \textbf{7759}$$

Then, subtract the correct term (Step 7) from the above value.

$$7759 - 7755 = \textbf{4} = SS_{expectancy}$$

Step 11. Computation of the effects of the second experimental condition on overall performance (the effects of failure versus success feedback—$SS_{feedback}$): First, sum the scores of the two failure and success groups (Step 3), disregarding the high versus low expectancy.

$$\begin{array}{rl} 206 = & \text{sum for Group 1 (high expectancy/failure)} \\ + \ 170 = & \text{sum for Group 3 (low expectancy/failure)} \\ \hline \textbf{376} = & \text{sum for all subjects who ''failed''} \end{array}$$

$$\begin{array}{rl} 249 = & \text{sum for Group 2 (high expectancy/success)} \\ + \ 307 = & \text{sum for Group 4 (low expectancy/success)} \\ \hline \textbf{556} = & \text{sum for all subjects who ''succeeded''} \end{array}$$

Then, square the sums of these two experimental conditions, and divide the squared values by the number of measures added to get each sum.

(Here, each sum was obtained by adding 56 scores.) Then, add the quotients.

$$\frac{376^2}{56} + \frac{556^2}{56} = 8045$$

Then, subtract the correction term (Step 7) from the above value.

$$8045 - 7755 = 290 = SS_{feedback}$$

Step 12. Computation of the interactive effects of the two experimental conditions: (In this example, we are attempting to determine whether the effects of failure or success feedback on learning are influenced by the expectancy of the subject—$SS_{expectancy \times feedback}$.) First, square the sum of each of the experimental groups (Step 3); then, divide these products by the number of meaures added to get each. (Here, 28 scores were added to get each.) Then add the quotients.

$$\frac{206^2}{28} + \frac{249^2}{28} + \frac{170^2}{28} + \frac{307^2}{28} = 8128$$

Then, from this value, subtract the correction term (Step 7), the $SS_{expectancy}$ (Step 10), and the $SS_{feedback}$ (Step 11).

$$8128 - 7755 - 4 - 290 = 79 = SS_{expectancy \times feedback}$$

Step 13. Computation of the between-subjects error term (error$_b$): Simply subtract the $SS_{expectancy}$ (Step 10), $SS_{feedback}$ (Step 11), and $SS_{expectancy \times feedback}$ (Step 12) from the SS_b (Step 9).

$$576 - 4 - 290 - 79 = 203 = error_b$$

Within-Subjects Effects (Steps 14–19)

Step 14. Computation of the within-subjects sum of squares (SS_w): Simply subtract the SS_b (Step 9) from the SS_t (Step 8).

$$1657 - 576 = 1081 = SS_w$$

Step 15. Computation of the sum of squares for trials (SS_{trials}): First, add the sums of each trial for *all* experimental groups (Step 2).

$$
\begin{aligned}
34 + 33 + 34 + 34 &= 135 = \text{sum of Trial 1} \\
44 + 52 + 41 + 63 &= 200 = \text{sum of Trial 2} \\
56 + 72 + 45 + 89 &= 262 = \text{sum of Trial 3} \\
72 + 92 + 50 + 121 &= 335 = \text{sum of Trial 4}
\end{aligned}
$$

Then, square each of these sums, divide each square by the number of measures on which that sum was based, and add the quotients.

$$\frac{135^2}{28} + \frac{200^2}{28} + \frac{262^2}{28} + \frac{335^2}{28} = \mathbf{8539}$$

Then, subtract the correction term (Step 7) from the above value.

$$8539 - 7755 = \mathbf{784} = SS_{trials}$$

Step 16. Computation of the sum of squares for the trials-by-first-condition interaction (here, the trials-by-expectancy interaction—$SS_{trials \times expectancy}$): First, sum the scores of the two expectancy groups on each trial (Step 2), disregarding the failure or success feedback.

	Trial 1	Trial 2	Trial 3	Trial 4
Group 1 (high expectancy/failure)	34	44	56	72
Group 2 (high expectancy/success)	33	52	72	92
Sums of all high-expectancy S_s =	**67**	**96**	**128**	**164**
Group 3 (low expectancy/failure)	34	41	45	50
Group 4 (low expectancy/success)	34	63	89	121
Sum of all low-expectancy S_s =	**68**	**104**	**134**	**171**

Then, square the sums of each trial, divide by the number of measures on which each of the sums was based, and add the quotients together.

$$\frac{67^2}{14} + \frac{96^2}{14} + \frac{128^2}{14} + \frac{164^2}{14} + \frac{68^2}{14} + \frac{104^2}{14} + \frac{134^2}{14} + \frac{171^2}{14} = \mathbf{8544}$$

Then, from this value, subtract the correction term (Step 7), $SS_{expectancy}$ (Step 10), and SS_{trials} (Step 15).

$$8544 - 7755 - 4 - 784 = \mathbf{1} = SS_{trials \times expectancy}$$

Step 17. Computation of the sum of squares for the trials-by-second-condition interaction (here, the trials-by-feedback interaction—$SS_{trials \times feedback}$): First, sum the scores of the two failure or success groups on each trial (Step 2), disregarding high and low expectancy.

	Trial 1	Trial 2	Trial 3	Trial 4
Group 1 (high expectancy/failure)	34	44	56	72
Group 3 (low expectancy/failure)	34	41	45	50
Sums for all "failure" S_s =	**68**	**85**	**101**	**122**
Group 2 (high expectancy/success)	33	52	72	92
Group 4 (low expectancy/success)	34	63	89	121
Sums for all "success" S_s =	**67**	**115**	**161**	**213**

Then, square the sums of each trial, divide by the number of measures on which each of the sums was based, and add the quotients together.

$$\frac{68^2}{14} + \frac{85^2}{14} + \frac{101^2}{14} + \frac{122^2}{14} + \frac{67^2}{14} + \frac{115^2}{14} + \frac{161^2}{14} + \frac{213^2}{14} = \mathbf{8996}$$

Then, from this value, subtract the correction term (Step 7), $SS_{feedback}$ (Step 11), and SS_{trials} (Step 15).

$$8996 - 7755 - 290 - 784 = \mathbf{167} = SS_{trials \times feedback}$$

Step 18. Computation of the sum of squares for the trials-by-first-condition-by-second-condition interaction (here, the interaction of trials by expectancy by feedback—$SS_{trials \times expectancy \times feedback}$): First, square the sums of each trial in each of the experimental groups (Step 2). Divide each squared value by the number of measures on which each sum was based, and then add the quotients together.

$$\frac{34^2}{7} + \cdots + \frac{72^2}{7} + \frac{33^2}{7} + \cdots + \frac{92^2}{7} + \frac{34^2}{7}$$
$$+ \cdots + \frac{50^2}{7} + \frac{34^2}{7} + \cdots + \frac{121^2}{7} = \frac{63,902}{7} = \mathbf{9129}$$

Then, from this value, subtract the correction term (Step 7), $SS_{expectancy}$ (Step 10), $SS_{feedback}$ (Step 11), $SS_{expectancy \times feedback}$ (Step 12), SS_{trials} (Step 15), $SS_{trials \times expectancy}$ (Step 16), and $SS_{trials \times feedback}$ (Step 17).

$$9129 - 7755 - 4 - 290 - 79 - 784 - 1 - 167$$
$$= \mathbf{49} = SS_{trials \times expectancy \times feedback}$$

Step 19. Computation of the within-subjects error term (error$_w$): Subtract SS_{trials} (Step 15), $SS_{trials \times expectancy}$ (Step 16), $SS_{trials \times feedback}$ (Step 17), and $SS_{trials \times expectancy \times feedback}$ (Step 18) from SS_w (Step 14).

$$1081 - 784 - 1 - 167 - 49 = \mathbf{80} = error_w$$

Step 20. All computations have now been completed.

SS_t	=	1657 (Step 8)
SS_b	=	576 (Step 9)
$SS_{expectancy}$	=	4 (Step 10)
$SS_{feedback}$	=	290 (Step 11)
$SS_{expectancy \times feedback}$	=	79 (Step 12)
SS_{error_b}	=	203 (Step 13)
SS_w	=	1081 (Step 14)
SS_{trials}	=	784 (Step 15)

$$SS_{trials \times expectancy} \qquad = \ 1 \ (Step \ 16)$$
$$SS_{trials \times feedback} \qquad = \ 167 \ (Step \ 17)$$
$$SS_{trials \times expectancy \times feedback} \ = \ 49 \ (Step \ 18)$$
$$SS_{error_w} \qquad\qquad = \ 80 \ (Step \ 19)$$

However, since the tests of significance (*F*-tests) are ratios of *mean* squares, we must compute these values. To do this, the degrees of freedom (*df*) for each of the components must be determined, as follows.

df for SS_t = the total number of measures recorded minus 1.

$$112 - 1 = 111$$

df for SS_b = the total number of subjects minus 1.

$$28 - 1 = 27$$

df for $SS_{expectancy}$ = the number of expectancy groups minus 1.

$$2 - 1 = 1$$

df for $SS_{feedback}$ = the number of feedback groups minus 1.

$$2 - 1 = 1$$

df for $SS_{expectancy \times feedback}$ = the *df* for expectancy times the *df* for feedback.

$$1 \times 1 = 1$$

df for SS_{error_b} = the *df* for SS_b minus the *df*s for expectancy feedback and expectancy by feedback.

$$27 - 1 - 1 - 1 = 24$$

df for SS_w = the *df* for the SS_t minus the *df* for SS_b.

$$111 - 27 = 84$$

df for SS_{trials} = the number of trials given each subject minus 1.

$$4 - 1 = 3$$

df for $SS_{trials \times expectancy}$ = the *df* for SS_{trials} times the *df* for $SS_{expectancy}$.

$$3 \times 1 = 3$$

df for $SS_{trials \times feedback}$ = the *df* for SS_{trials} times the *df* for $SS_{feedback}$.

df for $SS_{trials \times expectancy \times feedback}$ = the *df* for SS_{trials} times the *df* for $SS_{feedback}$ times the *df* for $SS_{expectancy}$.

$$3 \times 1 \times 1 = 3$$

df for SS_{error_w} = the df for SS_w minus the dfs for SS_{trials}, $SS_{trials \times expectancy}$, $SS_{trials \times feedback}$, and $SS_{trials \times expectancy \times feedback}$.

$$84 - 3 - 3 - 3 - 3 = 72$$

Step 21. The mean squares are then computed as SS/df.

ms_t (This value is not needed.)

ms_b (This value is not needed)

$$ms_{expectancy} = \frac{(Step\ 10)}{1} = \frac{4}{1} = 4$$

$$ms_{feedback} = \frac{(Step\ 11)}{1} = \frac{290}{1} = 290$$

$$ms_{expectancy \times feedback} = \frac{(Step\ 12)}{1} = \frac{79}{1} = 79$$

$$ms_{error_b} = \frac{(Step\ 13)}{24} = \frac{203}{24} = 8.46$$

ms_w (This value is not needed.)

$$ms_{trials} = \frac{(Step\ 15)}{3} = \frac{784}{8} = 231.33$$

$$ms_{trials \times expectancy} = \frac{(Step\ 16)}{3} = \frac{1}{3} = .33$$

$$ms_{trials \times feedback} = \frac{(Step\ 17)}{3} = \frac{167}{3} = 55.67$$

$$ms_{trials \times expectancy \times feedback} = \frac{(Step\ 18)}{3} = \frac{49}{3} = 16.33$$

$$ms_{error_w} = \frac{(Step\ 19)}{72} = \frac{80}{72} = 1.11$$

Step 22. Table the final analysis as follows. The tests of significance (*F* ratios) are equal to the mean squares of the experimental conditions divided by their appropriate error terms.

Source	SS	df	ms	F	p
Total	1657	111	—	—	—
Between subjects	576	27	—	—	—
Expectancy (high/low)	4	1	4	<1	—

Source	SS	df	ms	F	p
Feedback (success/failure)	290	1	290	34.27	<.001
Expectancy × feedback	79	1	79	9.34	<.001
Error$_b$	203	24	8.46	—	
Within subjects	1081	84	—	—	—
Trials	784	3	261.43	235.43	<.001
Trials × expectancy	1	3	.33	<1	—
Trials × feedback	167	3	55.67	50.15	<.001
Trials × expectancy × feedback	49	3	16.33	14.71	<.001
Error$_w$	80	72	1.11	—	—

Hence, it is concluded (see Appendix E) that:

1. The type of feedback given the subjects (failure versus success) affects the overall performance level.
2. The effect of the feedback, however, is dependent upon (interacts with) the expectancy of the subjects.
3. All subjects did learn as a function of practice (trials).
4. The rate of learning is dependent upon the type of feedback.
5. The rate of learning is dependent upon both the expectancy level of the subjects and the feedback they receive.

SECTION 2.9

Three-Factor Mixed Design: Repeated Measures on Two Factors

This design is basically a combination of the treatments-by-treatments-by-subjects design and the completely randomized design. Instead of only one group of subjects receiving all the treatment conditions (as in the treatments-by-treatments-by-subjects design), two or more experimental groups are compared (as in the completely randomized design). Thus, this design permits the comparison of several groups' performance and the evaluation of the variations in performance shown by the subjects under all treatment conditions within both factors presented.

Points to Consider When Using the Three-Factor Mixed Design

1. Each subject in a particular experimental group is tested under each treatment condition within both of the factors.
2. The number of subjects used in each experimental group is arbi-

trary. Again, while it is not necessary, it is best to have an equal number of subjects in each group. In most experiments, ten to fifteen subjects are assigned to each group.

3. The number of treatments given within each of the factors is arbitrary. However, it is rare that more than four or five treatments within either or both of the factors are compared.

Example

Consider a situation where two groups of subjects practice a coordination task using a pursuit rotor apparatus. All subjects in *both* groups practice for four 60-second trials with the rotor moving clockwise. Then all subjects practice for four 60-second trials with the rotor moving counterclockwise. (*Note:* If the experimenter decided to counterbalance and have half the subjects practice under the counterclockwise condition first, this would then be a Latin square design.)

The two experimental groups are formed as follows. One group practices under a schedule where the four practice trials (under both the clockwise and counterclockwise conditions) consist of 60 seconds of practice with 60 seconds of rest between each practice trial. The other group practices for 60 seconds with only 10 seconds of rest between each trial. Thus, the present experiment permits evaluation of (1) the effects of the amount of rest between practice trials, (2) the effects of the direction of rotation, and (3) changes across practice trials. The response measured for all subjects under all treatments is time on target during each 60-second trial.

Step 1. After the data have been collected, they are tabled as follows.

| | GROUP 1 (60-SECOND PRACTICE/60-SECOND REST) | | | | | | | |
| | Clockwise | | | | Counterclockwise | | | |
Subject	Trial 1	Trial 2	Trial 3	Trial 4	Trial 1	Trial 2	Trial 3	Trial 4
S_1	21	26	35	43	2	9	20	24
S_2	18	28	41	46	5	10	14	15
S_3	25	24	45	48	10	15	21	21
S_4	19	20	38	37	1	3	8	7
S_5	14	18	25	33	4	7	12	20
S_6	17	16	28	29	8	10	18	25
S_7	12	18	29	35	6	13	24	28
S_8	15	17	27	30	1	7	17	26
S_9	20	28	26	29	2	4	16	14
S_{10}	18	20	31	38	9	23	36	39
	GROUP 2 (60-SECOND PRACTICE/10-SECOND REST)							
S_{11}	15	17	26	30	11	14	17	20
S_{12}	6	9	17	24	15	14	19	22
S_{13}	22	23	28	26	10	18	16	13

	GROUP 2 (60-SECOND PRACTICE/10-SECOND REST)							
	Clockwise				Counterclockwise			
	Trial 1	Trial 2	Trial 3	Trial 4	Trial 1	Trial 2	Trial 3	Trial 4
S_{14}	18	26	29	32	16	18	15	21
S_{15}	11	16	15	18	2	1	6	9
S_{16}	14	22	22	32	7	11	10	14
S_{17}	19	21	24	29	9	12	11	17
S_{18}	16	17	17	22	13	17	21	20
S_{19}	12	16	25	23	4	9	14	19
S_{20}	10	12	16	17	8	10	15	15

Step 2. Add the scores for each treatment in both experimental groups. (*Note:* If you are using a calculator, Step 2 and Step 3 can be done at the same time.)

	GROUP 1							
	Clockwise				Counterclockwise			
Subject	Trial 1	Trial 2	Trial 3	Trial 4	Trial 1	Trial 2	Trial 3	Trial 4
S_1	21	26	35	43	2	9	20	24
S_2	18	28	41	46	5	10	14	15
⋮	⋮	⋮	⋮	⋮	⋮	⋮	⋮	⋮
Sums:	**179**	**215**	**325**	**368**	**48**	**101**	**186**	**219**
	GROUP 2							
S_{11}	15	17	26	30	11	14	17	20
S_{12}	6	9	17	24	15	14	19	22
⋮	⋮	⋮	⋮	⋮	⋮	⋮	⋮	⋮
Sums:	**143**	**179**	**219**	**253**	**95**	**124**	**144**	**170**

Step 3. Square each number in the entire table (Step 1), and then add the squares.

$$21^2 + 18^2 + \cdots + 15^2 = \mathbf{69{,}810}$$

Step 4. Add all the sums obtained in Step 2 to get the grand sum for the entire table.

$$179 + \cdots + 48 + \cdots + 143 + \cdots + 95 + \cdots + 170 = \mathbf{2968}$$

Step 5. Square the grand sum obtained in Step 4, and divide it by the total number of measures recorded—i.e., the total number of subjects times the number of measures recorded for each subject (in this example, $20 \times 8 = 160$). This yields the correction term.

$$\frac{2968^2}{160} = \frac{8{,}809{,}024}{160} = \mathbf{55{,}056.4}$$

Step 6. Computation of the total sum of squares (SS_t): Simply subtract the correction term (Step 5) from the sum of the squared values in Step 3.

$$69,810 - 55,056.4 = \mathbf{14,753.6} = SS_t$$

Between-Subjects Effects (Steps 7–9)

Step 7. Computation of the effects of the experimental conditions on overall performance (in the present example, the effects of 60- versus 10-second rest periods—SS_{groups}): First, add all the trial sums obtained in Step 2 for Group 1 and for Group 2.

$$179 + 215 + 325 + 368 + 48 + 101 + 186 + 219 =$$
$$\mathbf{1641} = \text{sum for Group 1}$$
$$143 + 179 + 219 + 253 + 95 + 124 + 144 + 170 =$$
$$\mathbf{1327} = \text{sum for Group 2}$$

Then, square each of these totals, divide each square by the number of scores added to get the total (80 in this example), and add the quotients.

$$\frac{1641^2}{80} + \frac{1327^2}{80} = \frac{4,453,810}{80} = \mathbf{55,672.6}$$

Then, subtract the correction term (Step 5) from the above value.

$$55,672.6 - 55,056.4 = \mathbf{616.2} = SS_{groups}$$

Step 8. Computation of the between-subjects sum of squares (SS_b): First, add the scores for each subject in each group.

Subject	Trial 1		Trial 2		Trial 3		Trial 4		Trial 1		Trial 2		Trial 3		Trial 4	Sums
						Group 1										
S_1	21	+	26	+	35	+	43	+	2	+	9	+	20	+	24	180
S_2	18	+	28	+	41	+	46	+	5	+	10	+	14	+	15	177
$\vdots$	$\vdots$		$\vdots$		$\vdots$		$\vdots$		$\vdots$		$\vdots$		$\vdots$		$\vdots$	$\vdots$
S_{10}	18	+	20	+	31	+	38	+	9	+	23	+	36	+	39	214
						Group 2										
S_{11}	15	+	17	+	26	+	30	+	11	+	14	+	17	+	20	150
S_{12}	6	+	9	+	17	+	24	+	15	+	14	+	19	+	22	126
$\vdots$	$\vdots$		$\vdots$		$\vdots$		$\vdots$		$\vdots$		$\vdots$		$\vdots$		$\vdots$	$\vdots$
S_{20}	10	+	12	+	16	+	17	+	8	+	10	+	15	+	15	103

Then, square the sum of each subject's scores, and add these squares.

$$180^2 + 177^2 + \cdots + 103^2 = \mathbf{460,482}$$

Then, divide the above value by the number of scores which were added to get the sum for each subject. (In the present example, 8 scores were added to get each of the above sums.)

$$\frac{460,482}{8} = 57,560.25$$

Then, subtract the correction term (Step 5) from the above value.

$$57,560.25 - 55,056.4 = \mathbf{2503.8} = SS_b$$

Step 9. Computation of the between-subjects error term (SS_{error_b}): Simply subtract the sum of squares for groups (Step 7) from the sum of squares for between subjects (Step 8).

$$2503.8 - 616.2 = \mathbf{1887.6} = SS_{error_b}$$

Within-Subjects Effects (Steps 10–20)

Step 10. Computation of the within-subjects sum of squares (SS_w): Simply subtract the between-subjects sum of squares (Step 8) from the total sum of squares (Step 6).

$$14,753.6 - 2503.8 = \mathbf{12,249.8} = SS_w$$

Step 11. Computation of the sum of squares for the first within-subjects factor (here, the clockwise versus counterclockwise rotation—$SS_{rotation}$): First, add the sums for the clockwise and for the counterclockwise conditions (Step 2), disregarding the trials and the rest-period dimensions.

$$170 + 215 + 325 + 368 + 143 + 179 + 219 + 253$$
$$= 1881 = \text{sum for clockwise}$$
$$48 + 101 + 186 + 219 + 95 + 124 + 144 + 170$$
$$= 1087 = \text{sum for counterclockwise}$$

Then, square each of these sums, and divide each square by the number of scores added to get each sum. (In this example 80 scores were added to obtain each sum.) Then, add the quotients.

$$\frac{1881^2}{80} + \frac{1087^2}{80} = \frac{4,719,730}{80} = \mathbf{58,996.6}$$

Then, subtract the correction term (Step 5) from the above value.

$$58,996.6 - 55,056.4 = \mathbf{3940.2} = SS_{rotation}$$

Step 12. Computation of the sum of squares for the second within-subjects factor (here, the effects of trials—SS_{trials}): First, sum the scores for each of the four trials (Step 2), disregarding the rotational-direction and rest-period dimensions:

$$170 + 143 + 48 + \ \ 95 = \ \ \textbf{465} = \text{sum for Trial 1}$$
$$215 + 179 + 101 + 124 = \ \ \textbf{619} = \text{sum for Trial 2}$$
$$325 + 219 + 186 + 144 = \ \ \textbf{874} = \text{sum for Trial 3}$$
$$368 + 253 + 219 + 170 = \textbf{1010} = \text{sum for Trial 4}$$

Then, square these sums, and divide each square by the number of scores added to get each of the above sums. (In this example, each of the above sums represents a total of 40 scores, since 10 subjects' scores were added to get each of the sums in Step 2.) Then, add the quotients.

$$\frac{465^2}{40} + \frac{619^2}{40} + \frac{874^2}{40} + \frac{1010^2}{40} + \frac{2,383,362}{40} = \textbf{59,584.0}$$

Then, subtract the correction term (Step 5) from the above value.

$$59,584.0 - 55,056.4 = \textbf{4527.6} = SS_{trials}$$

Step 13. Computation of the interaction between experimental groups (rest interval) and the first factor (rotational direction)— $SS_{groups \times rotation}$: First, add the sums for each treatment for each of the two groups (Step 2). In this example, disregard the trials dimension and obtain the sums for the clockwise treatment in Group 1 and the counterclockwise treatment in Group 1. Do the same for Group 2.

$$179 + 215 + 325 + 368 = \textbf{1087} = \text{sum for Group 1, clockwise}$$
$$143 + 179 + 219 + 253 = \textbf{794} \ \ = \text{sum for Group 2, clockwise}$$
$$\ \ 48 + 101 + 186 + 219 = \textbf{554} \ \ = \text{sum for Group 1, counterclockwise}$$
$$\ \ 95 + 124 + 144 + 170 = \textbf{533} \ \ = \text{sum for Group 2, counterclockwise}$$

Then, square these sums, and divide by the number of scores added to obtain each sum. (In the present example, each of the above sums represents a total of 40 scores.) Then add the quotients.

$$\frac{1087^2}{40} + \frac{794^2}{40} + \frac{554^2}{40} + \frac{533^2}{40} = \frac{2,403,010}{40} = \textbf{60,075.2}$$

Then, subtract the correction term (Step 5), the sum of squares for groups (Step 7), and the sum of squares for rotational direction (Step 11), from the above value.

$$60,075.2 - 55,056.4 - 616.2 - 3940.2 = \textbf{462.4} = SS_{groups \times rotation}$$

Step 14. Computation of the interaction between experimental groups (rest interval) and the second factor (trials)—$SS_{groups \times trials}$: First, add the sums for each trial for each of the two groups (Step 2). In the present example, disregard the rotational-direction dimension and obtain the sums for Trials 1, 2, 3, and 4 in Group 1. Do the same for Group 2.

$$179 + 48 = \mathbf{227} = \text{sum for Group 1, Trial 1}$$
$$215 + 101 = \mathbf{316} = \text{sum for Group 1, Trial 2}$$
$$325 + 186 = \mathbf{511} = \text{sum for Group 1, Trial 3}$$
$$368 + 219 = \mathbf{587} = \text{sum for Group 1, Trial 4}$$
$$143 + 95 = \mathbf{238} = \text{sum for Group 2, Trial 1}$$
$$179 + 124 = \mathbf{303} = \text{sum for Group 2, Trial 2}$$
$$219 + 144 = \mathbf{363} = \text{sum for Group 2, Trial 3}$$
$$253 + 170 = \mathbf{423} = \text{sum for Group 2, Trial 4}$$

Then, square each of these sums, and divide by the number of scores added to obtain each sum. (In the present example, each sum represents a total of 20 scores: 10 subjects and 2 scores per subject.) Then, add the quotients.

$$\frac{227^2}{20} + \frac{316^2}{20} + \cdots + \frac{423^2}{20} = \frac{1{,}216{,}226}{20} = \mathbf{60{,}811.3}$$

Then, subtract the correction (Step 5), the sum of squares for groups (Step 7), and the sum of squares for trials (Step 12) from the above value.

$$60{,}811.3 - 55{,}056.4 - 616.2 - 4527.6 = \mathbf{611.1} = SS_{\text{groups} \times \text{trials}}$$

Step 15. Computation of the interaction between the first factor (rotational direction) and the second factor (trials)—$SS_{\text{rotation} \times \text{trials}}$: First, add the sums for each trial for both rotational-direction conditions (Step 2). In the present example, disregard the groups effects and obtain the sum for Trials 1, 2, 3, and 4 under the clockwise condition. Then do the same for the counterclockwise condition.

$$179 + 143 = \mathbf{322} = \text{sum for Trial 1, clockwise}$$
$$215 + 179 = \mathbf{394} = \text{sum for Trial 2, clockwise}$$
$$325 + 219 = \mathbf{544} = \text{sum for Trial 3, clockwise}$$
$$368 + 253 = \mathbf{621} = \text{sum for Trial 4, clockwise}$$
$$48 + 95 = \mathbf{143} = \text{sum for Trial 1, counterclockwise}$$
$$101 + 124 = \mathbf{225} = \text{sum for Trial 2, counterclockwise}$$
$$186 + 144 = \mathbf{330} = \text{sum for Trial 3, counterclockwise}$$
$$219 + 170 = \mathbf{389} = \text{sum for Trial 4, counterclockwise}$$

Then, square each of these sums, and divide by the number of scores added to obtain each sum. (In the present example, each sum represents a total of 20 scores: 10 subjects and 2 scores per subject.) Then, add the quotients.

$$\frac{322^2}{20} + \frac{394^2}{20} + \cdots + \frac{389^2}{20} = \frac{1,271,792}{20} = \mathbf{63,589.6}$$

Then, subtract the correction term (Step 5), the sum of squares for rotation (Step 11), and the sum of squares for trials (Step 12) from the above value.

$$63,589.6 - 55,056.4 - 3940.2 - 4527.6 = \mathbf{65.4} = SS_{\text{rotation} \times \text{trials}}$$

Step 16. Computation of the interaction among groups, rotation, and trials $(SS_{\text{groups} \times \text{rotation} \times \text{trials}})$: First, square each sum obtained by Step 2, and divide each square by the number of scores added to obtain each sum. (In the present example, each sum represents a total of 10 scores.) Then add the quotients.

$$\frac{179^2}{10} + \frac{215^2}{10} + \cdots + \frac{170^2}{10} = \frac{652,874}{10} = \mathbf{65,287.4}$$

Then, from this value, subtract the correction term (Step 5), the sum of squares for groups (Step 7), the sum of squares for rotation (Step 11), the sum of squares for trials (Step 12), the sum of squares for groups by rotation (Step 13), the sum of squares for groups by trials (Step 14), and the sum of squares for rotation by trials (Step 15).

$$65,287.4 - 55,056.4 - 616.2 - 3940.2 - 4527.6$$
$$- 462.4 - 611.1 - 65.4 = \mathbf{8.1} = SS_{\text{groups} \times \text{rotation} \times \text{trials}}$$

Step 17. Computation of the total error for within subjects (SS_{error_w}): (*Note:* This error term is not used in any of the *F*-tests, but it is needed for later computations.) Subtract the final values of Steps 11, 12, 13, 14, 15, and 16 from the final value of Step 10.

$$12,249.8 - 3940.2 - 4527.6 - 462.4 - 611.1$$
$$- 65.4 - 8.1 = \mathbf{2635.0} = SS_{\text{error}_w}$$

Step 18. Computation of the error term for the rotation and for the rotation-by-groups interaction (SS_{error_1}): This requires that the data be retabled as follows. Sum the scores of each subject for the four trials within each rotational-direction condition.

	Group 1	
Subject	Clockwise	Counterclockwise
S_1	125 (21 + 26 + 35 + 43)	55 (2 + 9 + 20 + 24)
S_2	133 (18 + 28 + 41 + 46)	44 (5 + 10 + 14 + 15)
$\vdots$	$\vdots$	$\vdots$
S_{10}	107	107

	Group 2	
S_{11}	88	62
S_{12}	56	70
$\vdots$	$\vdots$	$\vdots$
S_{20}	55	48

Then, square each of these sums, and divide the squares by the number of scores added to obtain each sum. (In this example, each sum is based on 4 scores.) Then, add the quotients.

$$\frac{125^2}{4} + \frac{133^2}{4} + \cdots + \frac{48^2}{4} = \frac{254{,}140}{4} = \mathbf{63{,}535}$$

Then, from this value, subtract the correction term (Step 5), the sum of squares for between subjects (Step 8), the sum of squares for rotation (Step 11), and the sum of squares for rotation by groups (Step 13).

$$63{,}535 - 55{,}056.4 - 2503.8 - 3940.2 - 462.4 = \mathbf{1572.2} = SS_{error_1}$$

Step 19. Computation of the error term for trials and for the groups-by-trials interaction (SS_{error_2}): This requires that the data be retabled once again. Sum the scores of each subject on the four trials, disregarding the rotation-direction dimension.

	Group 1			
Subject	Trial 1	Trial 2	Trial 3	Trial 4
S_1	23(21 + 2)	35(26 + 9)	55(35 + 20)	67(43 + 24)
S_2	23(18 + 5)	38(28 + 10)	55(41 + 14)	61(46 + 15)
$\vdots$	$\vdots$	$\vdots$	$\vdots$	$\vdots$
S_{10}	27	43	67	77

	Group 2			
S_{11}	26	31	43	50
S_{12}	21	23	36	46
$\vdots$	$\vdots$	$\vdots$	$\vdots$	$\vdots$
S_{20}	18	22	31	32

Then, square all the sums in this table, and divide each square by the number of scores added to obtain each sum. (In the present example, 2 scores were added.) Then, add the quotients.

$$\frac{23^2}{2} + \frac{23^2}{2} + \cdots + \frac{32^2}{2} = \frac{126,418}{2} = \mathbf{63,209}$$

Then, from this value, subtract the correction term (Step 5), the sum of squares for between subjects (Step 8), the sum of squares for trials (Step 12), and the sum of squares for groups-by-trials (Step 14).

$$63,209 - 55,056.4 - 2503.8 - 4527.6 - 611.1 = \mathbf{510.1} = SS_{error_2}$$

Step 20. Computation of the error term for the trials-by-rotation interaction and for the trials-by-rotation-by-groups interaction (SS_{error_3}): Simply subtract the final values of Steps 18 and 19 from the final value of Step 17.

$$2635.0 - 1572.2 - 510.1 = \mathbf{552.7} = SS_{error_3}$$

Step 21. All the computations based directly on the data have been completed. However, to compute the F ratios, the degrees of freedom (df) are required. These are computed as follows:

df for SS_t	= the total number of scores entered in Step 1 minus 1.
	$160 - 1 = \mathbf{159}$
df for SS_b	= the total number of subjects used in the experiment minus 1.
	$20 - 1 = \mathbf{19}$
df for SS_{groups}	= the number of groups minus 1.
	$2 - 1 = \mathbf{1}$
df for SS_{error_b}	= the between-subjects df minus the groups df.
	$19 - 1 = \mathbf{18}$
df for SS_w	= the total df minus the between-subjects df.
	$159 - 19 = \mathbf{140}$
df for $SS_{rotation}$	= the number of different (rotation) conditions minus 1.
	$2 - 1 = \mathbf{1}$

df for SS$_{trials}$	= the number of trials given within each condition minus 1.
	$4 - 1 = \mathbf{3}$
df for SS$_{groups \times rotation}$	= the *df* for groups times the *df* for rotation.
	$1 \times 1 = \mathbf{1}$
df for SS$_{groups \times trials}$	= the *df* for groups times the *df* for trials.
	$1 \times 3 = \mathbf{3}$
df for SS$_{rotation \times trials}$	= the *df* for rotation times the *df* for trials.
	$1 \times 3 = \mathbf{3}$
df for SS$_{groups \times rotation \times trials}$	= the *df* for groups times the *df* for rotation times the *df* for trials.
	$1 \times 1 \times 3 = \mathbf{3}$
df for SS$_{error_w}$	= the *df* for within subjects minus the *dfs* for rotation, trials, groups-by-rotation, groups-by-trials, rotation-by-trials, and groups-by-rotation-by-trials.
	$140 - 1 - 3 - 1 - 3 - 3 - 3 = \mathbf{126}$
df for SS$_{error_1}$ (rotation and rotation-by-groups)	= the *df* for rotation, times the total number of subjects minus the number of groups.
	$1 \times (20 - 2) = \mathbf{18}$
df for SS$_{error_2}$ (trials and groups-by-trials)	= the *df* for trials, times the total number of subjects minus the number of groups.
	$3 \times (20 - 2) = \mathbf{54}$
df for SS$_{error_3}$ (rotation-by-trials and groups-by-rotation-by-trials	= the *df* for rotation, times the *df* for trials, times the total number of subjects minus the number of groups.
	$1 \times 3 \times (20 - 2) = \mathbf{54}$

Step 22. The mean squares are then computed as SS/df.

$$ms_{groups} = \frac{(Step\ 7)}{1} = \frac{616.2}{1} = \mathbf{616.2}$$

$$ms_{error_b} = \frac{(Step\ 9)}{18} = \frac{1887.6}{18} = \mathbf{104.9}$$

$$ms_{rotation} = \frac{(Step\ 11)}{1} = \frac{3940.2}{1} = \mathbf{3940.2}$$

$$ms_{trials} = \frac{(Step\ 12)}{3} = \frac{4527.6}{3} = \mathbf{1509.2}$$

$$ms_{groups \times rotation} = \frac{(Step\ 13)}{1} = \frac{462.4}{1} = \mathbf{462.4}$$

$$ms_{groups \times trials} = \frac{(Step\ 14)}{3} = \frac{611.1}{3} = \mathbf{203.7}$$

$$ms_{rotation \times trials} = \frac{(Step\ 15)}{3} = \frac{65.4}{3} = \mathbf{21.8}$$

$$ms_{groups \times rotation \times trials} = \frac{(Step\ 16)}{3} = \frac{8.1}{3} = \mathbf{2.7}$$

$$ms_{error_1} = \frac{(Step\ 18)}{18} = \frac{1572.2}{18} = \mathbf{87.3}$$

$$ms_{error_2} = \frac{(Step\ 19)}{54} = \frac{510.1}{54} = \mathbf{9.4}$$

$$ms_{error_3} = \frac{(Step\ 20)}{54} = \frac{552.7}{54} = \mathbf{10.2}$$

Step 23. The appropriate F ratios are as follows.

$$\text{Experimental groups: } \frac{ms_{groups}}{ms_{error_b}}$$

$$\text{Rotation direction: } \frac{ms_{rotation}}{ms_{error_1}}$$

$$\text{Trials: } \frac{ms_{trials}}{ms_{error_2}}$$

$$\text{Groups-by-rotation: } \frac{ms_{G \times R}}{ms_{error_1}}$$

$$\text{Groups-by-trials: } \frac{ms_{G \times T}}{ms_{error_2}}$$

$$\text{Rotation-by-trials: } \frac{ms_{R \times T}}{ms_{error_3}}$$

$$\text{Groups-by-rotation-by-trials: } \frac{ms_{G \times R \times T}}{ms_{error_3}}$$

Step 24. Table the results as follows.

Source	SS	df	ms	F	p
Total	14,753.6	159	—	—	—
Between subjects	2,503.8	19	—	—	—
Groups	616.2	1	616.2	5.88	<.05
Error$_b$	1,887.6	18	104.9	—	—
Within subjects	12,249.8	140	—	—	—
Rotation	3,940.2	1	3940.2	45.11	<.001
Trials	4,527.6	3	1509.2	159.77	<.001
G × R	462.4	1	462.4	5.29	<.05
G × T	611.1	3	203.7	21.56	<.001
R × T	65.4	3	21.8	2.13	n.s.
G × R × T	8.1	3	2.7	<1	n.s.
Error$_1$	1,572.2	18	87.3	—	—
Error$_2$	510.1	54	9.4	—	—
Error$_3$	552.7	54	10.2	—	—

Therefore, it is concluded (see Appendix E) that:

1. The overall performance of the groups was different.
2. The effects of directional change (rotation) were significant.
3. The subjects' performance changed across trials.
4. There was a significant groups-by-direction (rotation) interaction.
5. There was also a significant trials-by-groups interaction.

SECTION 2.10

Latin Square Design: Simple

This Latin square design is basically equivalent to the treatments-by-subjects design of Section 2.5, except that the effects of the order of treatment presentation can also be analyzed.

Points to Consider When Using the Simple Latin Square Design

1. All subjects in all the experimental subgroups receive the same experimental treatments. The only difference between the

subgroups is in the order of presentation of these treatments, which is systematically counterbalanced. For example, if three drugs were being tested, every subject would receive each of the drugs, but one third of the subjects would be given Drug 1 first; the second third, Drug 2 first; and the final third, Drug 3 first.

2. There should be an equal number of subjects in each of the experimental subgroups. In most experiments, five to fifteen subjects are assigned to each subgroup.

3. It is rare that more than four or five treatments are administered to each of the subjects. *Note:* The number of treatments is always equal to the number of experimental subgroups in a Latin square design.

Example

Assume that an experimenter wishes to determine the effects of three different shock intensities on the galvanic skin responses of a group of subjects. Since it seems reasonable to expect that the order in which the shocks are presented might affect the results, the experimenter decides to counterbalance the effects of the shock by using a Latin square design. To do this, one third of the subjects receive low-intensity shock followed by medium and then high. The second third of the subjects receive medium-intensity shock first followed by high and then low. The remaining third receive high, low, and then medium. The measure recorded is the number of responses above a certain criterion during the fifteen trials given under each shock condition.

In this experiment, the presentation of the shock intensities will be as follows. (The subgroups will hereafter be referred to simply as "groups.")

	Shock Intensity		
Order of Presentation	Low	Medium	High
First	Group 1	Group 2	Group 3
Second	Group 3	Group 1	Group 2
Third	Group 2	Group 3	Group 1

Or, rearranging the group and order-of-presentation dimensions, the above becomes:

	Shock Intensity		
Group	Low	Medium	High
Group 1	First	Second	Third
Group 2	Third	First	Second
Group 3	Second	Third	First

(*Note:* This rearrangement permits computations to be made more easily but does not affect the actual computational procedures.)

Step 1. After the experiment has been conducted, table the data as follows.

	Group 1 (Order: Low, Medium, High)		
Subject	Low Shock (First)	Medium Shock (Second)	High Shock (Third)
S_1	4	6	7
S_2	6	8	9
S_3	5	8	10
S_4	3	5	6
S_5	4	5	9
S_6	5	6	6
S_7	7	19	10

	Group 2 (Order: Medium, High, Low)		
	(Third)	(First)	(Second)
S_8	8	6	2
S_9	9	4	6
S_{10}	11	8	9
S_{11}	6	6	6
S_{12}	14	8	17
S_{13}	9	5	6
S_{14}	8	5	8

	Group 3 (Order: High, Low, Medium)		
	(Second)	(Third)	(First)
S_{15}	8	9	9
S_{16}	8	9	12
S_{17}	10	9	10
S_{18}	9	11	11
S_{19}	7	8	9
S_{20}	10	8	13
S_{21}	7	4	8

Step 2. Add the scores in each group for each shock intensity.

	Group 1		
Subject	Low (First)	Medium (Second)	High (Third)
S_1	4	6	7
S_2	6	8	9
⋮	⋮	⋮	⋮
Sums:	**34**	**57**	**57**

	Group 2		
Subject	(Third)	(First)	(Second)
S_8	8	6	2
S_9	9	4	6
$\vdots$	$\vdots$	$\vdots$	$\vdots$
Sums:	65	42	54

	Group 3		
	(Second)	(Third)	(First)
S_{15}	8	9	9
S_{16}	8	9	12
$\vdots$	$\vdots$	$\vdots$	$\vdots$
Sums:	59	58	72

Step 3. Obtain the sum for each group by adding the sums of the shock intensities.

$$34 + 57 + 57 = \mathbf{148} = \text{sum for Group 1}$$
$$65 + 42 + 54 = \mathbf{161} = \text{sum for Group 2}$$
$$59 + 58 + 72 = \mathbf{189} = \text{sum for Group 3}$$

Step 4. Add the scores for *each* subject in each group. (*Note:* If you are using a calculator, Step 5 can be done at the same time as Step 4.)

		Group 1			
Subject	Low		Medium	High	Sums
S_1	4	+	6	+ 7	17
S_2	6	+	8	+ 9	23
$\vdots$	$\vdots$		$\vdots$	$\vdots$	$\vdots$
S_7	7	+	19	+ 10	36
		Group 2			
S_8	8	+	6	+ 2	16
S_9	9	+	4	+ 6	19
$\vdots$	$\vdots$		$\vdots$	$\vdots$	$\vdots$
S_{14}	8	+	5	+ 8	21
		Group 3			
S_{15}	8	+	9	+ 9	26
S_{16}	8	+	9	+ 12	29
$\vdots$	$\vdots$		$\vdots$	$\vdots$	$\vdots$
S_{21}	7	+	4	+ 8	19

Step 5. Square each score in the entire table (Step 1), and add these squared values to get a grand sum of the squares. (Note that in the fol-

lowing example we are considering the scores as they go across—not down—the table.)

$$4^2 + 6^2 + 7^2 + 6^2 + 8^2 + 9^2 + \cdots + 7^2 + 4^2 + 8^2 = \mathbf{4500}$$

Step 6. Add the group sums (Step 3) to get the grand sum of the entire table.

$$148 + 161 + 189 = \mathbf{498}$$

Step 7. Square the grand sum and divide it by the total number of measures in the entire table—i.e., the number of subjects times the number of measures per subject ($21 \times 3 = 63$ in this example). This yields the correction term.

$$\frac{498^2}{63} = \frac{248{,}004}{63} = \mathbf{3936.571}$$

Step 8. Computation of the total sum of squares (SS_t): Simply subtract the correction term (Step 7) from the sum of the squares (Step 5).

$$4500 - 3936.571 = \mathbf{563.429} = SS_t$$

Between-Subjects Effects (Steps 9–11)

Step 9. Computation of the between-subjects sum of the squares (SS_b): First, square the sum of each subject's scores (Step 4), and add these squared values.

$$17^2 + 23^2 + \cdots + 19^2 = \mathbf{12{,}756}$$

Then, divide this value by the number of different shock intensities given each subject (i.e., the number of different orders).

$$\frac{12{,}756}{3} = \mathbf{4252}$$

Then, subtract the correction term (Step 7) from the above value.

$$4252 - 3936.571 = \mathbf{315.429} = SS_b$$

Step 10. Computation of the effects of groups—SS_G (i.e., the effects of order of presentation of the shock intensities on the overall group performance): First, square the sum of each of the experimental groups (Step 3), divide each squared value by the number of measures on which the sum was based, and then add these values.

$$\frac{148^2}{21} + \frac{161^2}{21} + \frac{189^2}{21} = \frac{83{,}546}{21} = \mathbf{3978.381}$$

Then, subtract the correction term (Step 7) from the above value.

$$3978.381 - 3936.571 = \textbf{41.810} = SS_G$$

Step 11. Computation of the between-subjects error term (SS_{error_b}): Simply subtract the SS_G (Step 10) from the SS_b (Step 9).

$$315.429 - 41.810 = \textbf{273.619} = SS_{error_b}$$

Within-Subjects Effects (Steps 12–16)

Step 12. Computation of the within-subjects sum of squares (SS_w): Simply subtract the SS_b (Step 9) from the SS_t (Step 8).

$$563.429 - 315.429 = \textbf{248.0} = SS_w$$

Step 13. Computation of the treatment effects (in this example, the effects of low- versus medium- versus high-intensity shock—SS_s): First, add the sums of all subjects when they received low shock (Step 2). Do the same for medium and high shock.

$$34 + 65 + 59 = \textbf{158} = \text{sum for low shock}$$
$$57 + 42 + 58 = \textbf{157} = \text{sum for medium shock}$$
$$57 + 54 + 72 = \textbf{183} = \text{sum for high shock}$$

Then, square each of these sums, divide by the number of measures on which each sum was based, and add the quotients.

$$\frac{158^2}{21} + \frac{157^2}{21} + \frac{183^2}{21} = \frac{83,102}{21} = \textbf{3957.238}$$

Then, subtract the correction term (Step 7) from the above value.

$$3957.238 - 3936.571 = \textbf{20.667} = SS_s$$

Step 14. Computation of the order effects—SS_O (i.e., the general effects of firstness versus secondness versus thirdness, irrespective of the particular shock intensity): First, add the sums of the shock intensities presented first, second, and third (Step 2).

$$34 + 42 + 72 = \textbf{148} = \text{sum of shocks presented first}$$
$$57 + 54 + 59 = \textbf{170} = \text{sum of shocks presented second}$$
$$57 + 65 + 58 = \textbf{180} = \text{sum of shocks presented third}$$

Then, square the sums, divide by the number of measures on which each sum was based, and then add the quotients.

$$\frac{148^2}{21} + \frac{170^2}{21} + \frac{180^2}{21} = \frac{83,204}{21} = \textbf{3962.095}$$

Then, subtract the correction term (Step 7) from the above value.

$$3962.095 - 3936.571 = \textbf{25.524} = SS_O$$

Step 15. Computation of the effects of a particular shock intensity being presented first versus second versus third—SS_{SO_w} (SS for shock $\times$ order within subjects): First, square the sums of each shock intensity in each of the nine experimental groups (Step 2), divide by the number of measures on which the sums were based (7 in this example), and then add the quotients.

$$\frac{34^2}{7} + \frac{57^2}{7} + \frac{57^2}{7} + \frac{65^2}{7} + \frac{42^2}{7} + \frac{54^2}{7} + \frac{59^2}{7}$$
$$+ \frac{58^2}{7} + \frac{72^2}{7} = \frac{28,588}{7} = \textbf{4084.0}$$

Then, from this value, subtract the correction term (Step 7).

$$4084 - 3936.571 = \textbf{147.429}$$

Then, from this value, subtract SS_G (Step 10), SS_S (Step 13), and SS_O (Step 14).

$$147.429 - 41.810 - 20.667 - 25.524 = \textbf{59.428} = SS_{SO_w}$$

Step 16. Computation of the within-subject error term (SS_{error_w}): Simply subtract SS_S (Step 13), SS_O (Step 14), and SS_{SO_w} (Step 15) from SS_w (Step 12).

$$248 - 20.667 - 25.524 - 59.428 = \textbf{142.381} = SS_{error_w}$$

Step 17. All computations based on the data have been completed:

$$
\begin{aligned}
SS_t &= 563.429 \text{ (Step 8)}\\
SS_b &= 315.429 \text{ (Step 9)}\\
SS_G &= 41.810 \text{ (Step 10)}\\
SS_{error_b} &= 273.619 \text{ (Step 11)}\\
SS_w &= 248.000 \text{ (Step 12)}\\
SS_S &= 20.667 \text{ (Step 13)}\\
SS_O &= 25.524 \text{ (Step 14)}\\
SS_{SO_w} &= 59.428 \text{ (Step 15)}\\
SS_{error_w} &= 142.381 \text{ (Step 16)}
\end{aligned}
$$

Since the F ratios are the ratios of mean squares (SS/df), the df must be determined, as follows.

df for SS_t = total number of measures recorded minus 1.

$$63 - 1 = \textbf{62}$$

df for SS_b = total number of subjects minus 1.

$$21 - 1 = \textbf{20}$$

df for SS_G = number of experimental subject groups minus 1.

$$3 - 1 = \textbf{2}$$

df for SS_{error_b} = df for SS_b minus the df for SS_G.

$$20 - 2 = \textbf{18}$$

df for SS_w = df for SS_t minus the df for SS_b.

$$62 - 20 = \textbf{42}$$

df for SS_S = number of treatment conditions minus 1.

$$3 - 1 = \textbf{2}$$

df for SS_O = number of different orders of presentation minus 1.

$$3 - 1 = \textbf{2}$$

df for SS_{SO_w} = number of treatment groups minus 2, times the number of orders minus 1.

$$(3 - 2) \times (3 - 1) = \textbf{2}$$

df for SS_{error_w} = df for SS_w minus the dfs for SS_S, SS_O, and SS_{SO_w}.

$$42 - 2 - 2 - 2 = \textbf{36}$$

Step 18. Compute the mean squares needed for the analysis (SS/df).

ms_t (This value is not needed.)

ms_b (This value is not needed.)

$$ms_G = \frac{(Step\ 10)}{2} = \frac{41.810}{2} = \textbf{20.905}$$

$$ms_{error_b} = \frac{(Step\ 11)}{18} = \frac{273.619}{18} = \textbf{15.201}$$

ms_w (This value is not needed.)

$$ms_S = \frac{(Step\ 13)}{2} = \frac{20.667}{2} = \textbf{10.333}$$

$$ms_O = \frac{(Step\ 14)}{2} = \frac{25.524}{2} = \textbf{12.762}$$

$$ms_{SO_w} = \frac{(\text{Step } 15)}{2} = \frac{59.428}{2} = 29.714$$

$$ms_{error_w} = \frac{(\text{Step } 16)}{36} = \frac{142.381}{36} = 3.955$$

Step 19. Compute the *F* ratios, and table the data as follows.

Source	SS	df	ms	F	p
Total	563.429	62	—	—	—
Between subjects	315.429	20	—	—	—
Groups (SO$_b$)	41.810	2	20.905	1.38	n.s.
Error$_b$	273.619	18	15.201	—	—
Within subjects	248.00	42	—	—	—
Shock (S)	20.667	2	10.333	2.61	n.s.
Order (O)	25.524	2	12.762	3.23	n.s.
Shock × order$_w$ (SO$_w$)	59.428	2	29.714	7.51	<.005*
Error$_w$	142.381	36	3.955	—	—

Therefore, it is concluded (see Appendix E) that the within-subjects shock and order effects were interactive. All other effects were nonsignificant.

SECTION 2.11
Latin Square Design: Complex

The complex Latin square design is essentially an extension of the simple Latin square. The basic difference is that one extra between-subjects dimension is added. The major advantage of this Latin square design is that it allows the use of many fewer separate groups of subjects to cover the treatment combinations of interest.

Points to Consider When Using the Complex Latin Square Design

1. Two or more main experimental subject groups are employed and are either *treated* differently or are *selected* on the basis of some

---

*The *p* value listed is taken from an *F* table for the appropriate degrees of freedom. However, according to Lindquist (1953, p. 280), in the instances where the source of variation is mixed (SO$_b$ and SO$_w$), each component should be tested at the X/2% level. In that case, it would be concluded that the SO$_w$ interaction is significant at the .025 rather than the .005 level.

criterion so that they differ (for example, high- versus medium-versus low-anxious subjects).

2. Within each of the main experimental subject groups, a number of subgroups are formed, corresponding to the size of the Latin square (i.e., if the square is 3×3, then each main group is divided into three subgroups).

3. An equal number of subjects should be assigned to each of the subgroups. In most experiments, five to fifteen subjects are assigned to each subgroup.

4. It is rare that more than four or five treatments are administered to each of the subjects. (*Note:* The number of treatments administered to each subject is always equal to the number of subgroups in a Latin square design.)

Example

Assume that an experimenter wishes to determine the effects of three different difficulties of verbal material on the learning ability of high-versus low-anxious subjects. However, she hypothesizes that if the three different difficulties of material were given to each subject, the order in which they were presented would affect the rate of learning. Therefore, instead of presenting the lists in the same order (for example, hard first, medium second, and easy last) to all the subjects in both experimental groups (high- versus low-anxious), she decides to counterbalance the difficulty effects by using a Latin square design. Consequently, within *both* the high- and low-anxiety conditions, the order of presentation of the lists is systematically varied so that an equal proportion of the subjects receive the hard, medium, and easy lists first, second, and third in counterbalanced order. The measure recorded by the experimenter is the number of trials needed to reach the criterion of one perfect recitation of each of the lists.

For the high- and low-anxious subjects, the presentation of the lists is as follows. (The subgroups will hereafter be referred to simply as "groups")

| Order of Presentation | Low-Anxious Subjects | | |
| | Type of List | | |
	Hard	Medium	Easy
First	Group 1	Group 2	Group 3
Second	Group 3	Group 1	Group 2
Third	Group 2	Group 3	Group 1
	High-Anxious Subjects		
First	Group 4	Group 5	Group 6
Second	Group 6	Group 4	Group 5
Third	Group 5	Group 6	Group 4

Or, rearranging the group and order-of-presentation dimensions for both the high- and low-anxious subjects, the above becomes:

Group	Low-Anxious Subjects		
		Type of List	
	Hard	Medium	Easy
Group 1	First	Second	Third
Group 2	Third	First	Second
Group 3	Second	Third	First
	High-Anxious Subjects		
Group 4	First	Second	Third
Group 5	Third	First	Second
Group 6	Second	Third	First

(*Note:* This rearrangement permits computations to be made more easily, but it in no way affects the actual computational procedures.)

Step 1. After the experiment has been completed, table the data as follows.

LOW-ANXIOUS SUBJECTS
Group 1 (Order: Hard, Medium, Easy)

Subject	Hard List (First)	Medium List (Second)	Easy List (Third)
S_1	20	16	9
S_2	14	15	7
S_3	18	10	11
S_4	16	12	10
S_5	18	9	12
S_6	19	14	10

Group 2 (Order: Medium, Easy, Hard)

	(Third)	(First)	(Second)
S_7	17	17	9
S_8	16	18	9
S_9	12	15	8
S_{10}	11	15	10
S_{11}	13	13	12
S_{12}	15	16	12

Group 3 (Order: Easy, Hard, Medium)

	(Second)	(Third)	(First)
S_{13}	19	9	11
S_{14}	14	8	13
S_{15}	13	6	14
S_{16}	14	12	10
S_{17}	16	10	15
S_{18}	18	16	13

HIGH-ANXIOUS SUBJECTS
Group 4 (Order: Hard, Medium, Easy)

Subject	Hard List (First)	Medium List (Second)	Easy List (Third)
S_{19}	26	16	7
S_{20}	23	16	6
S_{21}	21	11	8
S_{22}	19	11	6
S_{23}	24	10	8
S_{24}	27	12	10

Group 5 (Order: Medium, Easy, Hard)

	(Third)	(First)	(Second)
S_{25}	14	17	8
S_{26}	15	18	6
S_{27}	11	16	6
S_{28}	13	16	9
S_{29}	15	13	11
S_{30}	13	11	10

Group 6 (Order: Easy, Hard, Medium)

	(Second)	(Third)	(First)
S_{31}	19	10	10
S_{32}	15	9	11
S_{33}	14	7	6
S_{34}	13	12	10
S_{35}	15	10	9
S_{36}	17	11	8

Step 2. Add the scores in each group for each level of list difficulty.

LOW-ANXIOUS SUBJECTS
Group 1

Subject	Hard (First)	Medium (Second)	Easy (Third)
S_1	20	16	9
S_2	14	15	7
⋮	⋮	⋮	⋮
Sums:	**105**	**76**	**59**

Group 2

	(Third)	(First)	(Second)
S_7	17	17	9
S_8	16	18	9
⋮	⋮	⋮	⋮
Sums:	**84**	**94**	**60**

LOW-ANXIOUS SUBJECTS
Group 3

Subject	Hard (Second)	Medium (Third)	Easy (First)
S_{13}	19	9	11
S_{14}	14	8	13
$\vdots$	$\vdots$	$\vdots$	$\vdots$
Sums:	**94**	**61**	**76**

HIGH-ANXIOUS SUBJECTS
Group 4

Subject	Hard (First)	Medium (Second)	Easy (Third)
S_{19}	26	16	7
S_{20}	23	16	6
$\vdots$	$\vdots$	$\vdots$	$\vdots$
Sums:	**140**	**76**	**45**

Group 5

	(Third)	(First)	(Second)
S_{25}	14	17	8
S_{26}	15	18	6
$\vdots$	$\vdots$	$\vdots$	$\vdots$
Sums:	**81**	**91**	**50**

Group 6

	(Second)	(Third)	(First)
S_{31}	19	10	10
S_{32}	15	9	11
$\vdots$	$\vdots$	$\vdots$	$\vdots$
Sums:	**93**	**59**	**54**

Step 3. Obtain the sum for each group by adding the sums for each level of list difficulty.

$$105 + 76 + 59 = \mathbf{240} = \text{sum for Group 1}$$
$$84 + 94 + 60 = \mathbf{238} = \text{sum for Group 2}$$
$$94 + 61 + 76 = \mathbf{231} = \text{sum for Group 3}$$
$$140 + 76 + 45 = \mathbf{261} = \text{sum for Group 4}$$
$$81 + 91 + 50 = \mathbf{222} = \text{sum for Group 5}$$
$$93 + 59 + 54 = \mathbf{206} = \text{sum for Group 6}$$

Step 4. Add the scores for *each* subject in each group. (*Note:* If you are using a calculator, Step 5 can be done at the same time as Step 4.)

LOW-ANXIOUS SUBJECTS
Group 1

Subject	Hard		Medium		Easy	Sums
S_1	20	+	16	+	9	45
S_2	14	+	15	+	7	36
⋮	⋮		⋮		⋮	⋮
S_6	19	+	14	+	10	43

Group 2

S_7	17	+	17	+	9	43
S_8	16	+	18	+	9	43
⋮	⋮		⋮		⋮	⋮
S_{12}	15	+	16	+	12	43

Group 3

S_{13}	19	+	9	+	11	39
S_{14}	14	+	8	+	13	35
⋮	⋮		⋮		⋮	⋮
S_{18}	18	+	16	+	13	47

HIGH-ANXIOUS SUBJECTS
Group 4

Subject	Hard		Medium		Easy	Sums
S_{19}	26	+	16	+	7	49
S_{20}	23	+	16	+	6	45
⋮	⋮		⋮		⋮	⋮
S_{24}	27	+	12	+	10	49

Group 5

S_{25}	14	+	17	+	8	39
S_{26}	15	+	18	+	6	39
⋮	⋮		⋮		⋮	⋮
S_{30}	13	+	11	+	10	34

Group 6

S_{31}	19	+	10	+	10	39
S_{32}	15	+	9	+	11	35
⋮	⋮		⋮		⋮	⋮
S_{36}	17	+	11	+	8	36

Step 5. Square each score in the entire table (Step 1), and add these squared values to get a grand sum of the squared numbers.

$$20^2 + 16^2 + 9^2 + 14^2 + 15^2 + 7^2 + \cdots + 17^2 + 11^2 + 8^2 = \mathbf{20,106}$$

Step 6. Add the group totals (Step 3) to get the grand sum for the entire table.

$$240 + 238 + 231 + 261 + 222 + 206 = \mathbf{1398}$$

Step 7. Square the grand sum of Step 6, and divide it by the total number of measures recorded in Step 1—i.e., the number of subjects times the number of measures per subject ($36 \times 3 = 108$, in this example). This yields the correction term.

$$\frac{1398^2}{108} = \frac{1,954,404}{108} = \mathbf{18,096.333}$$

Step 8. Computation of the total sum of squares (SS_t): Simply subtract the correction term (Step 7) from the sum of squares obtained in Step 5.

$$20,106 - 18,096.333 = \mathbf{2009.667} = SS_t$$

Between-Subjects Effects (Steps 9–13)

Step 9. Computation of the between-subjects sum of squares (SS_b): First, square the sum of each subject's scores (Step 4), and add these squared values.

$$45^2 + 36^2 + \cdots + 36^2 = \mathbf{55,080}$$

Then, divide this value by the number of scores added to obtain each of the sums obtained in Step 4. (In the present example, each sum represents 3 scores.)

$$\frac{55,080}{3} = \mathbf{18,360}$$

Then, subtract the correction term from the above value.

$$18,360 - 18,096.333 = \mathbf{263.667} = SS_b$$

Step 10. Computation of the "pure" between-groups effects (SS_A): (The other "between" components in this design are partially between- and partially within-subjects sources of variance. In the present example, the overall effects of high versus low anxiety are not confounded.) First, add the group sums (Step 3) for the subjects in the low-anxious conditions (Groups 1, 2, and 3) and for subjects in the high-anxious conditions (Groups 4, 5, and 6).

$$240 + 238 + 231 = \mathbf{709} = \text{sum for low-anxious } Ss$$
$$261 + 222 + 206 = \mathbf{689} = \text{sum for high-anxious } Ss$$

Then, square each of these sums, divide by the total number of scores that make up each sum, and add the quotients.

$$\frac{709^2}{54} + \frac{689^2}{54} = \frac{977,402}{54} = \textbf{18,100.037}$$

Then, subtract the correction term (Step 7) from the above value.

$$18,100.037 - 18,096.333 = \textbf{3.704} = SS_A$$

Step 11. Computation of the between-subjects list-by-order interaction ($SS_{LO\ b}$): First add the sums for the groups that received the same order of presentation of the lists (Step 3), disregarding the effects of high versus low anxiety.

$$\begin{array}{rl} 240 = & \text{sum for Group 1} \\ +261 = & \text{sum for Group 4} \\ \hline 501 = & \text{sum for all } Ss \text{ who received the lists} \\ & \text{in the order: hard, medium, easy} \end{array}$$

$$\begin{array}{rl} 238 = & \text{sum for Group 2} \\ +222 = & \text{sum for Group 5} \\ \hline 460 = & \text{sum for all } Ss \text{ who received the lists} \\ & \text{in the order: easy, hard, medium} \end{array}$$

$$\begin{array}{rl} 231 = & \text{sum for Group 3} \\ +206 = & \text{sum for Group 6} \\ \hline 437 = & \text{sum for all } Ss \text{ who received the lists} \\ & \text{in the order: medium, easy, hard} \end{array}$$

Then, square the above sums, divide each squared value by the number of measures on which the sum was based, and then add the quotients.

$$\frac{501^2}{36} + \frac{460^2}{36} + \frac{437^2}{36} = \frac{653,570}{36} = \textbf{18,154.722}$$

Then, subtract the correction term (Step 7) from the above value.

$$18,154.722 - 18,096.333 = \textbf{58.389} = SS_{LOb}$$

Step 12. Computation of the between-subjects list-by-order-by-anxiety interaction (SS_{LOAb}): First, square the sum of each of the experimental groups (Step 3), divide each squared value by the number of measures on which the sum is based, and then add the quotients.

$$\frac{240^2}{18} + \frac{238^2}{18} + \frac{231^2}{18} + \frac{261^2}{18} + \frac{222^2}{18} + \frac{206^2}{18} = \frac{327,446}{18} = \textbf{18,191.444}$$

Then, from the above value, subtract the correction term (Step 7), the SS_A (Step 10), and the SS_{LO_b} (Step 11).

$$18{,}191.444 - 18{,}096.333 - 3.704 - 58.389 = \mathbf{33.018} = SS_{LOA_b}$$

Step 13. Computation of the between-subjects error term (SS_{error_b}): Simply subtract the SS_A (Step 10), SS_{LO_b} (Step 11), and SS_{LOA_b} (Step 12) from the SS_b (Step 9).

$$263.667 - 3.704 - 58.389 - 33.018 = \mathbf{168.556} = SS_{error_b}$$

Within-Subjects Effects (Steps 14–21)

Step 14. Computation of the within-subjects sum of squares (SS_w): Simply subtract the SS_b (Step 9) from the SS_t (Step 8)

$$2009.667 - 263.667 = \mathbf{1746.0} = SS_w$$

Step 15. Computation of the list effects (SS_L): (In this example, these are the effects of a hard versus medium versus easy list, disregarding the order and anxiety effects.) First, add the scores of all the subjects when they received the hard list (Step 2). Do the same for the medium and easy lists.

$$105 + 84 + 94 + 140 + 81 + 93 = \mathbf{597} = \text{sum for hard list}$$
$$76 + 94 + 61 + 76 + 91 + 59 = \mathbf{457} = \text{sum for medium list}$$
$$59 + 60 + 76 + 45 + 50 + 54 = \mathbf{344} = \text{sum for easy list}$$

Then, square each of these sums, divide by the number of measures on which each sum was based, and add the quotients.

$$\frac{597^2}{36} + \frac{457^2}{36} + \frac{344^2}{36} = \frac{683{,}594}{36} = \mathbf{18{,}988.722}$$

Then, subtract the correction term (Step 7) from the above value.

$$18{,}988.722 - 18{,}096.333 = \mathbf{892.389} = SS_L$$

Step 16. Computation of the order effects (SS_O): (In this example, these are the general effects of firstness versus secondness versus thirdness of list presentation, irrespective of anxiety level or the particular list presented.) First, add the sums of the lists presented first, second, and third (Step 2).

$$105 + 94 + 76 + 140 + 91 + 54 = \mathbf{560} = \text{sum of lists presented first}$$
$$76 + 60 + 94 + 76 + 50 + 93 = \mathbf{449} = \text{sum of lists presented second}$$
$$59 + 84 + 61 + 45 + 81 + 59 = \mathbf{389} = \text{sum of lists presented third}$$

Then, square these sums, divide by the number of measures on which each sum was based, and add the quotients.

$$\frac{560^2}{36} + \frac{449^2}{36} + \frac{389^2}{36} = \frac{666,522}{36} = 18,514.5$$

Then, subtract the correction term (Step 7) from the above value.

$$18,514.5 - 18,096.333 = 418.167 = SS_O$$

Step 17. Computation of the effects of each type of list being presented first versus second versus third, disregarding the effects of anxiety (SS_{LO_w}): First, add the sums of the high- and low-anxious groups for each of the treatment and order combinations (Step 2).

$$105 + 140 = 245 = \text{sum of hard list first}$$
$$76 + 76 = 152 = \text{sum of medium list second}$$
$$59 + 45 = 104 = \text{sum of easy list third}$$
$$94 + 91 = 185 = \text{sum of medium list first}$$
$$60 + 50 = 110 = \text{sum of easy list second}$$
$$84 + 81 = 165 = \text{sum of hard list third}$$
$$76 + 54 = 130 = \text{sum of easy list first}$$
$$94 + 93 = 187 = \text{sum of hard list second}$$
$$61 + 59 = 120 = \text{sum of medium list third}$$

Then, square each of the above sums, divide each square by the number of measures on which each sum is based, and then add the quotients.

$$\frac{245^2}{12} + \frac{152^2}{12} + \cdots + \frac{120^2}{12} = \frac{233,764}{12} = 19,480.333$$

Then, from this value, subtract the correction term (Step 7), the sum of squares for LO_b (Step 11), the sum of squares for lists (Step 15), and the sum of squares for order (Step 16).

$$19,480.333 - 18,096.333 - 58.389 - 892.389$$
$$- 418.167 = 15.055 = SS_{LO_w}$$

Step 18. Computation of the treatments-by-"pure"-between-groups interaction (in this example, the list-by-anxiety interaction—SS_{LA}): First, add the sums of the hard, medium, and easy lists within each of the anxiety conditions (Step 2), disregarding the effects of order of presentation of these lists.

$$105 + 84 + 94 = 283 = \text{sum of hard lists, low-anxious } Ss$$
$$76 + 94 + 61 = 231 = \text{sum of medium lists, low-anxious } Ss$$
$$59 + 60 + 76 = 195 = \text{sum of easy lists, low-anxious } Ss$$

$140 + 81 + 93 = \mathbf{314} = $ sum of hard lists, high-anxious Ss

$76 + 91 + 59 = \mathbf{226} = $ sum of medium lists, high-anxious Ss

$45 + 50 + 54 = \mathbf{149} = $ sum of easy lists, high-anxious Ss

Then, square the above sums, divide by the number of measures on which each sum was based, and add these quotients.

$$\frac{283^{24}}{18} + \frac{231^2}{18} + \frac{195^2}{18} + \frac{314^2}{18} + \frac{226^2}{18} + \frac{149^2}{18} = \frac{343,348}{18} = \mathbf{19,074.889}$$

Then, from this value, subtract the correction term (Step 7), the sum of squares for anxiety (Step 10), and the sum of squares for lists (Step 15).

$$19,074.889 - 18,096.333 - 3.704 - 892.389 - \mathbf{82.463} = SS_{LA}$$

Step 19. Computation of the order-by-"pure"-between-groups interaction (here, the order-by-anxiety interaction—SS_{OA}): First, add the sums of the lists presented first versus second versus third within each of the anxiety conditions (Step 2) disregarding the effects of list difficulty.

$105 + 94 + 76 = \mathbf{275} = $ sum of first lists, low-anxious Ss

$76 + 60 + 94 = \mathbf{230} = $ sum of second lists, low-anxious Ss

$59 + 84 + 61 = \mathbf{204} = $ sum of third lists, low-anxious Ss

$140 + 91 + 54 = \mathbf{285} = $ sum of first lists, high-anxious Ss

$76 + 50 + 93 = \mathbf{219} = $ sum of second lists, high-anxious Ss

$45 + 81 + 59 = \mathbf{185} = $ sum of third lists, high-anxious Ss

Then, square the above sums, divide by the number of measures on which each sum was based, and add these quotients.

$$\frac{275^2}{18} + \frac{230^2}{18} + \frac{204^2}{18} + \frac{285^2}{18} + \frac{219^2}{18} + \frac{185^2}{18} = \frac{333,552}{18} = \mathbf{18,530.667}$$

Then, from this value, subtract the correction term (Step 7), the sum of squares for anxiety (Step 10), and the sum of squares for order (Step 16).

$$18,530.667 - 18,096.333 - 3.704 - 418.167 = \mathbf{12.463} = SS_{OA}$$

Step 20. Computation of the within-subjects lists-by-order-by-anxiety interaction (SS_{LOA_w}): First, square the sums of each list within each experimental group (Step 2), divide by the number of measures on which each sum was based, and then add the quotients.

$$\frac{105^2}{6} + \frac{76^2}{6} + \frac{59^2}{6} + \frac{84^2}{6} + \frac{94^2}{6} + \frac{60^2}{6} + \cdots + \frac{93^2}{6}$$

$$+ \frac{59^2}{6} + \frac{54^2}{6} = \frac{117,896}{6} = \mathbf{19,649.333}$$

From the above value, subtract the correction term (Step 7), the sum of squares for anxiety (Step 10), the sum of squares LO_b (Step 11), the sum of squares LOA_b (Step 12), the sum of squares for lists (Step 15), the sum of squares for order (Step 16), the sum of squares LO_w (Step 17), the sum of squares LA (Step 18), and the sum of squares OA (Step 19).

$$19{,}649.333 - 18{,}096.333 - 3.704 - 58.389 - 33.018 - 892.389$$
$$- 418.167 - 15.055 - 82.463 - 12.463 = \mathbf{37.352} = SS_{LOA_w}$$

Step 21. Computation of the within-subjects error term (SS_{error_w}): From the within-subjects sum of squares (Step 14), subtract the sums of squares for lists (Step 15), order (Step 16), lists $\times$ order within (Step 17), lists $\times$ anxiety (Step 18), order $\times$ anxiety (Step 19), and list $\times$ order $\times$ anxiety (Step 20).

$$1746.000 - 892.389 - 418.167 - 15.055 - 82.463$$
$$- 12.463 - 37.352 = \mathbf{288.111} = SS_{error_w}$$

Step 22. All computations based on the data have been completed:

$$
\begin{aligned}
SS_t &= 2009.667 \text{ (Step 8)} \\
SS_b &= 263.667 \text{ (Step 9)} \\
SS_A &= 3.704 \text{ (Step 10)} \\
SS_{LO_b} &= 58.389 \text{ (Step 11)} \\
SS_{LOA_b} &= 33.018 \text{ (Step 12)} \\
SS_{error_b} &= 168.556 \text{ (Step 13)} \\
SS_w &= 1746.000 \text{ (Step 14)} \\
SS_L &= 892.389 \text{ (Step 15)} \\
SS_O &= 418.167 \text{ (Step 16)} \\
SS_{LO_w} &= 15.055 \text{ (Step 17)} \\
SS_{LA} &= 82.463 \text{ (Step 18)} \\
SS_{OA} &= 12.463 \text{ (Step 19)} \\
SS_{LOA_w} &= 37.352 \text{ (Step 20)} \\
SS_{error_w} &= 288.111 \text{ (Step 21)}
\end{aligned}
$$

Since the F ratios are ratios of mean squares (SS/df), the df must be determined, as follows.

df for SS_t = total number of measures recorded minus 1.

$$108 - 1 = \mathbf{107}$$

df for SS_b = total number of subjects minus 1.

$$36 - 1 = \mathbf{35}$$

df for SS_A = number of "pure" between-subjects groups minus 1.

$$2 - 1 = \mathbf{1}$$

df for SS_{LO_b} = number of treatments (lists) minus 1.

$$3 - 1 = \mathbf{2}$$

df for SS_{LOA_b} = number of treatments minus 1, times the number of "pure" between-subjects groups minus 1.

$$(3 - 1) \times (2 - 1) = \mathbf{2}$$

df for SS_{error_b} = df for SS_b minus the dfs for SS_A, SS_{LO_b}, and SS_{LOA_b}.

$$35 - 1 - 2 - 2 = \mathbf{30}$$

df for SS_W = df for SS_t minus the df for SS_b.

$$107 - 35 = \mathbf{72}$$

df for SS_L = number of treatment groups minus 1.

$$3 - 1 = \mathbf{2}$$

df for SO_O = number of orders of presentation minus 1.

$$3 - 1 = \mathbf{2}$$

df for SS_{LO_w} = number of treatments minus 1, times the number of orders minus 2.

$$(3 - 1) \times (3 - 2) = \mathbf{2}$$

df for SS_{LA} = number of treatments minus 1, times the number of "pure" between factors minus 1.

$$(3 - 1) \times (2 - 1) = \mathbf{2}$$

df for SS_{OA} = number of orders minus 1, times the number of "pure" between factors minus 1.

$$(3 - 1) \times (2 - 1) = \mathbf{2}$$

df for SS_{LOA_w} = number of treatments minus 1, times the number of "pure" between factors minus 1.

$$(3 - 1) \times (2 - 1) = \mathbf{2}$$

df for SS_{OA} = number of orders minus 1, times the number of "pure" between factors minus 1.

$$(3 - 1) \times (2 - 1) = \mathbf{2}$$

df for SS_{LOAw} = number of treatments minus 1, times the
number of orders minus 2, times the number of
"pure" between factors minus 1.

$$(3 - 1) \times (3 - 2) \times (2 - 1) = 2$$

df for SS_{errorw} = df for SS_w minus the dfs for SS_L, SS_O,
SS_{LOw}, SS_{LA}, SS_{OA}, and SS_{LOAw}.

$$72 - 2 - 2 - 2 - 2 - 2 - 2 = 60$$

Step 23. Computation of the mean squares needed for the analysis
$(ms = SS/df)$:

$$ms_t \qquad \text{(This value is not needed.)}$$

$$ms_b \qquad \text{(This value is not needed.)}$$

$$ms_A = \frac{(\text{Step } 10)}{1} = \frac{3.704}{1} = 3.704$$

$$ms_{LOb} = \frac{(\text{Step } 11)}{2} = \frac{58.389}{2} = 29.194$$

$$ms_{LOAb} = \frac{(\text{Step } 12)}{2} = \frac{33.018}{2} = 16.509$$

$$ms_{errorb} = \frac{(\text{Step } 13)}{30} = \frac{168.556}{30} = 5.619$$

$$ms_w \qquad \text{(This value is not needed.)}$$

$$ms_L = \frac{(\text{Step } 15)}{2} = \frac{892.389}{2} = 446.194$$

$$ms_O = \frac{(\text{Step } 16)}{2} = \frac{418.167}{2} = 209.083$$

$$ms_{LOw} = \frac{(\text{Step } 17)}{2} = \frac{15.055}{2} = 7.527$$

$$ms_{LA} = \frac{(\text{Step } 18)}{2} = \frac{82.463}{2} = 41.231$$

$$ms_{OA} = \frac{(\text{Step } 19)}{2} = \frac{12.463}{2} = 6.231$$

$$ms_{LOAw} = \frac{(\text{Step } 20)}{2} = \frac{37.352}{2} = 18.676$$

$$\mathrm{ms_{error_w}} = \frac{(\text{Step 21})}{60} = \frac{288.111}{60} = \mathbf{4.802}$$

Step 24. Compute the F ratios, and table the data as follows.

Source	SS	df	ms	F	p
Total	2009.667	107	—	—	—
Between subjects	263.667	35	—	—	—
Anxiety	3.704	1	3.704	.66	n.s.
List $\times$ order$_b$	58.389	2	29.194	5.20	<.025*
List $\times$ order$_b$ $\times$ anxiety$_b$	33.018	2	16.509	2.94	n.s.
Error$_b$	168.556	30	5.619	—	—
Within subjects	1746.000	72	—	—	—
List	892.389	2	446.194	92.92	<.001
Order	418.167	2	209.083	43.54	<.001
List $\times$ order$_w$	15.055	2	7.527	1.57	n.s.
List $\times$ anxiety	82.463	2	41.231	8.59	<.001
Order $\times$ anxiety	12.463	2	6.231	1.30	n.s.
List $\times$ order$_w$ $\times$ anxiety$_w$	37.352	2	18.676	3.89	<.05
Error$_w$	288.111	60	4.802	—	—

Therefore, it is concluded (see Appendix E) that:

1. The effects of list difficulty and order of presentation interacted.
2. The overall effects of list difficulty were significant.
3. The overall effects of order of presentation were significant.
4. The effects of anxiety and list difficulty interacted.

SECTION 2.12
Analysis of Variance for Dichotomous Data

This section and the one that follows are not designs of experiments but, rather, represent special situations in which the data recorded are dichotomous or rank ordered. The actual computational steps are the same as those presented in earlier sections. The only real difference is in the data collection and recording.

---

*The p values listed are those taken from F tables for the appropriate degrees of freedom. However, according to Lindquist (1953, p. 280), in those instances where the sources of variation are mixed (LO$_b$ and LO$_w$, and LOA$_b$ and LOA$_w$), each component should be tested at the $X/2\%$ level. In that case, it would be concluded that only the LO$_b$ of the mixed sources is significant (at the .05 rather than the .025 level).

Points to Consider When Analyzing Dichotomous Data

1. The design used and the analyses undertaken depend on the type of problem and the experimental treatments, or groups, employed.
2. The data recorded for each subject will be a "1" or a "0."

Example

The design used to demonstrate the analysis of dichotomous data will be the Repeated Measures Design presented in Section 2.5.

For this example, assume that three different automobiles are being evaluated by automotive experts. Several different questions are asked, but the question of particular interest is the final one, "If you had the necessary funds, would you purchase this car?" *Yes* responses are given a value of "1"; *No* responses are recorded as "0."

Step 1. Table the data as follows.

Expert Judges	Car #1	Car #2	Car #3
S_1	1	1	0
S_2	1	1	0
S_3	0	0	1
S_4	1	1	0
S_5	0	0	1
S_6	0	0	1
S_7	1	1	0
S_8	0	1	0
S_9	1	1	0
S_{10}	1	0	0

Turn to Section 2.5, and follow the instructions in Steps 2 through 16.

SECTION 2.13
Analysis of Variance for Rank-Order Data

This section also deals with the type of data recorded rather than focusing on experimental design. For the example in this section, the data represent rank ordering of preferences. As with analysis of dichotomous data, the computational steps are the same as those presented in earlier designs.

Points to Consider When Analyzing Rank-Order Data

1. The experimental design used and the analysis undertaken depend on the type of problem under investigation and the experimental treatments, or groups, employed.
2. The data recorded for each subject will be a rank or a series of ranks.

Example

The example used to demonstrate analysis of rank-order data will be the Repeated Measures Design of Section 2.5.

Assume that an experimenter wishes to determine whether there is a difference in taste preference for three brands of cola soft drinks. Appropriate experimental controls (randomized presentation, double blind, etc.) are used, and the data recorded are the rankings of the three drinks in order of preference. (*Note*: If the presentation of the cola drinks for tasting were systematically counterbalanced, the appropriate analysis would be the Latin Square presented in Section 2.10.)

Step 1. Table the data as follows.

Subject	Cola #1	Cola #2	Cola #3
S_1	3	2	1
S_2	2	3	1
S_3	1	2	3
S_4	2	3	1
S_5	3	2	1
S_6	2	1	3
S_7	2	3	1
S_8	3	1	2
S_9	3	2	1
S_{10}	2	3	1
S_{11}	1	2	3
S_{12}	3	2	1

Turn to Section 2.5, and follow Steps 2 through 16. (*Note:* An analysis of rank-order data will always result in a sum of squares for subjects [Step 10] equal to zero.)

Part 3

Supplemental Computations for Analysis of Variance

The next eleven sections present supplemental computations that are often needed to make a more complete and accurate analysis of data than is possible with the test in Parts 1 and 2. The eleven tests specifically treated in these sections are grouped below into three general types; a more detailed description of each test follows. This introductory material should help you to decide which test or tests will satisfy the demands of your particular data analysis.

1. Tests for equality of variances
2. Tests for "simple effects"—usually used when an interaction is found to be significant
3. Trend tests—used to determine whether linear, quadratic, and higher complex relationships are present in the data

Tests for Equality of Variances (Sections 3.1, 3.2, and 3.3)

Most statistical tests have a clearly stated requirement that population variances must be equal. Sections 3.1, 3.2, and 3.3 present simple tests to determine whether this is true. Any one of these tests may also be used to determine whether two experimental treatments produce an effect in terms of different variances.

Section 3.1 presents a test for equality of two independent variances. Section 3.2 is concerned with a test for equality of two related variances. (Related variances are variances computed from two measures of the same people or measures of matched people.) Section 3.3 presents a test for equality of several independent variances.

Tests for Simple Effects (Sections 3.4, 3.5, 3.6, 3.7, 3.8, 3.9, and 3.10)

When the more complex analyses, such as those presented in Part 2, are completed, a researcher often wishes to determine which specific means differ from each other. Sections 3.4–3.9 present several ways of dealing with such "simple effects."

There is, however, considerable controversy regarding which of several tests is most appropriate. The only point where there seems to be agreement is that in order to use t-tests for multiple comparisons, you must have stated the specific hypotheses regarding which means were expected to differ *before* the experiment was carried out. If you wait until the experiment has been completed, look at the data, and then do t-tests on only a few of the most widely different values, you are guilty of a severe violation of probability theory since the tabled probability values hold only when you have planned from the beginning of the data-gathering phase to compare specific means.

On the other hand, if you choose specific differences to compare after the data have been collected, you must use a test that will correct the probability for your choosing to state your comparisons after the fact. For a discussion of the complex issues regarding which of the several tests available gives the accurate probability estimate for the comparisons, the reader is referred to the textbooks listed in the Textbook Reference Chart.

The computational procedures of all of the tests are highly similar. The t-test (Section 3.4), Tukey (Section 3.7), and the Scheffé (Section 3.8) tests all use a single critical difference for making all comparisons, but the last two tests use tables which correct for the total number of means being compared. The Duncan's (Section 3.5) and Newman-Keuls' (Section 3.6), on the other hand, employ the so-called layer approach where the critical difference is determined by the range of the specific two means being compared. The Dunnett test (Section 3.9) is a specialized test for comparing experimental group means only with the control group rather than making all possible comparisons.

The same example will be used in Sections 3.4–3.9 to demonstrate the computation of the various tests. A table following Section 3.9 will summarize the results and the differences in conclusions that would be drawn from use of the various tests.

When the very complex analyses of Sections 2.6–2.11 are used, there are several complete factorial or completely randomized analyses that may be computed *before* the more simple comparisons of two means are made. Section 3.10 presents a series of questions that might be asked and subsequently answered in such complex research situations.

Tests for Trend (Section 3.11)

Occasionally, a researcher is confronted with a problem which requires describing the shape of the data points, or the type of equation that best fits the several points. In this case, the analyses presented in Section 3.11 would best serve this need.

SECTION 3.1

Test for Difference Between Variances of Two Independent Samples (Test for Homogeneity of Independent Variances)

There are two cases in which it is desirable to do a significance test of the difference between two independent variances: (1) when an experimental hypothesis is concerned with the variability of the samples and (2) when there is doubt concerning the requirement of equality of variances in a mean-difference test. In either case, the following procedure may be followed in testing the hypothesis of equal variances.

Step 1. Presumably, you already have the two variances (or the two variance estimates, i.e., mean squares). If you do not, go to Section 1.2, and follow the instructions in the first six steps. The value arrived at in Step 6 will be the variance for one of the samples. When you have both variances, go on to Step 2 of this section.

Step 2. Obtain the F ratio by placing the larger variance of Step 1 over the smaller. For example, suppose the two variances are

$$\text{variance}_1 = 126.81$$
$$\text{variance}_2 = 35.27$$

Then,

$$F = \frac{126.81}{35.27} = 3.60$$

Step 3. Before the F value can be evaluated for significance, the degrees of freedom associated with each variance must be determined. In all cases, the degrees of freedom for a variance (actually a population-variance estimate) will be one less than the number of cases on which the variance is based. For example, suppose variance_1 of Step 2 had been computed from a sample of 41 cases. Then,

$$df \text{ for variance}_1 = 41 - 1 = 40$$

In the same way, suppose variance_2 had been computed from a sample of 31 cases. Then,

$$df \text{ for variance}_2 = 31 - 1 = 30$$

Step 4. The F value of Step 2 (3.60 in this example) has the two degrees of freedom 40 and 30 associated with it. From tabled F values (see Appendix E), we find that when $df = 40/30$,* F values larger than 2.47 are significant at the .01 level. Since the ratio of the variances in the present example equals 3.60, it is concluded that the variances are not homogeneous (equal).

However, there is one slight complication due to the fact that we selected *only* the larger variance as the numerator of the F ratio. A correction is needed to take care of the selection of only the larger variance as the numerator of F. In all cases, this correction is accomplished very simply by multiplying the probability value by 2. Thus, an F value of 2.47, with degrees of freedom equal to 24/30, is considered significant at the .02 level rather than at the .01 level.

SECTION 3.2

Test for Difference Between Variances of Two Related Samples (Test for Homogeneity of Related Variances)

There are two cases in which it is desirable to do a significance test of the difference between two related (correlated) variances: (1) when an experimental hypothesis is concerned with the variability of the samples and (2) when there is doubt concerning the requirement of equality of variances in a mean-difference test. In either case, the following procedure may be followed in testing the hypothesis of equal variances.

Step 1. Presumably, you already have the two variances (or the two variance estimates, i.e., mean squares). If you do not, go to Section 1.2 and follow the instructions in the first six steps. The value arrived at in Step 6 will be the variance for one of the samples. When you have both variances, go on to Step 2 of this section.

Step 2. Since these samples are related (correlated), the degree of the relationship must be determined. Compute the correlation between the pairs of scores. This is done by following the directions from Step 1 through Step 18 in Section 4.1. For the present example, suppose the computed correlation is +.64.

---

*Since the df of 40/30 are not listed, go to the next closest df values. In this instance, it would be 24/30.

Step 3. Subtract the smaller variance of Step 1 from the larger variance. For example, suppose

$$\text{variance}_1 = 65.27$$
$$\text{variance}_2 = 243.61$$

Then,

$$243.61 - 65.27 = \textbf{178.34}$$

Step 4. Multiply the two variances together.

$$243.61 \times 65.27 = \textbf{15,900.425}$$

Step 5. There were two groups of scores from which the two variances of Step 1 were computed. Subtract 2 from the number of cases in one group. For example, suppose each group is composed of 36 scores.

$$36 - 2 = \textbf{34}$$

Step 6. Take the square root of the final value of Step 5.

$$\sqrt{34} = \textbf{5.831}$$

Step 7. Multiply the final value of Step 6 by the final value of Step 3.

$$5.831 \times 178.34 = \textbf{1039.9}$$

Step 8. Subtract the result of Step 2 from the number 1. (*Note:* The number 1 is always used.)

$$1 - .64 = \textbf{.36}$$

Step 9. Multiply the final value of Step 4 by the number 4. (*Note:* The number 4 is always used.)

$$4 \times 15,900.425 = \textbf{63,601.7}$$

Step 10. Multiply the value obtained in Step 9 by the result of Step 8.

$$63,601.7 \times .36 = \textbf{22,896.612}$$

Then, take the square root of the product.

$$\sqrt{22,896.612} = \textbf{151.316}$$

Step 11. Divide the final value of Step 7 by the final value of Step 10. This yields the *t* statistic.

$$t = \frac{1039.9}{151.316} = \textbf{6.87}$$

Step 12. The degrees of freedom for the t is the result of Step 5 (34 in this example). A t with 34 degrees of freedom that is larger than 2.042 (see Appendix B) is significant at the .05 level using a two-tailed test. If the computed t value is large enough to be significant, this means that the variances are very likely really different.

SECTION 3.3

Test for Differences Among Several Independent Variances (*F*-Maximum Test for Homogeneity of Variances)*

There are two cases in which it is desirable to do a significance test of the differences among several independent variances: (1) when an experimental hypothesis is concerned with the variability of the samples and (2) when there is doubt concerning the requirement of equality of variances in a mean-difference test. In either case, the following procedure may be followed in testing the hypothesis of equal variances. There is one restriction on the use of this F_{max} test: the sizes of all the samples *must be the same.*

Step 1. Presumably, you already have the several variances (or the variance estimates, i.e., mean squares). If you do not, go to Section 1.2, and follow the instructions in the first six steps. The value arrived at in Step 6 will be the variance for one of the samples. When you have all the variances, go on to Step 2 of this section.

Step 2. Obtain the F_{max} ratio by placing the largest variance of Step 1 over the smallest. For example, suppose we had computed five variances in Step 1:

$$\text{variance}_1 = 18.99$$

$$\text{variance}_2 = 27.62$$

$$\text{variance}_3 = 92.63$$

$$\text{variance}_4 = 77.38$$

$$\text{variance}_5 = 26.40$$

---

*A Master's thesis completed at Ohio University in 1967 by Mr. Henry Winkler, titled "An empirical sampling study testing the robustness of several tests of homogeneity of variance and Type I errors in the simple randomized analysis of variance," dealt with the power of the F_{max}, Bartlett's, Cochran's, and Levine's two tests for homogeneity of variance. His results indicate that the F_{max} test and Bartlett's test are the most robust and are to be preferred over the others. Because of its simplicity, the F_{max} is probably the best choice.

Then,

$$F_{\max} = \frac{92.63}{18.99} = 4.88$$

Step 3. The number of degrees of freedom is always one less than the number of cases in each sample. Suppose that the size of each sample is 18. Then,

$$df = 18 - 1 = 17$$

Step 4. The significance of the $F_{\max}$ ratio computed in Step 2 is obtained from a table of $F_{\max}$ values (see Appendix I). We must know both the number of degrees of freedom (17 in this example) and the number of variances being considered (5 in this example). From the table of $F_{\max}$ values, we see that when $df = 17$ and the number of variances = 5, a value larger than 4.37 is significant at the .05 level. Since the value computed in Step 2 above (4.88) is larger than 4.37, we conclude that these variances are not homogeneous (i.e., they are different).

SECTION 3.4
The *t*-Test for Differences Among Several Means

The *t*-test is used to determine which specific means differ significantly from each other *if* these differences have been hypothesized *prior to the collection of the experimental data.*

Example

The set of data used to demonstrate the computational procedures is that used for the simple randomized design presented in Section 2.1. The summary table presented in the final analysis was:

Source	SS	df	ms	F	p
Total	1132	47	—	—	—
Between groups	567	3	189.00	14.71	<.001
Within groups	565	44	12.84	—	—

The basic computational formula for computing the critical difference is

$$\text{C. diff.} = t \sqrt{\frac{2 \ ms_{\text{within gp. error}}}{n \ (\text{per gp.})}}$$

where t is obtained from the t tables presented in Appendix B.

The significance of the overall F indicates only that out of the means of the four groups, at least two differ. The problem is to determine *which* pairs of means are significantly different.

Step 1. Obtain the mean score for each group. (In this example, there are 12 scores in each group.)

Sum of Group 1 = 111 $\dfrac{111}{12}$ = **9.25**   = mean of Group 1

Sum of Group 2 = 88 $\dfrac{88}{12}$ = **7.33**   = mean of Group 2

Sum of Group 3 = 156 $\dfrac{156}{12}$ = **13.00** = mean of Group 3

Sum of Group 4 = 195 $\dfrac{195}{12}$ = **16.25** = mean of Group 4

Step 2. Derive the standard error of the difference among means by

$$\sqrt{\dfrac{2\ ms_{within}}{n\ (per\ gp.)}}$$

For this example,

$$\sqrt{\dfrac{2 \times 12.84}{12}} = \sqrt{2.14} = 1.463$$

If there are unequal numbers of cases in some groups, you must account for those unequal *ns*. If the *ns* are not too disparate (as a rough rule of thumb, if there are not more than twice as many in the largest group as in the smallest group), you may use $\overline{n}$, the harmonic mean of the numbers, where

$$\overline{n} = \dfrac{number\ of\ groups}{\dfrac{1}{n_1} + \dfrac{1}{n_2} + \dfrac{1}{n_3} + \cdots + \dfrac{1}{n_i}}$$

For example, suppose that $n_1 = 16$, $n_2 = 12$, $n_3 = 8$, and $n_4 = 12$. Then,

$$\overline{n} = \dfrac{4}{\frac{1}{16} + \frac{1}{12} + \frac{1}{8} + \frac{1}{12}} = \dfrac{4}{.353} = 11.33$$

So the standard error of the differences between means would be

$$\sqrt{\dfrac{2(12.84)}{11.33}} = \sqrt{2.266} = 1.51$$

If the *ns* are widely different, the standard error must be computed separately for each comparison as

$$\sqrt{ms_{within}\left(\frac{1}{n_1} + \frac{1}{n_2}\right)}$$

For example, if $n_1 = 18$, $n_2 = 6$, and $ms_{within} = 12.84$, we would have

$$\sqrt{12.84(\tfrac{1}{18} + \tfrac{1}{6})} = \sqrt{\frac{12.84}{18} + \frac{12.84}{6}}$$

$$= \sqrt{.713 + 2.14} = \sqrt{2.853} = \mathbf{1.69}$$

A new standard error must then be computed for each new *t*-test.

Step 3. The needed *t* value is obtained from Appendix B ($p = .05$). The table is entered at the row headed by the *df* for the mean square within groups. In the present example, the *df* for ms_{within} is 44. Since 44 is not listed, go to the next closest *df*, which is 40. Reading into the table, for 40 *df*, at the .05 level,

$$t = \mathbf{2.02}$$

Step 4. Now multiply the *t* value of Step 3 by the value obtained in Step 2. This gives the critical difference against which the several mean differences will be compared.

$$2.02 \times 1.463 = \mathbf{2.955} = \text{C. diff.}$$

Step 5. Testing the differences between the various means: In all instances, if the difference between any two means (see Step 1) is larger than the critical difference (in this case, 2.955), then the means are assumed to be significantly different.

Group 1 vs. Group 4 (C. diff. = 2.955):

$16.25 - 9.25 = \mathbf{7.00}$ (significant)

Group 1 vs. Group 3 (C. diff. = 2.955):

$13.00 - 9.25 = \mathbf{3.75}$ (significant)

Group 1 vs. Group 2 (C. diff. = 2.955):

$9.25 - 7.33 = \mathbf{1.92}$ (nonsignificant)

Group 2 vs. Group 4 (C. diff. = 2.955):

$16.25 - 7.33 = \mathbf{8.92}$ (significant)

Group 2 vs. Group 3 (C. diff. = 2.955):

$$13.00 - 7.33 = \textbf{5.67} \text{ (significant)}$$

Group 3 vs. Group 4 (C. diff. = 2.955):

$$16.25 - 13.00 = \textbf{3.25} \text{ (significant)}$$

SECTION 3.5
Duncan's Multiple-Range Test

The Duncan's multiple-range test involves the so-called layer or stair-step approach to the making of multiple comparisons. Instead of making all comparisons in relation to a single critical difference (as in the *t*-test), the size of the critical difference is adjusted depending upon whether the two means being compared are adjacent, or whether one or more other means fall between those being compared.

Example

The data used to demonstrate the computational procedures are those used for the simple randomized design presented in Section 2.1. Recall that the summary table was as follows:

Source	SS	df	ms	F	p
Total	1132	47	—	—	—
Between groups	567	3	189.00	14.71	<.001
Within groups	565	44	12.84	—	—

The basic computational formula for the Duncan's multiple-range test is

$$\text{C. diffs.} = k_r \sqrt{\frac{ms_{\text{within gp. error}}}{n \text{ (per gp.)}}}$$

where the several k values are obtained from tables presented in Appendix J.

Step 1. Obtain the mean score for each group. (In this example, there are 12 scores per group.)*

---

*For unequal ns, see C. Y. Kramer, Extension of multiple range tests to group means with unequal numbers of replications, *Biometrics*, 1956, **12**, 307–310.

Sum of Group 1 = 111 $\dfrac{111}{12}$ = **9.25**   = mean of Group 1

Sum of Group 2 =   88 $\dfrac{88}{12}$ = **7.33**   = mean of Group 2

Sum of Group 3 = 156 $\dfrac{156}{12}$ = **13.00** = mean of Group 3

Sum of Group 4 = 195 $\dfrac{195}{12}$ = **16.25** = mean of Group 4

Step 2. Derive the standard error of the means by

$$\sqrt{\frac{\text{ms}_{\text{within}}}{n \text{ (per gp.)}}}$$

For this example,

$$\sqrt{\frac{12.84}{12}} = \sqrt{1.07} = 1.03$$

If there are unequal numbers of cases in some groups, you must account for those unequal *n*s. If the *n*s are not too disparate (as a rough rule of thumb, if there are not more than twice as many in the largest group as in the smallest group), you may use $\bar{n}$, the harmonic mean of the numbers, where

$$\bar{n} = \frac{\text{number of groups}}{\dfrac{1}{n_1} + \dfrac{1}{n_2} + \dfrac{1}{n_3} + \cdots + \dfrac{1}{n_i}}$$

For example, suppose that $n_1 = 16$, $n_2 = 12$, $n_3 = 8$, and $n_4 = 12$. Then,

$$\bar{n} = \frac{4}{\frac{1}{16} + \frac{1}{12} + \frac{1}{8} + \frac{1}{12}} = \frac{4}{.353} = \textbf{11.33}$$

So the standard error of the means would be

$$\sqrt{\frac{12.84}{11.33}} = \sqrt{1.13} = \textbf{1.06}$$

Step 3. Go to Appendix J ($p = .05$) to obtain the "significant studentized ranges." The table is entered at the row headed by the *df* for the mean square within groups. In the present example, the *df* for ms$_{\text{within}}$ is 44. Since 44 *df* is not listed, use the next closest *df*, which is 40. Reading to the right in this row, for the different ranges of means (r) to be compared, the values are

$$r = 2: \quad k = 2.86$$
$$r = 3: \quad k = 3.01$$
$$r = 4: \quad k = 3.10$$

Step 4. Multiply each significant range by the value obtained in Step 2. The values obtained are the minimum critical differences for the given ranges of comparisons.

$$r = 2: \quad \text{C. diff.}_2 = 2.86 \times 1.03 = \textbf{2.95}$$
$$r = 3: \quad \text{C. diff.}_3 = 3.01 \times 1.03 = \textbf{3.10}$$
$$r = 4: \quad \text{C. diff.}_4 = 3.10 \times 1.03 = \textbf{3.19}$$

Step 5. Testing the differences between the various means: The means must first be *ranked* from *smallest to largest* in order that the appropriate critical difference for each comparison will be used. In this example, the means are ranked as: Gp. 2 = 7.33, Gp. 1 = 9.25, Gp. 3 = 13.00, Gp. 4 = 16.25. In each instance, if the difference between the means (see Step 1) is larger than the minimum for that range (Step 4), it is considered to be significant.

Group 2 vs. Group 4 (C. diff.$_4$ = 3.19)

$$16.25 - 7.33 = \textbf{8.92} \text{ (significant)}$$

Group 2 vs. Group 3 (C. diff.$_2$ = 3.10)

$$13.00 - 7.33 = \textbf{5.67} \text{ (significant)}$$

Group 2 vs. Group 1 (C. diff.$_2$ = 2.95)

$$9.25 - 7.33 = \textbf{1.92} \text{ (nonsignificant)}$$

Group 1 vs. Group 4 (C. diff.$_3$ = 3.10)

$$16.25 - 9.25 = \textbf{7.00} \text{ (significant)}$$

Group 1 vs. Group 3 (C. diff.$_2$ = 2.95)

$$13.00 - 9.25 = \textbf{3.75} \text{ (significant)}$$

Group 3 vs. Group 4 (C. diff.$_2$ = 2.95)

$$16.25 - 13.00 = \textbf{3.25} \text{ (significant)}$$

Therefore, it is concluded that Groups 2 and 4, 2 and 3, 1 and 4, 1 and 3, and 3 and 4 differ significantly in terms of their overall performance.

SECTION 3.6
The Newman-Keuls' Multiple-Range Test

A somewhat more stringent layer approach for making multiple comparisons is the Newman-Keuls' multiple-range test. The computational
steps for the Newman-Keuls' are the same as for the Duncan's, except
that the tabled values that determine the critical ranges are different.
Many statisticians argue that the mathematical bases for the Newman-
Keuls' tables are more defensible than those for the Duncan's.

Example

Again, the data from the simple randomized design of Section 2.1 are
used:

Source	SS	df	ms	F	p
Total	1132	47	—	—	—
Between groups	567	3	189.00	14.71	<.001
Within groups	565	44	12.84	—	—

The basic computational formula for the Newman-Keuls' is

$$\text{C. diffs.} = q_r \sqrt{\frac{\text{ms}_{\text{within gp. error}}}{n \text{ (per gp.)}}}$$

where the several q values are obtained from tables presented in Appendix K.

Step 1. Obtain the mean score for each group. (In this example, there
are 12 scores per group.)*

$$\text{Sum of Group 1} = 111 \quad \frac{111}{12} = 9.25 = \text{mean of Group 1}$$

$$\text{Sum of Group 2} = 88 \quad \frac{88}{12} = 7.33 = \text{mean of Group 2}$$

$$\text{Sum of Group 3} = 156 \quad \frac{156}{12} = 13.00 = \text{mean of Group 3}$$

$$\text{Sum of Group 4} = 195 \quad \frac{195}{12} = 16.25 = \text{mean of Group 4}$$

---

*For unequal ns, see C. Y. Kramer, Extension of multiple range tests to group means
with unequal numbers of replications, *Biometrics*, 1956, **12**, 307–10.

Step 2. Derive the standard error of the means by

$$\sqrt{\frac{ms_{within}}{n \text{ (per gp.)}}}$$

For this example,

$$\sqrt{\frac{12.84}{12}} = \sqrt{1.07} = \mathbf{1.03}$$

If there are unequal numbers of cases in some groups, you must account for those unequal *ns*. If the *ns* are not too disparate (as a rough rule of thumb, if there are not more than twice as many in the largest group as in the smallest group), you may use $\bar{n}$, the harmonic mean of the numbers, where

$$\bar{n} = \frac{\text{number of groups}}{\dfrac{1}{n_1} + \dfrac{1}{n_2} + \dfrac{1}{n_3} + \cdots + \dfrac{1}{n_i}}$$

For example, suppose that $n_1 = 16$, $n_2 = 12$, $n_3 = 8$, and $n_4 = 12$. Then,

$$\bar{n} = \frac{4}{\frac{1}{16} + \frac{1}{12} + \frac{1}{8} + \frac{1}{12}} = \frac{4}{.353} = \mathbf{11.33}$$

So the standard error of the means would be

$$\sqrt{\frac{12.84}{11.33}} = \sqrt{1.13} = \mathbf{1.06}$$

Step 3. Go to Appendix K ($p = .05$) to obtain the "significant studentized ranges." The table is entered at the row headed by the *df* for the error mean square within groups. In the present example, the *df* for ms_{within} is 44. Since 44 *df* is not listed, use the next closest *df*, which is 40. Reading to the right in this row, for the different ranges of means (r) to be compared, the values are

$$
\begin{aligned}
r = 2: &\quad q = 2.86 \\
r = 3: &\quad q = 3.44 \\
r = 4: &\quad q = 3.79
\end{aligned}
$$

Step 4. Multiply each significant range by the value obtained in Step 2. The values obtained are the minimum critical differences for the given ranges of comparisons.

$$
\begin{aligned}
r = 2: &\quad \text{C. diff.}_2 = 2.86 \times 1.03 = \mathbf{2.95} \\
r = 3: &\quad \text{C. diff.}_3 = 3.44 \times 1.03 = \mathbf{3.54} \\
r = 4: &\quad \text{C. diff.}_4 = 3.79 \times 1.03 = \mathbf{3.90}
\end{aligned}
$$

Step 5. Testing the differences between the various means: The means must first be ranked from smallest to largest in order that the range (r value) for each difference can be determined. In this example, the means are ranked as: Gp. 2 = 7.33, Gp. 1 = 9.25, Gp. 3 = 13.00, Gp. 4 = 16.25. In each instance, if the difference between the means (see Step 1) is larger than the minimum for that range (Step 4), it is considered to be significant.

Group 2 vs. Group 4 (C. diff.$_4$ = 3.90)

16.25 − 7.33 = **8.92** (significant)

Group 2 vs. Group 3 (C. diff.$_3$ = 3.54)

13.00 − 7.33 = **5.67** (significant)

Group 2 vs. Group 1 (C. diff.$_2$ = 2.95)

9.25 − 7.33 = **1.92** (nonsignificant)

Group 1 vs. Group 4 (C. diff.$_3$ = 3.54)

16.25 − 9.25 = **7.00** (significant)

Group 1 vs. Group 3 (C. diff.$_2$ = 2.95)

13.00 − 9.25 = **3.75** (significant)

Group 3 vs. Group 4 (C. diff.$_2$ = 2.95)

16.25 − 13.00 = **3.25** (significant)

Therefore, it is concluded that Groups 2 and 4, 2 and 3, 1 and 4, 1 and 3, and 3 and 4 differ significantly in terms of their overall performance.

SECTION 3.7
The Tukey Test

Tukey's test is similar to the t-test in that a single critical difference is computed for making all the comparisons between the means, but it is also similar to the Duncan's and Newman-Keuls' in that the size of this critical difference is dependent upon the number of comparisons which are being made. The computational procedures are basically the same as those for the Newman-Keuls' multiple-range test.

Example

The set of data used to demonstrate the computational procedures is again that used for the simple randomized design presented in Section 2.1. Data from the final analysis of that example are:

Source	SS	df	ms	F	p
Total	1132	47	—	—	—
Between groups	567	3	189.00	14.71	<.001
Within groups	565	44	12.84	—	—

The basic computational formula for Tukey's test is

$$\text{C. diff.} = q_r \sqrt{\frac{\text{ms}_{\text{within gp. error}}}{n \text{ (per gp.)}}}$$

where the needed q value is obtained from tables presented in Appendix K.

Step 1. Obtain the mean score for each group. (In this example, there are 12 scores per group.)*

Sum of Group 1 = 111 $\dfrac{111}{12}$ = **9.25** = mean of Group 1

Sum of Group 2 = 88 $\dfrac{88}{12}$ = **7.33** = mean of Group 2

Sum of Group 3 = 156 $\dfrac{156}{12}$ = **13.00** = mean of Group 3

Sum of Group 4 = 195 $\dfrac{195}{12}$ = **16.25** = mean of Group 4

Step 2. Derive the standard error of the means by

$$\sqrt{\frac{\text{ms}_{\text{within}}}{n \text{ (per gp.)}}}$$

For this example,

$$\sqrt{\frac{12.84}{12}} = \sqrt{1.07} = \textbf{1.03}$$

---

*For unequal ns, see C. Y. Kramer, Extension of multiple range tests to group means with unequal numbers of replications, *Biometrics*, 1956, **12**, 307–10.

If there are unequal numbers of cases in some groups, you must account for those unequal ns. If the ns are not too disparate (as a rough rule of thumb, if there are not more than twice as many in the largest group as in the smallest group), you may use $\bar{n}$, the harmonic mean of the numbers, where

$$\bar{n} = \frac{\text{number of groups}}{\frac{1}{n_1} + \frac{1}{n_2} + \frac{1}{n_3} + \cdots + \frac{1}{n_i}}$$

For example, suppose that $n_1 = 16$, $n_2 = 12$, $n_3 = 8$, and $n_3 = 12$. Then,

$$\bar{n} = \frac{4}{\frac{1}{16} + \frac{1}{12} + \frac{1}{8} + \frac{1}{12}} = \frac{4}{.353} = \mathbf{11.33}$$

So the standard error of the means would be

$$\sqrt{\frac{12.84}{11.33}} = \sqrt{1.13} = \mathbf{1.06}$$

Step 3. The needed q value is obtained from Appendix K ($p = .05$). The table is entered at the column headed by the number of means to be compared and the row headed by the df for the mean square within groups. In the present example, the df for ms_{within} is 44. Since 44 is not listed, go to the next closest df, which is 40. At the .05 level for four means ($r = 4$) and 40 df, we find:

$$r = 4: \qquad q = 3.79$$

Step 4. Now multiply the q value of Step 3 by the value obtained in Step 2. This gives the critical difference against which the several mean differences will be compared.

$$3.79 \times 1.03 = \mathbf{3.90} = \text{C. diff.}$$

Step 5. Testing the differences between the various means: In all instances, if the difference between any two means (see Step 1) is larger than the critical difference (in this case 3.90), then the means are assumed to be significantly different.

Group 1 vs. Group 4 (C. diff. $= 3.90$)

$$16.25 - 9.25 = \mathbf{7.00} \text{ (significant)}$$

Group 1 vs. Group 3 (C. diff. $= 3.90$)

$$13.00 - 9.25 = \mathbf{3.75} \text{ (nonsignificant)}$$

Group 1 vs. Group 2 (C. diff. $= 3.90$)

$$9.25 - 7.33 = \mathbf{1.92} \text{ (nonsignificant)}$$

Group 2 vs. Group 4 (C. diff. = 3.90)

$$16.25 - 7.33 = \textbf{8.92} \text{ (significant)}$$

Group 2 vs. Group 3 (C. diff. = 3.90)

$$13.00 - 7.33 = \textbf{5.67} \text{ (significant)}$$

Group 3 vs. Group 4 (C. diff. = 3.90)

$$16.25 - 13.00 = \textbf{3.25} \text{ (nonsignificant)}$$

SECTION 3.8
Scheffé's Test

The Scheffé test is similar to the Tukey test in that a single critical difference is computed regardless of whether the means to be compared are immediately adjacent, or whether several other means fall between those being compared. The major computational difference is that Scheffé's test makes use of F tables versus the studentized range tables for the other tests. The Scheffé test is also more stringent than the Tukey, and thus the probability of Type I error is lower. The computational steps of the Scheffé are basically the same as those for the t-test.

Example

The data used to demonstrate the computational procedures are those from the simple randomized design of Section 2.1. The summary table of that analysis was as follows:

Source	SS	df	ms	F	p
Total	1132	47	—	—	—
Between groups	567	3	189.00	14.71	<.001
Within groups	565	44	12.84	—	—

The basic computational formula is

$$\text{C. diff.} = \sqrt{(a-1)F_{df_a-1, df_{error}}} \sqrt{\frac{2ms_{\text{within gps. error}}}{n \text{ (per gp.)}}}$$

where a = number of groups to be compared,

F = the tabled F value (see Appendix E) for the appropriate df.

Step 1. Obtain the mean score for each group. (In this example, there are 12 scores in each group.)

Sum of Group 1 = 111 $\dfrac{111}{12}$ = 9.25 = mean of Group 1

Sum of Group 2 = 88 $\dfrac{88}{12}$ = 7.33 = mean of Group 2

Sum of Group 3 = 156 $\dfrac{156}{12}$ = 13.00 = mean of Group 3

Sum of Group 4 = 195 $\dfrac{195}{12}$ = 16.25 = mean of Group 4

Step 2. Derive the standard error of the difference among means by

$$\sqrt{\frac{2\ \mathrm{ms_{within}}}{n\ (\text{per gp.})}}$$

For this example,

$$\sqrt{\frac{2 \times 12.84}{12}} = \sqrt{2.14} = 1.46$$

If there are unequal numbers of cases in some groups, you must account for those unequal *ns*. If the *ns* are not too disparate (as a rough rule of thumb, if there are not more than twice as many of the largest group as in the smallest group), you may use $\bar{n}$, the harmonic mean of the numbers, where

$$\bar{n} = \frac{\text{number of groups}}{\dfrac{1}{n_1} + \dfrac{1}{n_2} + \dfrac{1}{n_3} + \cdots + \dfrac{1}{n_i}}$$

For example, suppose that $n_1 = 16$, $n_2 = 12$, $n_3 = 8$, and $n_4 = 12$. Then,

$$\bar{n} = \frac{4}{\frac{1}{16} + \frac{1}{12} + \frac{1}{8} + \frac{1}{12}} = \frac{4}{.353} = 11.33$$

So the standard error of the differences between means would be

$$\sqrt{\frac{2(12.84)}{11.33}} = \sqrt{2.266} = 1.51$$

If the *ns* are widely different, the standard error must be computed separately for each comparison as

$$\sqrt{\mathrm{ms_{within}}\left(\frac{1}{n_1} + \frac{1}{n_2}\right)}$$

For example, if $n = 16$, $n = 6$, and $ms_{within} = 12.84$, we would have

$$\sqrt{12.84(\tfrac{1}{18} + \tfrac{1}{6})} = \sqrt{\frac{12.84}{18} + \frac{12.84}{6}}$$

$$= \sqrt{.713 + 2.14} = \sqrt{2.853} = \textbf{1.69}$$

Step 3. The needed F value is obtained from Appendix E $(p = .05)$ for df equal to the number of groups minus 1, and the df for mean square$_{within}$. In this example, the appropriate df would be 3 and 44. However, since 44 is not listed, go to the next smallest tabled df, which would be 40.

$$F(3, 40) = 2.84$$

Step 4. Multiply this value by the number of groups minus 1; i.e.,

$$4 - 1 = 3 \times 2.84 = \textbf{8.52}$$

Step 5. Take the square root of the above value and multiply that answer by the value obtained in Step 2. The final result is the critical difference needed for comparison of the several means.

$$\sqrt{8.52} = 2.92$$

Then,

$$2.92 \times 1.46 = \textbf{4.26} = \text{C. diff.}$$

Step 6. Testing the differences between the various means: In all instances, if the difference between any two means (see Step 1) is larger than the critical difference (in this case, 4.09), then the means are assumed to be significantly different.

Group 1 vs. Group 4 (C. diff. = 4.26)

$$16.25 - 9.25 = \textbf{7.00} \text{ (significant)}$$

Group 1 vs. Group 3 (C. diff. = 4.26)

$$13.00 - 9.25 = \textbf{3.75} \text{ (nonsignificant)}$$

Group 1 vs. Group 2 (C. diff. = 4.26)

$$7.33 - 9.25 = \textbf{1.92} \text{ (nonsignificant)}$$

Group 2 vs. Group 4 (C. diff. = 4.26)

$$16.25 - 7.33 = \textbf{8.92} \text{ (significant)}$$

Group 2 vs. Group 3 (C. diff. = 4.26)

$$13.00 - 7.33 = \textbf{5.67} \text{ (significant)}$$

Group 3 vs. Group 4 (C. diff. = 4.26)

$$16.25 - 13.00 = \textbf{3.25} \text{ (nonsignificant)}$$

SECTION 3.9
Dunnett's Test

The Dunnett is a specialized multiple-comparison test that is used in those instances when the only comparisons desired are between the control group and the various experimental groups (i.e., all possible comparisons involving all experimental groups are not made). The computational procedures of the Dunnett test are basically the same as those for the t-test and Scheffé's. As in those tests, a single critical difference is calculated against which all mean comparisons are made.

Example

The data used to demonstrate the computational procedures are those used for the simple randomized design of Section 2.1. For this example, however, it will be assumed that Group 1 was the control group and Groups 2, 3, and 4 were experimental groups.

Source	SS	df	ms	F	p
Total	1132	47	—	—	—
Between groups	567	3	189.00	14.71	<.001
Within groups	565	44	12.84	—	—

The computational formula for Dunnett's test is

$$\text{C. diff.} = d_r \sqrt{\frac{2 \; ms_{\text{within gps. error}}}{n \; (\text{per gp.})}}$$

where the needed d value is obtained from the tables presented in Appendix L.

Step 1. Obtain the mean score for each group. (In this example, there are 12 scores in each group.)

$$\text{Sum of Group 1} = 111 \qquad \frac{111}{12} = 9.25 \; = \text{mean of Group 1}$$

$$\text{Sum of Group 2} = 88 \qquad \frac{88}{12} = 7.33 \; = \text{mean of Group 2}$$

$$\text{Sum of Group 3} = 156 \qquad \frac{156}{12} = 13.00 = \text{mean of Group 3}$$

$$\text{Sum of Group 4} = 195 \qquad \frac{195}{12} = 16.25 = \text{mean of Group 4}$$

Step 2. Derive the standard error of the difference among means by

$$\sqrt{\frac{2 \ ms_{within}}{n \ (per \ gp.)}}$$

For this example,

$$\sqrt{\frac{2 \times 12.84}{12}} = \sqrt{2.14} = 1.46$$

If there are unequal numbers of cases in some groups, you must account for those unequal ns. If the ns are not too disparate (as a rough rule of thumb, if there are not more than twice as many in the largest group as in the smallest group), you may use $\overline{n}$, the harmonic mean of the numbers, where

$$\overline{n} = \frac{\text{number of groups}}{\dfrac{1}{n_1} + \dfrac{1}{n_2} + \dfrac{1}{n_3} + \cdots + \dfrac{1}{n_i}}$$

For example, suppose that $n_1 = 16$, $n_2 = 12$, $n_3 = 8$, and $n_4 = 12$. Then,

$$\overline{n} = \frac{4}{\frac{1}{16} + \frac{1}{12} + \frac{1}{8} + \frac{1}{12}} = \frac{4}{.353} = 11.33$$

So the standard error of the differences between means would be

$$\sqrt{\frac{2(12.84)}{11.33}} = \sqrt{2.266} = 1.51$$

If the ns are widely different, the standard error must be computed separately for each comparison as

$$\sqrt{ms_{within} \left(\frac{1}{n_1} + \frac{1}{n_2} \right)}$$

For example, if $n = 18$, $n = 6$, and $ms_{within} = 12.84$, we would have

$$\sqrt{12.84(\tfrac{1}{18} + \tfrac{1}{6})} = \sqrt{\frac{12.84}{18} + \frac{12.84}{6}}$$

$$= \sqrt{.713 + 2.14} = \sqrt{2.853} = 1.69$$

Step 3. The needed d value is obtained from Appendix L $(p = .05)$. The table is entered at the column headed by the number of means to be compared and the row headed by the df for the error mean square within groups. In the present example, the df for ms_{within} is 44, but since that value is not listed, read the table at the next smallest value, which is 40. Thus for 4 groups and 40 *df*, the *d* value is:

$$r = 4: \qquad d = 2.44$$

Step 4. Multiply the value obtained in Step 2 by the value of Step 3 to obtain the critical difference.

$$1.46 \times 2.44 = \mathbf{3.56} = \text{C. diff.}$$

Step 5. Testing the differences between the control and experimental groups: In all instances, if the difference between the means is larger than the critical difference obtained in Step 4 (3.56 in this example), the means are assumed to be significantly different.

Group 1 vs. Group 2 (C. diff. = 3.56)

$9.25 - 7.33 = \mathbf{1.92}$ (nonsignificant)

Group 1 vs. Group 3 (C. diff. = 3.56)

$13.00 - 9.25 = \mathbf{3.75}$ (significant)

Group 1 vs. Group 4 (C. diff. = 3.56)

$16.25 - 9.25 = \mathbf{7.00}$ (significant)

See page 133 for a summary of the differences in the several test results.

SECTION 3.10
F-Tests for Simple Effects

The *F*-test for simple effects is most commonly used when one of the factorial or mixed designs yields a significant interaction.

Example

To demonstrate some of the more commonly used tests of simple effects, a complete analysis of the two-factor mixed design of Section 2.7 will be presented. Although the presentation in this section represents a considerably more detailed analysis than is usually carried out, the several operations are presented for the sake of completeness.

SUMMARY TABLE

	t-test	Duncan's	Newman-Keuls'	Tukey's	Scheffé's	Dunnett's
	C. diff. = 2.94	C. diff. = $2.94_{(r=2)}$ $3.10_{(r=3)}$ $3.19_{(r=4)}$	C. diff. = $2.94_{(r=2)}$ $3.54_{(r=3)}$ $3.90_{(r=4)}$	C. diff. = 3.90	C. diff. = 4.26	C. diff. = 3.56
Gp 2 vs. 4 7.33 vs. 16.25	Significant	Significant	Significant	Significant	Significant	—
Gp 2 vs. 3 7.33 vs. 13.00	Significant	Significant	Significant	Significant	Significant	—
Gp 2 vs. 1 7.33 vs. 9.25	Nonsignificant	Nonsignificant	Nonsignificant	Nonsignificant	Nonsignificant	Nonsignificant
Gp 1 vs. 4 9.25 vs. 16.25	Significant	Significant	Significant	Nonsignificant	Nonsignificant	Significant
Gp 1 vs. 3 9.25 vs. 13.00	Significant	Significant	Significant	Significant	Significant	Significant
Gp 3 vs. 4 13.00 vs. 16.25	Significant	Significant	Significant	Nonsignificant	Nonsignificant	—

First, recall the final analysis of the mixed design of Section 2.7:

Source	SS	df	ms	F	p
Total	467	98	—	—	—
Between subjects	181	32	—	—	—
Conditions	39	2	19.50	4.12	<.05
Error$_b$	142	30	4.73	—	—
Within subjects	286	66	—	—	—
Trials	170	2	85.00	52.46	<.001
Trials × Conditions	19	4	4.75	2.93	<.05
Error$_w$	97	60	1.62	—	—

The conclusions drawn from this analysis were:

1. Meaningfulness of material significantly affected overall amount learned.
2. The overall tendency of the subjects was to change (learn) as a function of practice (trials).
3. The subjects in the various groups learned at different rates.

The specific questions to be answered by the analyses of simple effects relate to the differences noted above and can be stated as follows.

First Questions. The first two questions relate to the significant main effects of "Conditions" and "Trials." In the case of "Conditions," the significant F ratio merely means that of the three groups, at least two differ significantly. The problem is to determine which specific pairs differ. Similarly, the significant overall F for "Trials" means that on at least two trials a change in performance took place. Thus, further tests are needed to determine on which specific trials a significant change occurred. In both instances, the Newman-Keuls' range test will be applied to analyze the difference.* The graphical representation of the group means (Question 1a) being compared is presented in Figure 1. (*Note:* In most instances where there is a significant interaction, as is the case in this example, analyses of the main effects would not be undertaken and the researcher would proceed directly to the third question, listed below.)

Question 1a. Which of the three groups (conditions) differ significantly from each other in terms of overall performance?

Step 1. Obtain the overall mean for each of the three groups. The sums obtained in Step 3 of the original analysis (Section 2.7) must be divided by the number of subjects in each of the experimental groups

---

*Obviously, Scheffé's, etc., tests could also be used.

Figure 1

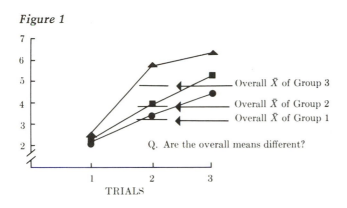

times the number of scores recorded for each subject (in this case, 33, since 33 measures were recorded in each group).

Sum of Group 1 = 111 $\frac{111}{33}$ = **3.36** = mean of Group 1

Sum of Group 2 = 124 $\frac{124}{33}$ = **3.76** = mean of Group 2

Sum of Group 3 = 160 $\frac{160}{33}$ = **4.85** = mean of Group 3

Step 2. Derive the standard error by

$$\sqrt{\frac{ms_{error_b}}{n \text{ (measures per group)}}} = \sqrt{\frac{4.73}{33}} = \sqrt{.1433} = \mathbf{.38}$$

Step 3. The significant studentized ranges are obtained from Appendix K of Newman-Keuls' multiple-range values ($p = .05$). The table is entered at the row headed by the *df* for ms_{error_b}, which is 30. Reading to the right in this row, for the different ranges of means (*r*) to be compared, the values are

$$r = 2: \quad q = 2.89$$
$$r = 3: \quad q = 3.49$$

Step 4. Multiply each significant range by the value obtained in Step 2. This gives the minimum critical differences against which the actual differences in the means will be compared.

$$r = 2: \quad \text{C. diff.} = 2.89 \times .38 = \mathbf{1.10}$$
$$r = 3: \quad \text{C. diff.} = 3.49 \times .38 = \mathbf{1.32}$$

Step 5. Testing the difference between the various means: First, rank the means from smallest to largest. In each instance, if the difference between the means (see Step 1) is larger than the minimum for that range (Step 4), it is considered to be significant.

Group 3 vs. Group 1 $(R_3 = 1.32)$

$4.85 - 3.36 = $ **1.49** (significant)

Group 3 vs. Group 2 $(R_2 = 1.10)$

$4.85 - 3.76 = $ **1.09** (nonsignificant)

Group 2 vs. Group 1 $(R_2 = 1.10)$

$3.76 - 3.36 = $ **.40** (nonsignificant)

Therefore, it is concluded that only Groups 1 and 3 differ significantly in terms of their overall performance.

Question 1b. On which of the three trials is there a significant change in the average performance of the three groups?

Step 1. Obtain the overall mean for each of the three trials. The sums obtained in Step 13 of the original analysis (Section 2.7) must be divided by the number of subjects in each group times the number of groups (in this case, 33 since there were 11 subjects in each of the three groups).

Sum of Trial 1 = 73 $\dfrac{73}{33} = $ **2.21** = overall mean of Trial 1

Sum of Trial 2 = 146 $\dfrac{146}{33} = $ **4.42** = overall mean of Trial 2

Sum of Trial 3 = 176 $\dfrac{176}{33} = $ **5.33** = overall mean of Trial 3

Step 2. Derive the standard error by:

$$\sqrt{\frac{ms_{error_w}}{n(\text{total number of measures per trial})}}$$

$$= \sqrt{\frac{1.62}{33}} = \sqrt{.049} = \textbf{.22}$$

Step 3. The significant studentized ranges are obtained from Appendix K. The table is entered at the row headed by the *df* for ms_{error_w}, which is 60.

$$r = 2: \quad q = 2.83$$
$$r = 3: \quad q = 3.40$$

Step 4. Multiply each significant range by the value obtained in Step 2.

$$r = 2: \quad \text{C. diff.} = 2.83 \times .22 = \textbf{.62}$$
$$r = 3: \quad \text{C. diff.} = 3.40 \times .22 = \textbf{.75}$$

Step 5. Testing the difference between the various means: First, rank the means from smallest to largest. In each instance, if the difference (Step 1) is larger than the minimum for that range (Step 4), it is considered to be significant.

$$\text{Trial 3 vs. Trial 1} \quad (R_3 = .75)$$
$$5.33 - 2.21 = \textbf{3.12} \text{ (significant)}$$
$$\text{Trial 2 vs. Trial 1} \quad (R_2 = .62)$$
$$4.42 - 2.21 = \textbf{2.21} \text{ (significant)}$$
$$\text{Trial 3 vs. Trial 2} \quad (R_2 = .62)$$
$$5.33 - 4.42 = \textbf{.91} \text{ (significant)}$$

Therefore, it is concluded that the overall means of the three trials differ significantly from each other.

Second Question: Although the overall tendency of the groups was to change (learn) as a function of trials (practice), did each of the groups actually improve? (*Note:* In actuality, most researchers would not be interested in this question until the third question had been answered. However, since this is merely an example, the various significant effects will be treated in order.) The graphical representation of the trials effect being analyzed is presented in Figure 2.

Step 1. From Step 2 of the original analysis (Section 2.7), obtain the sums of the trials for Group 1.

$$\text{Sum for Trial 1} = \textbf{24}$$
$$\text{Sum for Trial 2} = \textbf{38}$$
$$\text{Sum for Trial 3} = \textbf{49}$$

Figure 2

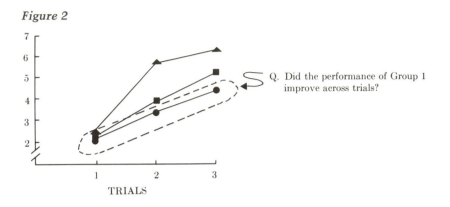

Q. Did the performance of Group 1 improve across trials?

TRIALS

Step 2. Square each of these values, divide by the number of scores on which each sum was based (in this case, 11), and add the quotients.

$$\frac{24^2}{11} + \frac{38^2}{11} + \frac{49^2}{11} = \frac{4421}{11} = \textbf{401.91}$$

Step 3. Obtain the grand sum for Group 1.

$$24 + 38 + 49 = \textbf{111}$$

Then, square this sum and divide by the total number of scores recorded in Group 1.

$$\frac{111^2}{33} = \frac{12{,}321}{33} = \textbf{373.36}$$

Step 4. Subtract the value of Step 3 from that of Step 2. This yields the Group 1 sum of squares for trials.

$$401.91 - 373.36 = \textbf{28.55} = SS_{trials} \text{ for Group 1}$$

Step 5. The *df* for this component is equal to the number of trials minus 1.

$$3 - 1 = 2$$

Step 6. Computation of the mean square for the simple trials effects: Merely divide the SS_{trials} for Group 1 (Step 4) by the *df* (Step 5).

$$\frac{28.55}{2} = \textbf{14.27} = ms_{trials} \text{ for Group 1}$$

Step 7. The test of significance (*F*-test) of the trials effect for Group 1 is $ms_{trials(Gp.1)}/ms_{errorw(total)}$.

$$F = \frac{14.27}{1.62} = 8.81$$

From Appendix E, it is found that this value, with *df* 2 and 60, is significant ($p < .001$). Therefore, it is concluded that performance of the subjects in Group 1 changed as a function of practice. The same procedures should be repeated for Groups 2 and 3. In each instance, the *F* ratio equals $\text{ms}_{\text{trials (Gp. 2 or 3)}}/\text{ms}_{\text{errorw(total)}}$.

Alternate Analysis to Answer Question 2. The analysis presented above merely indicates that there was an overall tendency for the performance of subjects in Group 1 to change over trials. By use of the Newman-Keuls' analysis presented below, it will be possible to determine which specific trials within Group 1 (or any other group to which the analysis might be applied) actually differ significantly from each other.

Step 1. From Step 2 of the original analysis, obtain the mean score for each trial in Group 1.

Sum for Trial 1 = 24 $\quad \frac{24}{11} = 2.18 =$ mean for Group 1/Trial 1

Sum for Trial 2 = 38 $\quad \frac{38}{11} = 3.45 =$ mean for Group 1/Trial 2

Sum for Trial 3 = 49 $\quad \frac{49}{11} = 4.45 =$ mean for Group 1/Trial 3

Step 2. Derive the standard error as:

$$\sqrt{\frac{\text{ms}_{\text{errorw}}}{n(\text{number of measures per trial})}} = \sqrt{\frac{1.62}{11}} = \sqrt{.147} = .38$$

Step 3. The significant studentized ranges are obtained from Appendix K. The table is entered at the row headed by the *df* for $\text{ms}_{\text{errorw}}$.

$$r = 2: \quad q = 2.83$$
$$r = 3: \quad q = 3.40$$

Step 4. Multiply each significant range by the value obtained in Step 2.

$$r = 2: \quad \text{C. diff.} = 2.83 \times .38 = 1.08$$
$$r = 3: \quad \text{C. diff.} = 3.40 \times .38 = 1.29$$

Step 5. Testing the difference between the various means. First, rank the means from the smallest to largest. In each instance, if the difference (Step 1) is larger than the minimum for that range (Step 4), it is considered to be significant.

$$\text{Trial 3 vs. Trial 1 } (R_3 = 1.29)$$

$$4.45 - 2.18 = 2.27 \text{ (significant)}$$

$$\text{Trial 2 vs. Trial 1 } (R_2 = 1.08)$$

$$3.45 - 2.18 = 1.27 \text{ (significant)}$$

$$\text{Trial 3 vs. Trial 2 } (R_2 = 1.08)$$

$$4.45 - 3.45 = 1.00 \text{ (nonsignificant)}$$

Therefore, it is concluded that trials 1 and 2, and 1 and 3 in Group 1 differ significantly from each other.

Third Question: The trials-by-conditions interaction indicates that at least two of the groups of subjects learned at different rates: Which groups were they? Comparisons must be made between Groups 1 and 2, Groups 1 and 3, and Groups 2 and 3. In the present example, the first comparison will be made between Groups 1 and 2. The graphical representation of the comparison being made is presented in Figure 3.

Step 1. From Step 2 of the original analysis (Section 2.7), obtain the sums of each of the trials for Group 1 and Group 2.

	Trial 1	Trial 2	Trial 3
Sums for Group 1 =	24	38	49
Sums for Group 2 =	24	43	57

Figure 3

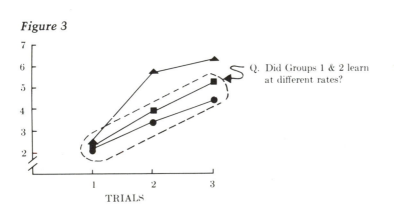

Q. Did Groups 1 & 2 learn at different rates?

TRIALS

Step 2. Obtain the overall sum for each of the groups.

$$24 + 38 + 49 = 111 = \text{sum for Group 1}$$
$$24 + 43 + 57 = 124 = \text{sum for Group 2}$$

Then, square these sums, divide by the number of scores on which each sum was based, and add the quotients.

$$\frac{111^2}{33} + \frac{124^2}{33} = \frac{27,697}{33} = 839.30$$

Step 3. Obtain the grand sum for both groups.

$$111 + 124 = 235$$

Then, square this value, and divide it by the total number of scores recorded in Groups 1 and 2. This yields the correction term for the Group 1 versus Group 2 comparison.

$$\frac{235^2}{66} = \frac{55,225}{66} = 836.74$$

Step 4. Computation of the Group 1 versus Group 2 effects: Merely subtract the correction term (Step 3) from the value obtained in Step 2.

$$839.30 - 836.74 = 2.56 = \text{SS}_{\text{groups}} \text{ for Groups 1 and 2}$$

Step 5. Computation of the effects of trials for Groups 1 and 2: First, obtain the sum of Trials 1, 2, and 3 for both experimental groups (see Step 1).

$$24 + 24 = 48 = \text{sum for Trial 1}$$
$$38 + 43 = 81 = \text{sum for Trial 2}$$
$$49 + 57 = 106 = \text{sum for Trial 3}$$

Then, square each of these sums, divide by the number of scores on which each was based, and add the quotients.

$$\frac{48^2}{22} + \frac{81^2}{22} + \frac{106^2}{22} = \frac{20,101}{22} = 913.68$$

Then, subtract the correction term (Step 3) from the above value.

$$913.68 - 836.74 = 76.94 = \text{SS}_{\text{trials}} \text{ for Groups 1 and 2}$$

Step 6. Computation of interaction of trials by conditions (groups) for Groups 1 and 2: First, square each of the sums for each trial for Groups 1 and 3 (Step 1). Divide each value by the number of scores on which each sum was based, and then add the quotients.

$$\frac{24^2}{11} + \frac{38^2}{11} + \frac{49^2}{11} + \frac{24^2}{11} + \frac{43^2}{11} + \frac{57^2}{11} = \frac{10,095}{11} = \mathbf{917.73}$$

Then, subtract the correction term (Step 3), the SS_{groups} (Step 4), and SS_{trials} (Step 5) from the above value.

$917.73 - 836.74 - 2.56 - 76.94$

$$= \mathbf{1.48} = SS_{trials \times groups} \text{ for Groups 1 and 2}$$

Step 7. The *df* for this component is equal to the number of trials minus 1, times the number of groups minus 1.

$$(3 - 1)(2 - 1) = \mathbf{2}$$

Step 8. Computation of the ms for the trials-by-groups interaction for Groups 1 and 2: Merely divide the $SS_{trials \times groups}$ (Step 6) by the *df* (Step 7).

$$\frac{1.48}{2} = \mathbf{.74} = ms_{trials \times groups} \text{ for Groups 1 and 2}$$

Step 9. The test of significance (*F*-test) is computed as

$$F = \frac{ms_{trials \times groups \text{ (Gps. 1 and 2)}}}{ms_{error_{w}(total)}}$$

Thus,

$$F = \frac{.74}{1.62} = \mathbf{.46}$$

From Appendix E, it is found that this *F* value, with $df = 2$ and 60, is nonsignificant. Therefore, it is concluded that the subjects in Groups 1 and 2 learned at the same rate. The remaining tests of simple interactions are carried out in exactly the same way, except that Groups 1 and 3 are compared and then Groups 2 and 3. In all instances, the test of significance is $F = ms_{trials \times groups}/ms_{error_w}$.

Alternate Approaches to Question Three. Since both the overall conditions effect and the trials-by-conditions interaction were significant in the present example, it may be of interest to determine exactly where (on which trials) the groups differed significantly. One of the most common techniques is to compare the performance of the several groups on each trial. Although many analyses are used (Newman-Keuls', Tukey's, etc.), the *F*-test is employed most often and is a series of completely randomized designs for each trial. The *F*-test of group differences for

Figure 4

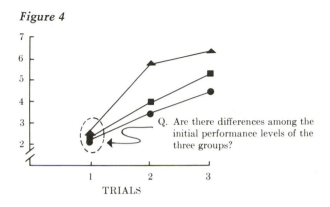

TRIALS

Q. Are there differences among the initial performance levels of the three groups?

Trial 1* is presented below. The graphical representation of this analysis is shown in Figure 4.

Step 1. From Step 1 of the original analysis (Section 2.7), obtain the scores for each subject in all three groups on Trial 1. Table these scores as follows.

		Trial 1			
Group 1		Group 2		Group 3	
Subject	Score	Subject	Score	Subject	Score
S_1	3	S_{12}	2	S_{23}	2
S_2	3	S_{13}	1	S_{24}	3
S_3	1	S_{14}	3	S_{25}	4
S_4	1	S_{15}	2	S_{26}	1
S_5	2	S_{16}	1	S_{27}	3
S_6	4	S_{17}	4	S_{28}	1
S_7	4	S_{18}	1	S_{29}	1
S_8	1	S_{19}	1	S_{30}	2
S_9	1	S_{20}	2	S_{31}	3
S_{10}	2	S_{21}	3	S_{32}	1
S_{11}	2	S_{22}	4	S_{33}	4

Step 2. Turn to the completely randomized design as presented in Section 2.1, and continue with Step 2 of that analysis. The final *F* value (Step 12) will indicate whether there are significant differences among

---

*There are occasions when this analysis might be applied even if the trials-by-conditions interaction were not significant. Although the problem is not present in this example, sometimes there is a question as to whether the groups might have differed significantly on the first trial (sampling error). If this were the case, any differences noted in the overall analysis would be reflecting this error rather than any real differences due to treatment effects.

the group means on this particular trial. Repeat the same procedure for the remaining trials.

Note: If, when the above analyses have been completed, significant conditions (groups) effects are found, one of the multiple comparisons tests may be used to determine which specific pairs of groups differ.

The second alternative to answering Question Three proceeds in nearly the same fashion as the method presented above, except that the comparisons between the groups are made against a single, pooled error term. It should be noted that this approach is recommended by many authors but is considered dubious by others. This pooled error term is computed as follows*:

Step 1. Secure the needed mean square$_{within cell}$ which will serve as the error term for all comparisons. The sums of squares and *df* values are obtained from the summary table of the original analysis.

$$\text{ms}_{within\ cells} = \frac{SS_{error_b} + SS_{error_w}}{df_{error_b} + df_{error_w}}$$

$$= \frac{142 + 97}{30 + 60}$$

$$= \frac{239}{90}$$

$$= \mathbf{2.66} \text{ with } df = 90$$

Step 2. Proceed to compute the mean square$_{conditions}$ for each trial. First, secure the sums for Groups 1, 2, and 3 on the first trial, and also obtain the overall sum for that trial.

Sum for Group 1 on Trial 1 = **24**

Sum for Group 2 on Trial 1 = **24**

Sum for Group 3 on Trial 1 = **25**

Overall Sum for Trial 1 = **73**

Step 3. Square each sum for Trial 1, and divide that value by the number of scores on which the sum was based. Add those values.

$$\frac{24^2}{11} + \frac{24^2}{11} + \frac{25^2}{11} = \frac{1777}{11} = \mathbf{161.54}$$

---

*See Winer (1971) pp. 530–31 for a discussion of an alternate method to calculate the *df* for ms$_{within cells}$.

Step 4. Square the overall sum for Trial 1, and divide that value by the number of scores added to obtain that sum.

$$\frac{73^2}{33} = \frac{5329}{33} = \textbf{161.48}$$

Step 5. Computation of $SS_{conditions}$ for Trial 1: Subtract the value of Step 4 from that of Step 3.

$$161.54 - 161.48 = \textbf{.06}$$

Step 6. Computation of the mean square$_{conditions \ for \ Trial \ 1}$: The *df* are equal to the number of groups minus 1, and the mean square is computed as $SS_{conditions}/df_{conditions}$.

$$ms_{conditions} = \frac{.06}{2} = \textbf{.02}$$

Step 7. The *F* ratio to test for differences among the groups of Trial 1 is as follows:

$$F = \frac{ms_{conditions \ at \ Trial \ 1}}{ms_{within \ cells}} = \frac{.03}{2.66} = \textbf{.01}$$

For *df* = 2 an 90, an *F* ratio of approximately 3.15 is needed to achieve significance. Obviously, the computed *F* ratio of .01 falls far short of the value required.

Step 8. Repeat steps 2, 3, 4, 5, and 6 for the remaining trials. In all cases, the error term for the *F* ratio will be the value computed in Step 1 ($ms_{within \ cells}$).

SECTION 3.11
Use of Orthogonal Components in Tests for Trend

In general, after an *F* ratio has been found to be significant, the only additional tests are those for simple effects (Sections 3.4, 3.5, 3.6, 3.7, 3.8, 3.9, and 3.10). Occasionally, however, it is of interest to determine the shape of the curves that best fit the data points of the various experimental groups. If the treatments form an increasing or decreasing continuum, such as trials or variations in shock intensity or anxiety level, the treatment effects can be further subdivided into trend components by use of orthogonal polynomials.

Three examples will be presented: The first is based on the data of Section 2.1 (the completely randomized design), the second on data presented in Section 2.7 (the two-factor mixed design), and the third on data from Section 2.8 (the three-factor mixed design). These examples represent the most frequent instances where this type of trend analysis is used, but they by no means exhaust the possibilities of application since the analysis can be used with any design. It is important to note that this type of trend analysis merely gives information regarding the shape of the performance curves. The nature of the data and the purpose of the experiment determine entirely whether these analyses are worthwhile.

Example 1

This example is based on the data and results of the completely randomized analysis presented in Section 2.1. First, recall the final analysis of that section:

Source	SS	df	ms	F	p
Total	1132	47	—	—	—
Between groups	567	3	189.00	14.71	<.001
Within groups	565	44	12.84	—	—

Question. What degree polynomial best approximates the form of the curve that connects the data points (group means) of the four groups?

Step 1. Turn to Appendix N and obtain the appropriate coefficients for the re-analysis. Since there are four groups, the coefficients for $k = 4$ are

Linear:	−3	−1	+1	+3
Quadratic:	+1	−1	−1	+1
Cubic:	−1	+3	−3	+1

Step 2. The linear component will be analyzed first. To do this, multiply the *sum* of each group (Step 2 of the original analysis) by its appropriate linear coefficient.

Sum of Group 1 = 111	$111 \times (-3) = -333$
Sum of Group 2 = 88	$88 \times (-1) = -88$
Sum of Group 3 = 156	$156 \times (+1) = +156$
Sum of Group 4 = 195	$195 \times (+3) = +585$

Step 3. Get the algebraic sum of the values obtained in Step 2.

$$(-333) + (-88) + 156 + 585 = +320$$

Then, square that sum.

$$320^2 = 102,400$$

Step 4. Square each of the linear coefficients (Step 1), and sum these squared values.

$$(-3)^2 + (-1)^2 + 1^2 + 3^2 = 20$$

Then, multiply the sum by the number of subjects in each group (12 in the present example).

$$20 \times 12 = 240$$

Step 5. Divide the value of Step 3 by that of Step 4.

$$\frac{102,400}{240} = 426.67 = SS_{linear}$$

Step 6. The test for linear trend is an *F* ratio that is equal to the value of Step 5 divided by the mean square within groups of the original analysis. (In the present example, $ms_{within} = 12.84$.) Compute this *F*, and record the value for future use.

$$F_{linear} = \frac{426.67}{12.84} = 33.22$$

The degrees of freedom (*df*) for this *F* ratio are always equal to 1 and the *df* for ms_{within} of the original analysis (44 in the present example).

After the linear computations are completed, the same basic steps must be repeated to analyze the quadratic component.

Step 1. From Step 1 of the linear analysis (or Appendix N), obtain the coefficients for the quadratic component. In the present example, where $k = 4$, these are:

Quadratic: +1   −1   −1   +1

Step 2. This time multiply the sum of each group by its appropriate quadratic coefficient.

Sum of Group 1 = 111 $111 \times (+1) = +111$
Sum of Group 2 = 88 $88 \times (-1) = -88$
Sum of Group 3 = 156 $156 \times (-1) = -156$
Sum of Group 4 = 195 $195 \times (+1) = +195$

Step 3. Get the algebraic sum of the values obtained in Step 2.

$$111 + (-88) + (-156) + 195 = +\mathbf{62}$$

Then, square this sum.

$$+62^2 = \mathbf{3844}$$

Step 4. Square the quadratic coefficients, and sum these squared values.

$$1^2 + (-1)^2 + (-1)^2 + 1^2 = \mathbf{4}$$

Then, multiply this sum by the number of subjects in each group. (In the present example, the number of subjects per group is 12.)

$$4 \times 12 = \mathbf{48}$$

Step 5. Divide the value of Step 3 by that of Step 4.

$$\frac{3844}{48} = \mathbf{80.08} = SS_{quadratic}$$

Step 6. The test for quadratic trend is an *F* ratio that is equal to the value of Step 5 divided by the mean square within groups of the original analysis. Compute this *F*, and record the value for future use.

$$F_{quadratic} = \frac{80.08}{12.84} = \mathbf{6.23}$$

The *df*s for this *F* ratio are always equal to 1 and the *df* for ms_{within} of the original analysis (44 in this example).

Once again the computations must be repeated; this time, analyze the cubic component.

Step 1. From Step 1 of the linear analysis (or Appendix N), obtain the cubic coefficients. In the present example, where $k = 4$, these are

$$\text{Cubic:} \quad -1 \quad +3 \quad -3 \quad +1$$

Step 2. Multiply the sum of each group by the appropriate cubic coefficient.

Sum of Group 1 = 111	$111 \times (-1) = \mathbf{-111}$
Sum of Group 2 = 88	$88 \times (+3) = \mathbf{+264}$
Sum of Group 3 = 156	$156 \times (-3) = \mathbf{-468}$
Sum of Group 4 = 195	$195 \times (+1) = \mathbf{+195}$

Step 3. Get the algebraic sum of the values obtained in Step 2.

$$(-111) + 264 + (-468) + 195 = \mathbf{-120}$$

Then, square this sum.

$$-120^2 = \textbf{14,400}$$

Step 4. Square each of the cubic coefficients, and sum these squared values.

$$(-1)^2 + 3^2 + (-3)^2 + 1^2 = \textbf{20}$$

Then, multiply that sum by the number of subjects in each group (12 in this example).

$$20 \times 12 = \textbf{240}$$

Step 5. Divide the value of Step 3 by that of Step 4.

$$\frac{14,400}{240} = \textbf{60.00} = SS_{cubic}$$

Step 6. The test of significance for cubic trend is again an *F* ratio that is equal to the value of Step 5 divided by the mean square within groups of the original analysis. Compute this *F*, and record the value for future use.

$$F_{cubic} = \frac{60.00}{12.84} = \textbf{4.67}$$

The *dfs* are equal to 1 and the *df* for ms_{within} (44 in this example).

Step 7. Final check for accuracy: Since the sum of squares between groups was broken into its several trend components, the sums of the linear, quadratic, and cubic components must equal (within rounding error) the sum of squares between groups. (In the present example, SS_b = 567.)

$$426.67 + 80.08 + 60.00 = \textbf{566.75}$$

Note that if the sum of the trend components does not equal the sum of squares between groups, an error in computation has been made.

Step 8. Final tabling of the data: Although there are several ways to do this, the following is most often used.

Source	SS	df	ms	F	p
Total	1132	47	—	—	—
Between groups	567	3	189.00	14.71	<.001
Linear	426.67	1	426.67	33.22	<.001
Quadratic	80.08	1	80.08	6.23	<.025
Cubic	60.00	1	60.00	4.67	<.05
Within groups	565	44	12.84	—	—

The results are interpreted as showing that the data can be best described mathematically by a complex equation. The significant linear component shows that there is some tendency for the data to lie along a straight line. The significant quadratic component shows that there is a tendency for the data to deviate from a straight line in an arcing manner. A quadratic function is represented by a curving line, such as a parabolic arc. The significant cubic component shows that there is a tendency for the data to crisscross a straight line in a regular fashion. A cubic function is similar to two connected parabolic arcs, each peaking in opposite directions.

Example 2

First, recall the two-factor mixed analysis as presented in Section 2.7. The final table of the results was as follows:

Source	SS	df	ms	F	p
Total	467	98	—	—	—
Between subjects	181	32	—	—	—
Conditions	39	2	19.50	4.12	<.05
Error$_b$	142	30	4.73	—	—
Within subjects	286	66	—	—	—
Trials	170	2	85.00	52.46	<.001
Trials by conditions	19	4	4.75	2.93	<.05
Error$_w$	97	60	1.62	—	—

Question. What level of polynomial functions best describes the form of the data points of the various within-subjects effects?

Step 1. Turn to Appendix N, and obtain the appropriate coefficients for the re-analysis. In the present example, each subject was measured three times (three trials). Therefore, the coefficients for $k = 3$ are:

Linear:	-1	0	$+1$
Quadratic:	$+1$	-2	$+1$

Step 2. The linear component will be analyzed first. Go back to the data recorded in Step 1 of the original analysis (Section 2.7), and multiply each score of each subject by the appropriate linear coefficient. In the present example, the retabled data would appear as follows.

| GROUP 1 (LOW-MEANINGFUL MATERIAL) | | |
| Trial Block 1 | Trial Block 2 | Trial Block 3 |
Score $\times -1$	Score $\times 0$	Score $\times +1$

Subject	Score $\times -1$	Score $\times 0$	Score $\times +1$
S_1	-3	0	4
S_2	-3	0	4
S_3	-1	0	4
S_4	-1	0	3
S_5	-2	0	5
S_6	-4	0	6
S_7	-4	0	6
S_8	-1	0	5
S_9	-1	0	5
S_{10}	-2	0	3
S_{11}	-2	0	4

GROUP 2 (MEDIUM-MEANINGFUL MATERIAL)

S_{12}	-2	0	6
S_{13}	-1	0	5
S_{14}	-3	0	7
S_{15}	-2	0	6
S_{16}	-1	0	3
S_{17}	-4	0	5
S_{18}	-1	0	4
S_{19}	-1	0	4
S_{20}	-2	0	4
S_{21}	-3	0	6
S_{22}	-4	0	7

GROUP 3 (HIGH-MEANINGFUL MATERIAL)

S_{23}	-2	0	8
S_{24}	-3	0	3
S_{25}	-4	0	7
S_{26}	-1	0	4
S_{27}	-3	0	12
S_{28}	-1	0	2
S_{29}	-1	0	3
S_{30}	-2	0	6
S_{31}	-3	0	6
S_{32}	-1	0	12
S_{33}	-4	0	7

Step 3. Obtain the algebraic sum of the "weighted scores" for each subject in all groups. (*Note:* All subsequent linear computations will be based on these algebraic sums.)

		Group 1				
Subject	Trial Block 1		Trial Block 2		Trial Block 3	Sum
S_1	-3	$+$	0	$+$	4	$+1$
S_2	-3	$+$	0	$+$	4	$+1$
$\vdots$	$\vdots$		$\vdots$		$\vdots$	$\vdots$
S_{11}	-2	$+$	0	$+$	4	$+2$
		Group 2				
S_{12}	-2	$+$	0	$+$	6	$+4$
S_{13}	-1	$+$	0	$+$	5	$+4$
$\vdots$	$\vdots$		$\vdots$		$\vdots$	$\vdots$
S_{22}	-4	$+$	0	$+$	7	$+3$
		Group 3				
S_{23}	-2	$+$	0	$+$	8	$+6$
S_{24}	-3	$+$	0	$+$	3	0
$\vdots$	$\vdots$		$\vdots$		$\vdots$	$\vdots$
S_{33}	-4	$+$	0	$+$	7	$+3$

Step 4. Add the subject sums obtained in Step 3 to get the algebraic sum for each experimental group. (*Note:* These group sums may be negative.)

$$1 + 1 + \cdots + 2 = +25 = \text{sum of Group 1}$$
$$4 + 4 + \cdots + 3 = +33 = \text{sum of Group 2}$$
$$6 + 0 + \cdots + 3 = +45 = \text{sum of Group 3}$$

Step 5. Square each of the subject sums obtained in Step 3, and add these squared values to get the overall sum of squares.

$$1^2 + 1^2 + \cdots + 4^2 + 4^2 + \cdots + 6^2 + 0^2 + \cdots + 3^2 = 473$$

Step 6. Add the group sums obtained in Step 4 to get the overall algebraic sum. (*Note:* Again, this may be a negative value.)

$$25 + 33 + 45 = +103$$

Step 7. Square each of the linear coefficients (Step 1), and add the squared values.

$$(-1)^2 + 0^2 + 1^2 = 2$$

Step 8. Computation of the within-subjects linear component: Divide the sum of the squared values obtained in Step 5 by the value of Step 7.

$$\frac{473}{2} = \textbf{236.5}$$

Step 9. Computation of the repeated-measure effect (trials, in the present example) for the linear component: Square the overall algebraic sum obtained in Step 6, and then divide by the value of Step 7 times the total number of subjects in the experiment (33, in the present example).

$$\frac{103^2}{2 \times 33} = \frac{10,609}{66} = \textbf{160.742}$$

Step 10. Computation of the trials-by-groups effect for the linear component: First, square each of the group sums obtained in Step 4, divide each by the value of Step 7 times the number of subjects in each group (11 in the present example), and add the quotients.

$$\frac{25^2}{2 \times 11} + \frac{33^2}{2 \times 11} + \frac{45^2}{2 \times 11} = \frac{3739}{22} = \textbf{169.955}$$

Then, from this sum, subtract the value obtained in Step 9.

$$169.955 - 160.742 = \textbf{9.213}$$

Step 11. Computation of the error term for the linear component: Simply subtract the values obtained in Steps 9 and 10 from the value of Step 8.

$$236.5 - 160.742 - 9.213 = \textbf{66.545}$$

Step 12. Computation of the *df* for each of the above components:

df for within subjects (see Step 8) = the total number of subjects used in the experiment.

No. of subjects = **33**

df for trials (see Step 9) = **1** in all instances.

df for the trials-by-groups interaction (see Step 10) = the number of experimental groups minus 1.

$$3 - 1 = \textbf{2}$$

df for the linear-component = df for within subjects
error term (see Step 11)   minus the df for trials and the df
for trials by groups.

$$33 - 1 - 2 = 30$$

Step 13. Computation of the required linear mean squares: Simply divide each sum of squares value as computed in Steps 8, 9, 10 and 11 by its appropriate degrees of freedom (Step 12). The test of significance in each instance is an F ratio of the appropriate linear component divided by the linear error term. Table the mean-square values and the F ratios temporarily as follows.

Source	SS	df	ms	F	p
Within$_{linear}$	236.50	33	(not needed)	—	—
Trials	160.742	1	160.742	72.47	<.001
Trials by groups	9.213	2	4.606	2.08	n.s.
Error$_{linear}$	66.545	30	2.218	—	—

After completing the linear-component computations, the same basic steps must be repeated to analyze the quadratic component.

Step 1. From Step 1 of the linear analysis (or Appendix N), obtain the coefficients for the quadratic component. In the present example, where $k = 3$, the coefficients are:

Quadratic: $+1$ -2 $+1$

Step 2. The data must again be retabled. The same basic procedure of Step 2 of the linear analysis is followed except that each score is multiplied by its appropriate quadratic coefficient.

GROUP 1

Subject	Trial Block 1 Score $\times$ +1	Trial Block 2 Score $\times$ -2	Trial Block 3 Score $\times$ +1
S_1	3	-6	4
S_2	3	-8	4
S_3	1	-6	4
S_4	1	-4	3
S_5	2	-6	5
S_6	4	-10	6
S_7	4	-10	6
S_8	1	-8	5
S_9	1	-8	5
S_{10}	2	-6	3
S_{11}	2	-4	4

GROUP 2

Subject	Trial Block 1 Score × +1	Trial Block 2 Score × −2	Trial Block 3 Score × +1
S_{12}	2	−6	6
S_{13}	1	−8	5
S_{14}	3	−12	7
S_{15}	2	−8	6
S_{16}	1	−4	3
S_{17}	4	−10	5
S_{18}	1	−6	4
S_{19}	1	−4	4
S_{20}	2	−6	4
S_{21}	3	−10	6
S_{22}	4	−12	7

GROUP 3

S_{23}	2	−8	8
S_{24}	3	−12	3
S_{25}	4	−14	7
S_{26}	1	−14	4
S_{27}	3	−14	12
S_{28}	1	−8	2
S_{29}	1	−10	3
S_{30}	2	−10	6
S_{31}	3	−12	6
S_{32}	1	−10	12
S_{33}	4	−18	7

Step 3. Obtain the algebraic sum of the "weighted scores" for each subject in all groups. (*Note:* All remaining quadratic computations will be based on these sums.)

Group 1

Subject	Trial Block 1		Trial Block 2		Trial Block 3	Sum
S_1	3	+	(−6)	+	4	+1
S_2	3	+	(−8)	+	4	−1
⋮	⋮		⋮		⋮	⋮
S_{11}	2	+	(−4)	+	4	+2

Group 2

S_{12}	2	+	(−6)	+	6	+2
S_{13}	1	+	(−8)	+	5	−2
⋮	⋮		⋮		⋮	⋮
S_{22}	4	+	(−12)	+	7	−1

			Group 3			
Subject	Trial Block 1		Trial Block 2		Trial Block 3	Sum
S_{23}	2	+	(-8)	+	8	+2
S_{24}	3	+	(-12)	+	3	-6
$\vdots$	$\vdots$		$\vdots$		$\vdots$	$\vdots$
S_{33}	4	+	(-18)	+	7	-7

Step 4. Add the subject sums obtained in Step 3 to get the algebraic sum for each group.

$$1 + (-1) + \cdots + 2 = -3 = \text{sum for Group 1}$$
$$2 + (-2) + \cdots + (-1) = -5 = \text{sum for Group 2}$$
$$2 + (-6) + \cdots + (-7) = -35 = \text{sum for Group 3}$$

Step 5. Square each of the subject sums obtained in Step 3, and add these squared values.

$$1^2 + (-1)^2 + \cdots + 2^2 + (-2)^2$$
$$+ \cdots + 2^2 + (-6)^2 + \cdots + (-7)^2 = 297$$

Step 6. Add the group sums obtained in Step 4 to get the overall algebraic sum.

$$(-3) + (-5) + (-35) = -43$$

Step 7. Square each of the quadratic coefficients (Step 1), and add the squared values.

$$1^2 + (-2)^2 + 1^2 = 6$$

Step 8. Computation of the within-subjects quadratic component: Divide the value of Step 5 by that of Step 7.

$$\frac{297}{6} = 49.50$$

Step 9. Computation of the trials effect for the quadratic component: Square the value of Step 6, and divide it by the value of Step 7 times the total number of subjects used in the experiment (33 in this example).

$$\frac{(-43)^2}{6 \times 33} = \frac{1849}{198} = 9.34$$

Step 10. Computation of the trials-by-groups effect for the quadratic component: First, square each sum obtained in Step 4, divide each by the value of Step 7 times the number of subjects in each group (11 in this example), and add the quotients.

$$\frac{(-3)^2}{6 \times 11} + \frac{(-5)^2}{6 \times 11} + \frac{(-35)^2}{6 \times 11} = \frac{1259}{66} = \textbf{19.08}$$

Then, from this sum subtract the value obtained in Step 9.

$$19.08 - 9.34 = \textbf{9.74}$$

Step 11. Computation of the error term for the quadratic component: Simply subtract the values of Step 9 and 10 from the value of Step 8.

$$49.50 - 9.34 - 9.74 = \textbf{30.42}$$

Step 12. Computation of the *df* for each quadratic component:

df for within subjects = total number of subjects used in the experiment.

No. of subjects = **33**

df for trials = **1** in all instances.

df for trials-by-group = the number of experimental groups minus 1. interaction

$$3 - 1 = \textbf{2}$$

df for the quadratic component error term = *df* for within subjects minus the *df* for trials and the *df* for trials by groups.

$$33 - 1 - 2 = \textbf{30}$$

Step 13. Computation of the quadratic mean squares: Simply divide each of the sums of squares of Steps 9, 10, and 11 by the appropriate *df* (Step 12). The test of significance in each case is an *F* ratio of the appropriate quadratic component divided by the quadratic error term. Table the data temporarily as follows.

Source	SS	df	ms	F	p
Within$_{quadratic}$	49.50	33	(not needed)	—	—
Trials	9.34	1	9.34	9.21	<.001
Trials by groups	9.74	2	4.87	4.80	<.05
Error$_{quadratic}$	30.42	30	1.01	—	—

Step 14. Final check for accuracy: Since each of the within-subjects sums of squares of the original analysis was broken into trend components, the sums of the components must equal (within rounding error) the original values.

Source	Linear		Quadratic		Sum	Original
Within subjects	236.500	+	49.500	=	**286.000**	286.00
Trials	160.742	+	9.340	=	**170.082**	170.00
Trials × groups	9.213	+	9.740	=	**18.953**	19.00
Error$_{within}$	66.545	+	30.420	=	**96.965**	97.00

Step 15. Final tabling of the analysis: Although there are several techniques, the following is used most often.

Source	SS	df	ms	F	p
Within subjects	286	66	—	—	—
Trials	170	2	85.000	52.46	<.001
Linear	160.742	1	160.742	72.47	<.001
Quadratic	9.340	1	9.340	9.21	<.001
Trials × groups	19	4	4.750	2.93	<.05
Linear	9.213	2	4.606	2.08	n.s.
Quadratic	9.740	2	4.870	4.80	<.05
Error	97	60	1.620	—	—
Linear	66.545	30	2.218	—	—
Quadratic	30.420	30	1.014	—	—

A significant linear component for the trials effect means that there is a tendency for the data to lie along a straight line. The significance of the quadratic component means that there is also a tendency for the data to follow a quadratic (arcing) function.

It is more difficult to interpret the quadratic component of the trials-by-groups interaction, since at least two sets of data points must be represented.

Example 3

This example is based on the three-factor mixed design presented in Section 2.8. First, recall the final analysis in that section:

Source	SS	df	ms	F	p
Total	1657	111	—	—	—
Between subjects	576	27	—	—	—
Expectancy (high/low)	4	1	4	<1	—
Feedback (success/failure)	290	1	290	34.27	<.001
Expectancy × feedback	79	1	79	9.34	<.001
Error$_b$	203	24	8.46	—	—

Source	SS	df	ms	F	p
Within subjects	1081	84	—	—	—
Trials	784	3	261.33	235.43	<.001
Trials × expectancy	1	3	.33	<1	—
Trials × feedback	167	3	55.67	50.15	<.001
Trials × expectancy × feedback	49	3	16.33	14.71	<.001
Error$_w$	80	72	1.11	—	—

Question: What level of polynomial functions best describes the form of the data points of the several within-subjects effects in this analysis?

Step 1. Obtain from Appendix N the appropriate coefficients for the re-analysis. In the present example, four measures were recorded for each subject (four trials). Therefore, the coefficients for $k = 4$ are:

Linear:	−3	−1	+1	+3
Quadratic:	+1	−1	−1	+1
Cubic:	−1	+3	−3	+1

Step 2. The linear component will be analyzed first. Multiply each score used in the original analysis by the appropriate linear coefficient, and retable the data as follows.

GROUP 1

(HIGH EXPECTANCY/FAILURE)

Subject	Trial 1 Score × −3	Trial 2 Score × −1	Trial 3 Score × +1	Trial 4 Score × +3
S_1	−12	−6	7	27
S_2	−18	−7	9	33
S_3	−15	−8	10	39
S_4	−9	−4	6	27
S_5	−12	−6	9	30
S_6	−15	−5	6	24
S_7	−21	−8	9	36

GROUP 2
(HIGH EXPECTANCY/SUCCESS)

S_8	−15	−8	9	36
S_9	−9	−5	10	42
S_{10}	−21	−10	12	39
S_{11}	−15	−8	11	45
S_{12}	−12	−9	13	45
S_{13}	−9	−5	9	36
S_{14}	−18	−7	8	33

GROUP 3
(LOW EXPECTANCY/FAILURE)

Subject	Trial 1 Score × −3	Trial 2 Score × −1	Trial 3 Score × +1	Trial 4 Score × +3
S_{15}	−9	−4	4	15
S_{16}	−15	−6	8	27
S_{17}	−15	−6	6	24
S_{18}	−15	−6	7	24
S_{19}	−9	−4	4	15
S_{20}	−21	−7	8	21
S_{21}	−18	−8	8	24

GROUP 4
(LOW EXPECTANCY/SUCCESS)

Subject				
S_{22}	−12	−8	10	42
S_{23}	−15	−10	14	54
S_{24}	−9	−7	12	54
S_{25}	−15	−8	11	51
S_{26}	−18	−11	15	57
S_{27}	−18	−8	10	45
S_{28}	−15	−11	17	60

Step 3. Obtain and record the algebraic sum of the "weighted scores" for each subject in all groups. (*Note:* All linear computations will be based on these sums.)

Group 1

Subject	Trial 1		Trial 2		Trial 3		Trial 4	Sum
S_1	(−12)	+	(−6)	+	7	+	27	+16
S_2	(−18)	+	(−7)	+	9	+	33	+17
⋮	⋮		⋮		⋮		⋮	⋮
S_7	(−21)	+	(−8)	+	9	+	36	+16

Group 2

S_8	(−15)	+	(−8)	+	9	+	36	+22
S_9	(−9)	+	(−5)	+	10	+	42	+38
⋮	⋮		⋮		⋮		⋮	⋮
S_{14}	(−18)	+	(−7)	+	8	+	33	+16

Group 3

S_{15}	(−9)	+	(−4)	+	4	+	15	+6
S_{16}	(−15)	+	(−6)	+	8	+	27	+14
⋮	⋮		⋮		⋮		⋮	⋮
S_{21}	(−18)	+	(−8)	+	8	+	24	+6

		Group 4					
Subject	Trial 1		Trial 2		Trial 3		Trial 4	Sum

Subject	Trial 1		Trial 2		Trial 3		Trial 4	Sum
S_{22}	(-12)	$+$	(-8)	$+$	10	$+$	42	$+32$
S_{23}	(-15)	$+$	(-10)	$+$	14	$+$	54	$+43$
$\vdots$	$\vdots$		$\vdots$		$\vdots$		$\vdots$	$\vdots$
S_{28}	(-15)	$+$	(-11)	$+$	17	$+$	60	$+51$

Step 4. Add the subject sums obtained in Step 3 to get the algebraic sum for each group. Record these sums and whether they are positive or negative; they will be used later.

$$16 + 17 + \cdots + 16 = +126 = \text{sum of Group 1}$$
$$22 + 38 + \cdots + 16 = +197 = \text{sum of Group 2}$$
$$6 + 14 + \cdots + 6 = + 52 = \text{sum of Group 3}$$
$$32 + 43 + \cdots + 51 = +287 = \text{sum of Group 4}$$

Step 5. Square each of the subject sums obtained in Step 3. Add these squared values, and record the sum for later use.

$$16^2 + 17^2 + \cdots + 51^2 = \mathbf{21{,}092}$$

Step 6. Add the group sums obtained in Step 4 to get the overall algebraic sum. Record this sum for later use.

$$126 + 197 + 52 + 287 = +662$$

Step 7. Square each of the linear coefficients (Step 1), and add the squared values.

$$(-3)^2 + (-1)^2 + 1^2 + 3^2 = \mathbf{20}$$

Step 8. Computation of the within-subjects linear component: Simply divide the overall sum of the squared values obtained in Step 5 by the value of Step 7.

$$\frac{21{,}092}{20} = \mathbf{1054.60}$$

Step 9. Computation of the repeated-measures (trials) effect for the linear component: Square the algebraic sum obtained in Step 6, and divide by the value of Step 7 times the total number of subjects used in the experiment (28 in the present example).

$$\frac{662^2}{20 \times 28} = \frac{438{,}244}{560} = \mathbf{782.58}$$

Step 10. Computation of the trials-by-first-factor (expectancy) linear component: First, obtain the algebraic sums for all subjects who were the same in terms of the first factor (expectancy), disregarding the second factor (feedback). (See Step 4 for group sums to be added.)

$$126 + 197 = \mathbf{323} = \text{sum of all high-expectancy subjects}$$
$$52 + 287 = \mathbf{339} = \text{sum of all low-expectancy subjects}$$

Then, square these sums, divide each by the value of Step 7 times the number of scores on which each sum was based (14 in this example), and add the quotients.

$$\frac{323^2}{20 \times 14} + \frac{339^2}{20 \times 14} = \frac{219{,}250}{280} = \mathbf{783.03}$$

Then, from this sum, subtract the value obtained in Step 9.

$$783.03 - 782.58 = \mathbf{0.45}$$

Step 11. Computation of the trials-by-second-factor (feedback) linear component: Repeat the computations of Step 10, except using the algebraic sums of subjects who were the same in terms of the second factor, disregarding the first factor (expectancy). (See Step 4 for the group sums to be added.) First,

$$126 + 52 = \mathbf{178} = \text{sum of all failure subjects}$$
$$197 + 287 = \mathbf{484} = \text{sum of all success subjects}$$

Then,

$$\frac{178^2}{20 \times 14} + \frac{484^2}{20 \times 14} = \frac{265{,}940}{280} = \mathbf{949.78}$$

Then,

$$949.78 - 782.58 = \mathbf{167.20}$$

Step 12. Computation of the repeated-measures-by-first-by-second-factor linear component (trials × expectancy × feedback, in the present example): Square the algebraic sum of each group (Step 4), divide each square by the value of Step 7 times the number of subjects in each group, and add the quotients.

$$\frac{126^2}{20 \times 7} + \frac{197^2}{20 \times 7} + \frac{52^2}{20 \times 7} + \frac{287^2}{20 \times 7} = \frac{139{,}758}{140} = \mathbf{998.27}$$

Then, from this sum, subtract the values obtained in Steps 9, 10, and 11.

$$998.27 - 782.58 - .45 - 167.20 = \mathbf{48.04}$$

Step 13. Computation of the linear-component error term: Simply subtract the value obtained in the first computation of Step 12 (998.27, in the present example) from the value of Step 8 (or subtract the values obtained in Steps 9, 10, 11, and 12 from Step 8).

$$1054.60 - 998.27 = \mathbf{56.33}$$

Step 14. Computation of the *df* for the linear components:

df for within$_{\text{linear}}$ = the total number
of subjects used in the experiment.

$$\text{No. of subjects} = \mathbf{28}$$

df for repeated measures (trials) = **1** (always).
df for trials by first factor (expectancy)
= the number of experimental conditions
within the first factor minus 1.

$$2 - 1 = \mathbf{1}$$

df for trials by second factor (feedback) = the number of
experimental conditions within the second factor minus 1.

$$2 - 1 = \mathbf{1}$$

df for trials by first by second factor
= the *df* for the first factor times the *df* for the second factor.

$$1 \times 1 = \mathbf{1}$$

df for error = *df* for within subjects minus *df* for trials, trials by first factor, trials by second factor, and trials by first by second factor.

$$28 - 1 - 1 - 1 - 1 = \mathbf{24}$$

Step 15. Computation of the mean squares for the linear components: Simply divide each sum-of-squares value as computed in Steps 8, 9, 10, 11, 12, and 13 by its appropriate degrees of freedom. The test of significance in each instance is an *F* ratio of the appropriate linear component divided by the linear-component error term. The mean squares and *F* ratios should be tabled as follows for future use.

Source	SS	df	ms	F
Within$_{\text{linear}}$	1054.60	28	(not needed)	—
Trials	782.58	1	782.58	333.01
T × Exp.	.45	1	.45	<1
T × Feed.	167.20	1	167.20	71.15
T × Exp. × Feed.	48.04	1	48.04	20.44
Error$_{\text{linear}}$	56.33	24	2.35	—

The computations for the linear components have been completed. The computations for the quadratic components must now be undertaken.

Step 1. Obtain the coefficients for the quadratic component from Step 1 of the linear analysis (or Appendix N). In the present example, where $k = 4$:

Quadratic: $+1$ -1 -1 $+1$

Step 2. The data must again be retabled. The same basic procedure of Step 2 of the linear analysis is followed, except that each score is multiplied by its appropriate quadratic coefficient.

GROUP 1
(HIGH EXPECTANCY/FAILURE)

Subject	Trial 1 Score × −3	Trial 2 Score × −1	Trial 3 Score × +1	Trial 4 Score × +3
S_1	4	−6	−7	9
S_2	6	−7	−9	11
S_3	5	−8	−10	13
S_4	3	−4	−6	9
S_5	4	−6	−9	10
S_6	5	−5	−6	8
S_7	7	−8	−9	12

GROUP 2
(HIGH EXPECTANCY/SUCCESS)

S_8	5	−8	−9	12
S_9	3	−5	−10	14
S_{10}	7	−10	−12	13
S_{11}	5	−8	−11	15
S_{12}	4	−9	−13	15
S_{13}	3	−5	−9	12
S_{14}	6	−7	−8	11

GROUP 3
(LOW EXPECTANCY/FAILURE)

S_{15}	3	−4	−4	5
S_{16}	5	−6	−8	9
S_{17}	5	−6	−6	8
S_{18}	5	−6	−7	8
S_{19}	3	−4	−4	5
S_{20}	7	−7	−8	7
S_{21}	6	−8	−8	8

GROUP 4
(LOW EXPECTANCY/SUCCESS)

Subject	Trial 1 Score × +1	Trial 2 Score × −1	Trial 3 Score × −1	Trial 4 Score × +1
S_{22}	4	−8	−10	14
S_{23}	5	−10	−14	18
S_{24}	3	−7	−12	18
S_{25}	5	−8	−11	17
S_{26}	6	−11	−15	19
S_{27}	6	−8	−10	15
S_{28}	5	−11	−17	20

Step 3. Using the data of Step 2, obtain and record the algebraic sum of the "weighted scores" of each subject in all groups. (*Note:* All quadratic computations will be based on these scores.)

Group 1

Subject	Trial 1		Trial 2		Trial 3		Trial 4	Sum
S_1	4	+	(-6)	+	(-7)	+	9	**0**
S_2	6	+	(-7)	+	(-9)	+	11	**+1**
⋮	⋮		⋮		⋮		⋮	⋮
S_7	7	+	(-8)	+	(-9)	+	12	**+2**

Group 2

Subject	Trial 1		Trial 2		Trial 3		Trial 4	Sum
S_8	5	+	(-8)	+	(-9)	+	12	**0**
S_9	3	+	(-5)	+	(-10)	+	14	**+2**
⋮	⋮		⋮		⋮		⋮	⋮
S_{14}	6	+	(-7)	+	(-8)	+	11	**+2**

Group 3

Subject	Trial 1		Trial 2		Trial 3		Trial 4	Sum
S_{15}	3	+	(-4)	+	(-4)	+	5	**0**
S_{16}	5	+	(-6)	+	(-8)	+	9	**0**
⋮	⋮		⋮		⋮		⋮	⋮
S_{21}	6	+	(-8)	+	(-8)	+	8	**−2**

Group 4

Subject	Trial 1		Trial 2		Trial 3		Trial 4	Sum
S_{22}	4	+	(-8)	+	(-10)	+	14	**0**
S_{23}	5	+	(-10)	+	(-14)	+	18	**−1**
⋮	⋮		⋮		⋮		⋮	⋮
S_{28}	5	+	(-11)	+	(-17)	+	(20)	**−3**

Step 4. Add the subject sums of Step 3 to get the algebraic sums for each group. Record these sums and whether they are positive or negative.

$$0 + 1 + \cdots + 2 = +6 = \text{sum of Group 1}$$

$$0 + 2 + \cdots + 2 = +1 = \text{sum of Group 2}$$
$$0 + 0 + \cdots + (-2) = -2 = \text{sum of Group 3}$$
$$0 + (-1) + \cdots + (-3) = +3 = \text{sum of Group 4}$$

Step 5. Square each of the subject sums of Step 3. Add these squared values, and record the sum for later use.

$$0^2 + 1^2 + \cdots + 0^2 + 2^2 + \cdots + 0^2 + 0^2$$
$$+ \cdots + 0^2 + (-1)^2 + \cdots + (-3)^2 = \mathbf{76}$$

Step 6. Add the subject sums of Step 4 to get the overall algebraic sum.

$$6 + 1 + (-2) + 3 = \mathbf{+8}$$

Step 7. Square and sum the quadratic coefficients (Step 1).

$$1^2 + (-1)^2 + (-1)^2 + 1^2 = \mathbf{4}$$

Step 8. Computation of the within-subjects quadratic component: Simply divide the value of Step 5 by that of Step 7.

$$\frac{76}{4} = \mathbf{19.00}$$

Step 9. Computation of the trials effect for the quadratic component: Square the value of Step 6, and then divide by the value of Step 7 times the total number of subjects used in the experiment (28 in this example).

$$\frac{8^2}{4 \times 28} = \frac{64}{112} = \mathbf{.57}$$

Step 10. Computation of the trials-by-first-factor (expectancy) quadratic component: First, obtain the algebraic sum of the subjects who were the same in terms of the first factor (expectancy in the present example), disregarding the second factor. (See Step 4 for the group sums to be added.)

$$6 + 1 = +7 = \text{sum for all high-expectancy subjects}$$
$$(-2) + 3 = +1 = \text{sum for all low-expectancy subjects}$$

Then, square these sums, divide each by the value of Step 7 times the number of scores on which each sum was based (14 in this example), and add the quotients.

$$\frac{7^2}{4 \times 14} + \frac{1^2}{4 \times 14} = \frac{50}{56} = \mathbf{.89}$$

Then, from this value, subtract the value of Step 9.

$$.89 - .57 = .32$$

Step 11. Computation of the trials-by-second-factor (feedback) quadratic component: Repeat the computations of Step 10, except using the algebraic sums of subjects who were the same in terms of the second factor, disregarding the first factor. (See Step 4 for the group sums to be added.) First,

$$6 + (-2) = +4 = \text{sum of all failure subjects}$$
$$1 + 3 = +4 = \text{sum of all success subjects}$$

Then,

$$\frac{4^2}{4 \times 14} + \frac{4^2}{4 \times 14} = \frac{32}{56} = .57$$

Then,

$$.57 - .57 = .00$$

Step 12. Computation of the trials-by-first-by-second-factor quadratic component: Square the sums of Step 4, divide each square by the value of Step 7 times the number of subjects in each group, and add the quotients.

$$\frac{6^2}{4 \times 7} + \frac{1^2}{4 \times 7} + \frac{(-2)^2}{4 \times 7} + \frac{3^2}{4 \times 7} = \frac{50}{28} = 1.79$$

Then, from this sum, subtract the values obtained in Steps 9, 10, and 11.

$$1.79 - .57 - .32 - .00 = .90$$

Step 13. Computation of the quadratic-component error term: Simply subtract the value obtained in the first computation of Step 12 from the value of Step 8.

$$19.00 - 1.79 = 17.21$$

Step 14. The *df* computations for the quadratic component are the same as for the linear component.

df for within$_{\text{quadratic}}$
= the total number of subjects used in the experiment.

$$\text{No. of subjects} = 28$$

df for trials = 1 (always).
df for trials by first factor
= number of first-factor conditions minus 1.

$$2 - 1 = 1$$

df for the trials by second factor
= number of second-factor conditions minus 1.

$$2 - 1 = 1$$

df for trials by first by second factor
= the df for the first factor times the df for the second factor.

$$1 \times 1 = 1$$

df for error$_{quadratic}$ = df for within subjects minus the df for trials, trials by first factor, trials by second factor, and trials by first and by second factor.

$$28 - 1 - 1 - 1 - 1 = 24$$

Step 15. Computation of the quadratic mean squares: Again, divide the values of Steps 8, 9, 10, 11, 12, and 13 by the appropriate degrees of freedom. Compute each F ratio by dividing the appropriate quadratic component by the quadratic error term. Table the data temporarily as follows.

Source	SS	df	ms	F
Within$_{quadratic}$	19.00	28	(not needed)	—
Trials	.57	1	.57	<1
T × Exp.	.32	1	.32	<1
T × Feed.	.00	1	.00	<1
T × Exp. × Feed.	.90	1	.90	1.25
Error$_{quadratic}$	17.21	24	.72	—

When the computations of the quadratic components have been completed, the computation of the cubic components must be undertaken.

Step 1. Obtain the coefficients for the cubic component from Step 1 of the linear analysis (or Appendix N). In the present example, where $k = 4$:

Cubic: -1 $+3$ -3 $+1$

Step 2. Once again, the data must be retabled. This time each score is multiplied by its appropriate cubic coefficient.

		GROUP 1		
	Trial 1	**Trial 2**	**Trial 3**	**Trial 4**
Subject	Score × −1	Score × +3	Score × −3	Score × +1
S_1	−4	18	−21	9
S_2	−6	21	−27	11
S_3	−5	24	−30	13
S_4	−3	12	−18	9
S_5	−4	18	−27	10
S_6	−5	15	−18	8
S_7	−7	24	−27	12
		GROUP 2		
S_8	−5	24	−27	12
S_9	−3	15	−30	14
S_{10}	−7	30	−36	13
S_{11}	−5	24	−33	15
S_{12}	−4	27	−39	15
S_{13}	−3	15	−27	12
S_{14}	−6	21	−24	11
		GROUP 3		
S_{15}	−3	12	−12	5
S_{16}	−5	18	−24	9
S_{17}	−5	18	−18	8
S_{18}	−5	18	−21	8
S_{19}	−3	12	−12	5
S_{20}	−7	21	−24	7
S_{21}	−6	24	−24	8
		GROUP 4		
S_{22}	−4	24	−30	14
S_{23}	−5	30	−42	18
S_{24}	−3	21	−36	18
S_{25}	−5	24	−33	17
S_{26}	−6	33	−45	19
S_{27}	−6	24	−30	15
S_{28}	−5	33	−51	20

Step 3. Using the data of Step 2, obtain the algebraic sum of each subject's "weighted scores." These sums will be used for all cubic computations.

		Group 1						
Subject	Trial 1		Trial 2		Trial 3		Trial 4	Sum
S_1	(−4)	+	18	+	(−21)	+	9	+2
S_2	(−6)	+	21	+	(−27)	+	11	−1
⋮	⋮		⋮		⋮		⋮	⋮
S_7	(−7)	+	24	+	(−27)	+	12	+2

			Group 2				
Subject	Trial 1		Trial 2		Trial 3	Trial 4	Sum
S_8	(-5)	$+$	24	$+$	(-27) $+$	12	$+4$
S_9	(-3)	$+$	15	$+$	(-30) $+$	14	-4
$\vdots$	$\vdots$		$\vdots$		$\vdots$	$\vdots$	$\vdots$
S_{14}	(-6)	$+$	21	$+$	(-24) $+$	11	$+2$
			Group 3				
S_{15}	(-3)	$+$	12	$+$	(-12) $+$	5	$+2$
S_{16}	(-5)	$+$	18	$+$	(-24) $+$	9	-2
$\vdots$	$\vdots$		$\vdots$		$\vdots$	$\vdots$	$\vdots$
S_{21}	(-6)	$+$	24	$+$	(-24) $+$	8	$+2$
			Group 4				
S_{22}	(-4)	$+$	24	$+$	(-30) $+$	14	$+4$
S_{23}	(-5)	$+$	30	$+$	(-42) $+$	18	$+1$
$\vdots$	$\vdots$		$\vdots$		$\vdots$	$\vdots$	$\vdots$
S_{28}	(-5)	$+$	33	$+$	(-51) $+$	20	-3

Step 4. Add the sums of Step 3 to get the algebraic sum for each group.

$$2 + (-1) + \cdots + 2 = +2 = \text{sum of Group 1}$$
$$4 + (-4) + \cdots + 2 = -1 = \text{sum of Group 2}$$
$$2 + (-2) + \cdots + 2 = +4 = \text{sum of Group 3}$$
$$4 + 1 + \cdots + (-3) = +9 = \text{sum of Group 4}$$

Step 5. Square the sums of Step 3. Add these squared values, and record them for later use.

$$2^2 + (-1)^2 + \cdots + 4^2 + (-4)^2 + \cdots + 2^2$$
$$+ (-2)^2 + \cdots + 4^2 + 1^2 + \cdots + (-3)^2 = 148$$

Step 6. Add the group sums of Step 4 to obtain the overall algebraic sum.

$$2 + (-1) + 4 + 9 = +14$$

Step 7. Square and sum the quadratic coefficients (Step 1).

$$(-1)^2 + 3^2 + (-3)^2 + 1^2 = 20$$

Step 8. Computation of the within-subjects cubic component: Simply divide Step 5 by Step 7.

$$\frac{148}{20} = 7.40$$

Step 9. Computation of the cubic-component trials effect: Square the value of Step 6. Then, divide by the value of Step 7 times the total number of subjects used in the experiment.

$$\frac{14^2}{20 \times 28} = \frac{196}{560} = .35$$

Step 10. Computation of the trials-by-first-factor (expectancy) cubic component: First, obtain the algebraic sum of the subjects who were the same in terms of the first factor, disregarding the second factor. (See Step 4 for the group sums to be added.)

$$2 + (-1) = +1 = \text{sum of high-expectancy subjects}$$
$$4 + 9 = +13 = \text{sum of low-expectancy subjects}$$

Then, square these sums, divide each by the value of Step 7 times the number of scores on which each sum was based, and add the quotients.

$$\frac{1^2}{20 \times 14} + \frac{13^2}{20 \times 14} = \frac{170}{280} = .61$$

Then, from this value, subtract the final value of Step 9.

$$.61 - .35 = .26$$

Step 11. Computation of the trials-by-second-factor (feedback) cubic component: Repeat the computations of Step 10, except that the algebraic sums of the second factor are used, disregarding the first factor. (See Step 4 for the group sums to be added.) First,

$$2 + 4 = 6 = \text{sum of failure subjects}$$
$$(-1) + 9 = 8 = \text{sum of success subjects}$$

Then,

$$\frac{6^2}{20 \times 14} + \frac{8^2}{20 \times 14} = \frac{100}{280} = .36$$

Then,

$$.36 - .35 = .01$$

Step 12. Computation of the trials-by-first-by-second-factor cubic component: Square the sums of Step 4, divide each by the value of Step 7 times the number of subjects in each group, and add the quotients.

$$\frac{2^2}{20 \times 7} + \frac{(-1)^2}{20 \times 7} + \frac{4^2}{20 \times 7} + \frac{9^2}{20 \times 7} = \frac{102}{140} = .73$$

Then, from this sum, subtract the values of Steps 9, 10, and 11.

$$.73 - .35 - .26 - .01 = .11$$

Step 13. Computation of the cubic-component error term: Simply subtract the value obtained in the initial computation of Step 12 from that of Step 8.

$$7.40 - .73 = \mathbf{6.67}$$

Step 14. The *df* computations for the cubic components are the same as for the linear and quadratic components. In the present example, these are

df for within$_{\text{cubic}}$ = **28**

df for trials = **1**

df for trials by first factor = **1**

df for trials by second factor = **1**

df for trials by first by second factor = **1**

df for error$_{\text{cubic}}$ = **24**

Step 15. Computation of the mean squares and *F* ratios for the cubic component: These are computed in the same manner as for the linear and quadratic components. Again, table the results temporarily as follows.

Source	SS	df	ms	F
Within$_{\text{cubic}}$	7.40	28	(not needed)	—
Trials	.35	1	.35	1.25
T × Exp.	.26	1	.26	<1
T × Feed.	.01	1	.01	<1
T × Exp. × Feed.	.11	1	.11	<1
Error$_{\text{cubic}}$	6.67	24	.28	—

Step 16. Final check for accuracy: Since each of the within-subjects sums of squares of the original analysis was broken into trend components, the sums of these components must equal (within rounding error) the original values. In the present example:

Source	Linear		Quadratic		Cubic		Sum	Original
Within subjects	1054.60	+	19.00	+	7.40	=	1081.00	1081
Trials	782.58	+	.57	+	.35	=	783.50	784
T × Exp.	.45	+	.32	+	.26	=	1.03	1
T × Feed.	167.20	+	.00	+	.01	=	167.21	167
T × Exp. × Feed.	48.04	+	.90	+	.11	=	49.05	49
Error$_{\text{within}}$	56.33	+	17.21	+	6.67	=	80.21	80

Step 17. Final tabling of the results: Although several procedures can be used, the following is probably used most often.

Source	SS	df	ms	F	p
Within subjects	1081	84	—	—	—
Trials	784	3	261.33	235.43	<.001
Linear	782.58	1	782.58	333.01	<.001
Quadratic	.57	1	.57	<1	—
Cubic	.35	1	.35	1.25	n.s.
Trials × Exp.	1	3	.33	<1	—
Linear	.45	1	.45	<1	—
Quadratic	.32	1	.32	<1	—
Cubic	.26	1	.26	<1	—
Trials × Feed.	167	3	55.67	50.15	<.001
Linear	167.20	1	167.20	71.15	<.001
Quadratic	.00	1	.00	<1	—
Cubic	.01	1	.01	<1	—
Trials × Exp. × Feed.	49	3	16.33	14.71	<.001
Linear	48.04	1	48.04	20.44	<.001
Quadratic	.90	1	.90	1.25	n.s.
Cubic	.11	1	.11	<1	—
Error$_{within}$	80	72	1.11	—	—
Linear	56.33	24	2.35	—	—
Quadratic	17	24	.72	—	—
Cubic	6.67	24	.28	—	—

The significant linear components mean that the data points tend to lie along a straight line.

If there had been significant quadratic components, it would have meant that there was a tendency for the data to follow quadratic functions (such as a parabola). Significant cubic components would have meant that there was a tendency for the data to follow an even more complex path with two inflections in the functions. The interpretation of linear, quadratic, and cubic components of interactions is very complex because at least two sets of data points are involved in each. Only inspection of the data or additional analyses of each individual set of data points (using the same procedures just outlined) can give an indication as to which of the sets is contributing to the significant findings.

Part 4

Correlation and Related Topics

The sixteen sections in this part present six rather different kinds of analyses. These six types are listed below, followed by a more detailed description of the different kinds of problems that can be handled with each analysis. This material should enable you to choose quickly the section that most clearly suits your individual needs.

1. Pearson product-moment correlation (the common, or usual, correlation)
2. Rank-order correlation, point-biserial correlation, and the correlation ratio *(eta)*
3. Partial and multiple correlation, simple and multiple regression
4. Analyses of covariance
5. Tests for reliability of measurement
6. Significance tests for deciding whether two correlation coefficients are different in magnitude

Pearson Product-Moment Correlation (Section 4.1)

Use the correlation coefficient described in Section 4.1 if your numbers represent amounts of some measurable quantity—such as height, age, IQ, grade points, test scores, etc. This analysis assumes that the two variables for which you have measures are linearly related. If you have ranked data or are reasonably sure that the variables are related in some complex curved way, use one of the correlation coefficients below.

Rank-Order Correlation, Point-Biserial Correlation, and Eta (Sections 4.2, 4.3, 4.4., 4.5)

If you have ranked data or wish to convert measures to ranks, use one of the rank-order correlations described in Sections 4.2 and 4.3. These

174

analyses make no assumptions about the shape of the relationship between variables.

If you have measures along one variable, such as grade points, and have another variable that is dichotomized, such as sex, use the point-biserial correlation described in Section 4.4.

If you are fairly certain that the variables are related in some complex curved manner or if you have already computed an F ratio, you can use the correlation ratio *(eta)*, described in Section 4.5, to express the relationship between the two variables.

Partial and Multiple Correlation, Simple and Multiple Regression (Sections 4.6, 4.7, 4.8, 4.9, and 4.10)

If you have three variables and wish to know how highly two of them are related when the mutual relationships with the third variable are taken out, use the partial correlation described in either Section 4.6 or Section 4.7.

If you have three variables and wish to know how highly two of them, taken together, are related to the third, use the multiple correlation described in Section 4.8.

Simple regression (Section 4.9) is used when a researcher wishes to use the scores on one variable to predict the scores on another variable.

Multiple regresssion (Section 4.10) is used if you wish to determine the weights which will give the highest possible correlation between the predicted and observed values of the criterion variable.

Covariance (Sections 4.11 and 4.12)

When you wish to make the kinds of tests of significance described in Sections 2.1 and 2.2 (the completely randomized design and the two-factor factorial design) but cannot select random groups that are essentially equal, you must make statistical corrections for such differences. For example, if one group is composed of grade-school children and another of high school students and you wish to measure the amount of verbal material learned in a certain interval of time, you would reasonably expect the groups to differ initially by a large amount because of their different levels of education. The analyses of covariance, described in Section 4.11 and 4.12, provide a technique whereby you can equate the two groups before training begins.

A major condition that must be met if the analysis of covariance is to be unequivocally interpreted is that the group regressions are homogeneous. Whether this condition is met or not can be tested by referring to the supplements at the end of Section 4.11 and 4.12.

Reliability of Measurement (Sections 4.13 and 4.14)

Whenever measures are gathered by some sort of testing, the validity of the measures depends upon the quality of the test being used. The tests for the reliability of measurement involve particular interpretations of a correlation coefficient. Sections 4.13 and 4.14 consider the topic of reliability of measurement.

Significance Tests of Differences Between Correlations (Sections 4.15 and 4.16)

Each correlation coefficient is regarded as being either (1) indicative of a real relationship or (2) due to chance variation. Each of Sections 4.1 through 4.8 presents a test to help you decide which of the two cases is more likely for a particular correlation value. However, it is often desirable to know whether two correlation coefficients differ by more than would be expected by chance. Sections 4.15 and 4.16 present tests to help evaluate differences between correlation values. If the two correlation coefficients are computed using completely different groups of people (that is, if the correlations are independent), use the test presented in Section 4.15. If the correlation values have been computed using the same people (that is, if the correlations are dependent), use the test presented in Section 4.16.

In this part, computational formulas will again be presented in those instances in which a complete analysis can be shown by a single formula or in which a standard set of terms and symbols is used in nearly all textbooks.

SECTION 4.1
Pearson Product-Moment Correlation

The Pearson product-moment correlation (r) is used to determine if there is a relationship between two sets of paired numbers. Generally, the paired numbers are either (1) two different measures on each of several objects or persons or (2) one measure on each of several pairs of objects or people where the pairing is based on some natural relationship, such as parent to child, or on some initial matching according to one specific varible, such as IQ score.

Example

Assume that an experimenter wishes to determine whether there is a relationship between the grade point averages (GPAs) and the scores on a reading-comprehension test of fifteen college freshmen.

The basic computational formula for the Pearson product-moment correlation is

$$r = \frac{N\Sigma XY - (\Sigma X)(\Sigma Y)}{\sqrt{[N\Sigma X^2 - (\Sigma X)^2][N\Sigma Y^2 - (\Sigma Y)^2]}}$$

where N = number of pairs of scores

ΣXY = sum of the products of the paired scores

ΣX = sum of scores on one variable

ΣY = sum of scores on the other variable

ΣX^2 = sum of the squared scores on the X variable

ΣY^2 = sum of the squared scores on the Y variable

Step 1. The scores must be paired in some meaningful way in order to use the Pearson r. In the present example, the two different scores— reading comprehension and grade point average—are paired and recorded for each of the fifteen students.

Student	Reading Score (X)	Freshman GPA (Y)
S_1	38	2.1
S_2	54	2.9
S_3	43	3.0
S_4	45	2.3
S_5	50	2.6
S_6	61	3.7
S_7	57	3.2
S_8	25	1.3
S_9	36	1.8
S_{10}	39	2.5
S_{11}	48	3.4
S_{12}	46	2.6
S_{13}	44	2.4
S_{14}	39	2.5
S_{15}	48	3.3

Step 2. Multiply the two numbers in each pair; then add the products.

$$(38 \times 2.1) + (54 \times 2.9) + \cdots + (48 \times 3.3) = \mathbf{1846.1}$$

Step 3. Multiply the number obtained in Step 2 by N, the number of *paired* scores (15 in this example).

$$1846.1 \times 15 = \mathbf{27{,}691.5}$$

Step 4. Square each number in the first column, and add the squared values.

$$38^2 + 54^2 + \cdots + 48^2 = \mathbf{31{,}327}$$

Step 5. Multiply the sum of Step 4 by the number of paired scores ($N = 15$ in this example).

$$31{,}327 \times 15 = \mathbf{469{,}905}$$

Step 6. Add all the scores in the first column (in this example, the reading-comprehension scores).

$$38 + 54 + \cdots + 48 = \mathbf{673}$$

Step 7. Square the value obtained in Step 6.

$$673^2 = \mathbf{452{,}929}$$

Step 8. Square each number in the second column, and add the squared values. (*Note:* The value of Step 10 can be obtained at the same time, as explained in Step 4.)

$$2.1^2 + 2.9^2 + \cdots + 3.3^2 = \mathbf{110.2}$$

Step 9. Multiply the result of Step 8 by the number of paired scores ($N = 15$ in this example).

$$110.2 \times 15 = \mathbf{1653}$$

Step 10. Add all the scores in the second column (in this example, the freshman GPAs).

$$2.1 + 2.9 + \cdots + 3.3 = \mathbf{39.6}$$

Step 11. Square the value obtained in Step 10.

$$39.6^2 = \mathbf{1568.16}$$

Step 12. Multiply the final value of Step 6 by that of Step 10.

$$673 \times 39.6 = \mathbf{26{,}650.8}$$

Step 13. The numerator of r is now computed by subtracting the value obtained in Step 12 from that obtained in Step 3. Be careful to note whether the answer is positive or negative.

$$27{,}691.5 - 26{,}650.8 = \mathbf{+1040.7}$$

Step 14. Subtract the value of Step 7 from that of Step 5.

$$469,905 - 452,929 = \mathbf{16,976}$$

Note: If the variance of these first-column scores is desired, divide the result of Step 14 by N^2, i.e., $16,976/15^2 = 16,976/225 = 75.45$. Or if an estimate of the variance of the population from which these scores were drawn is desired, divide by $N(N - 1)$, i.e., $16,976/[15(15 - 1)] = 16,976/210 = 80.84$.

Step 15. Subtract the final value of Step 11 from that of Step 9.

$$1653 - 1568.16 = \mathbf{84.84}$$

Note: The variance of these second-column scores can be obtained by dividing this result by N^2, i.e., $84.84/15^2 = 84.84/225 = .38$. For the estimate of the population variance, divide by $N(N - 1)$, i.e., $84.84/210 = .40$.

Step 16. Multiply the result of Step 14 by the result of Step 15.

$$16,976 \times 84.84 = \mathbf{1,440,243.84}$$

Step 17. Take the square root of the result of Step 16.

$$\sqrt{1,440,243.84} = \mathbf{1200.1}$$

Step 18. Divide the value of Step 13 by that of Step 17. This yields the value of the Pearson product-moment correlation.

$$r = \frac{+1040.7}{1200.1} = \mathbf{+.87}$$

Supplement

Testing the significance of r: Two different procedures are used to test the hypothesis that $r = 0$. If N (the number of pairs) is 30 or larger, a critical-ratio z-test can easily be done. If N is smaller than 30, a slightly more complicated t-test should be done.

Step 19. If N is 30 or larger, compute $z = r\sqrt{N - 1}$. For example, suppose $r = -.56$ and $N = 37$. Then,

$$z = (-.56)\sqrt{37 - 1} = (-.56)\sqrt{36} = -.56 \times 6 = \mathbf{-3.36}$$

If z is greater than ± 1.96, then r is significant at the .05 level using a two-tailed test (see Appendix A).

Step 20. If N is smaller than 30, compute $t = r\sqrt{(N-2)/(1-r^2)}$. Using the value of r from Step 18 and the value of N used to compute that r, we have $r = +.87$ and $N = 15$. Thus,

$$t = (+.87)\sqrt{\frac{15-2}{1-.87^2}} = .87\sqrt{\frac{13}{1-.76}}$$

$$= .87\sqrt{54.16} = .87 \times 7.36 = +6.40$$

The degrees of freedom for this t are $N - 2$.

$$df = 15 - 2 = 13$$

With 13 degrees of freedom, a t value larger than ± 2.16 (see Appendix B) is significant at the .05 level when a two-tailed test is used.

Note: Several critical values of r are given in Appendix G for various degrees of freedom.

SECTION 4.2
Spearman Rank-Order Correlation (*rho*)

Spearman's *rho* is used when an experimenter wishes to determine whether two sets of rank-ordered data are related.

Example

The experimenter first asked that twenty children in a class be ranked according to what their teacher believed their intelligence to be. Then the children were tested and their actual Wechsler IQ scores obtained. The data were as follows.

Child	Teacher's Ranking	IQ Score
A	1	116
B	2	111
C	3	97
D	4	122
E	5	116
F	6	105
G	7	108
H	8	95
I	9	124
J	10	98
K	11	116
L	12	109
M	13	103
N	14	103

Child	Teacher's Ranking	IQ Score
O	15	96
P	16	90
Q	17	134
R	18	87
S	19	96
T	20	91

The following formula for rho (ρ) describes the computational procedures:

$$rho = 1 - \frac{6\Sigma D^2}{N(N^2 - 1)}$$

where D = difference score between each X and Y pair

N = number of pairs of scores

Step 1. Rank the IQ scores so that both variables are ranked.

Note: When two children have the same IQ score, the same rank is given to each. But notice that the rank given is the mean value of the two ranks for the two tied scores. For example, Child M and Child N both have IQ scores of 103. These scores should fall into ranks 11 and 12. But since both children have the same IQ score, both are ranked as 11.5 and the next score (Child J, with an IQ score of 98) is ranked as 13.

Child	Teacher's Ranking	IQ Rank on the Basis of Test Score
A	1	5
B	2	7
C	3	14
D	4	3
E	5	5
F	6	10
G	7	9
H	8	17
I	9	2
J	10	13
K	11	5
L	12	8
M	13	11.5
N	14	11.5
O	15	15.5
P	16	19
Q	17	1
R	18	20
S	19	15.5
T	20	18

Step 2. Compute the difference between the two ranks for each child. The resulting value is called the *D value.* List these values in a column, making sure to note whether they are positive or negative.

Child	D Value	
A	−4	(1 − 5)
B	−5	(2 − 7)
C	−11	(3 − 14)
D	+1	(4 − 3)
E	0	(5 − 5)
F	−4	(6 − 10)
G	−2	(7 − 9)
H	−9	(8 − 17)
I	+7	(9 − 2)
J	−3	(10 − 13)
K	+6	(11 − 5)
L	+4	(12 − 8)
M	+1.5	(13 − 11.5)
N	+2.5	(14 − 11.5)
O	−0.5	(15 − 15.5)
P	−3	(16 − 19)
Q	+16	(17 − 1)
R	−2	(18 − 20)
S	+3.5	(19 − 15.5)
T	+2	(20 − 18)

Step 3. Square all the *D* values from Step 2, and add all the squared values.

$$(-4^2) + (-5^2) + \cdots + 2^2 = \mathbf{668}$$

Step 4. Multiply the result of Step 3 by the number 6. (*Note:* The number 6 is *always* used, regardless of the number of ranks, etc., involved.)

$$668 \times 6 = \mathbf{4008}$$

Step 5. Compute $N(N^2 - 1)$. (In our example $N = 20$.)

$$20(20^2 - 1) = 20(400 - 1) = 20(399) = \mathbf{7980}$$

Step 6. Divide the result of Step 4 by the result of Step 5.

$$\frac{4008}{7980} = \mathbf{.50}$$

Step 7. Subtract the final value of Step 6 from the number 1. (*Note:* The number 1 is also always used.) This yields the value of Spearman's *rho.* Be careful to record whether it is positive or negative.

$$rho = 1 - .50 = \mathbf{+.50}$$

Supplement

Testing the significance of *rho:* Two different procedures are needed to test the hypothesis that $rho = 0$. If N (the number of ranks) is 30 or larger, a critical-ratio z-test can easily be done. If N is between 10 and 30, a slightly more complicated t-test should be done. No test is presented here for the significance of *rho* where N is less than 10 since it is rare that so few ranks would be considered.

   Step 8. When N is 30 or greater, compute $z = rho \sqrt{N-1}$. For example, suppose $rho = -.21$ and $N = 50$. Then,

$$z = (-.21) \sqrt{50-1} = (-.21) \sqrt{49} = -.21 \times 7 = -1.47$$

If z is greater than ± 1.96, then *rho* is significant at the .05 level using a two-tailed test (see Appendix A).

   Step 9. When N is between 10 and 30, compute

$$t = rho \sqrt{\frac{N-2}{1-rho^2}}$$

Using the value of *rho* from Step 7 above and the value of N used to compute that *rho*, we have $rho = +.50$ and $N = 20$. Thus,

$$t = (+.50) \sqrt{\frac{20-2}{1-rho^2}} = (+.50) \sqrt{\frac{18}{1-.25}} = (+.50) \sqrt{\frac{18}{.75}}$$

$$= (+.50) \sqrt{24} = +.50 \times 4.9 = +2.45$$

The degrees of freedom for this t are $N - 2$. Thus,

$$df = 20 - 2 = 18$$

With 18 degrees of freedom, a t value larger than ± 2.10 (see Appendix B) is significant at the .05 level when a two-tailed test is used.

SECTION 4.3
Kendall Rank-Order Correlation *(tau)*

Kendall's *tau* is sometimes used in place of Spearman's *rho*, which was presented in Section 4.2. If you have a pair of ranks for each of several individuals, the *tau* statistic can be computed to express the degree of relationship between the ranks. Its significance can be tested as follows.

Example

Assume that an experimenter wishes to determine the degree of relationship between the ratings of two judges on ten contestants in a figure-skating contest. The measures recorded represent the ranks given to the contestants by each of the judges.

The usual formula for *tau* when there are no tied ranks is

$$tau = \frac{P = Q}{\frac{1}{2}N(N - 1)}$$

where P = total number of higher ranks

Q = total number of lower ranks

N = number of pairs of ranks

(*Note:* See the following for the explanation of "higher" and "lower.")

A more general computation, covering a case where some ranks are tied, is presented here. Steps 7 through 12 show how to correct for the tied ranks.

Step 1.   Table the data as follows. In computing *tau,* always arrange the first column of ranks so that they range from the lowest to the highest as you go down the column.

Contestant	Ranking by Judge A	Ranking by Judge B
A	1	3
B	2	5
C	3.5	1.5
D	3.5	4
E	5	1.5
F	6	6
G	8	8
H	8	7
I	8	10
J	10	9

Step 2. The ranks in the first column run from the lowest to the highest. The ranks in the second column are "mixed up." We must determine the extent of the "mix-up" in Column 2. In order to do so, count the number of ranks in Column 2 that are higher than that for Contestant A—i.e., the number of ranks that are higher in numerical value than the number 3. In this example, only the two 1.5's are lower than 3; therefore there are seven numbers higher than 3.

No. of higher ranks = 7

Then, count the number of ranks that are lower than 3.

No. of lower ranks = **2**

Step 3. Next, count the number of ranks in Column 2 that are higher than the rank for Contestant B—i.e., the number of ranks higher than 5. In this example, 6, 8, 7, 10, and 9 are higher than 5. Thus,

No. of higher ranks = **5**

Then count the number of ranks lower than 5. Do *not* include ranks that have already been considered (Contestant A). With the ranking for A excluded,

No. of lower ranks = **3**

Step 4. Now count the number of ranks that are higher than the rank for Contestant C—i.e., the number of ranks higher than 1.5. All ranks except the other 1.5 are higher, but ties are not counted, and we have already taken care of numbers 3 and 5 for Contestants A and B. Thus,

No. of higher ranks = **6**

Now count the number of ranks lower than 1.5.

No. of lower ranks = **0**

Step 5. Continue counting both the number of ranks higher and the number of ranks lower than the one under consideration for all the rest of the ranks in Column 2. Table the results as follows. (*Note:* The values for the bottom contestant are both zero since no ranks appear below them.)

Contestant	No. of Higher Ranks	No. of Lower Ranks
A	7	2
B	5	3
C	6	0
D	5	1
E	5	0
F	4	0
G	2	1
H	2	0
I	0	1
J	0	0

Step 6. Add the numbers in each of the columns in Step 5.

$7 + 5 + \cdots + 0 = $ **36** $ = $ sum of Column 1 (higher ranks)

$2 + 3 + \cdots + 0 = $ **8** $ = $ sum of Column 2 (lower ranks)

Then, subtract the sum of Column 2 from the sum of Column 1. Be careful to record whether the result is positive or negative.

$$36 - 8 = +28$$

Step 7. Go to the table in Step 1, and count the number of ranks that are tied in the first column (the ranks assigned by Judge A). Categorize each tie as being a set of two, three, four, etc., according to the number of contestants in that tie. Then, compute $x(x - 1)$ for each tie, where x is the number of contestants tied for that place—e.g., $2(2 - 1) = 2(1) = 2$; $5(5 - 1) = 5(4) = 20$; etc. Add these products, and divide by 2. (*Note:* The number 2 is always used.) In this example, Contestants C and D tied for the third and fourth positions (both were assigned to the rank of 3.5), and Contestants G, H, and I tied for the seventh, eighth, and ninth places (all were assigned to the rank of 8). In this example, then, there is one set of two and one set of three. Thus,

$$\frac{2(2 - 1) + 3(3 - 1)}{2} = \frac{2(1) + 3(2)}{2} = \frac{2 + 6}{2} = \frac{8}{2} = 4$$

Step 8. Repeat all the computations of Step 7 using the ranks in the second column of Step 1 (the ranks assigned by Judge B). Notice that Contestants C and E are tied at a rank of 1.5. Thus, we have one set of two, so

$$\frac{2(2 - 1)}{2} = \frac{2(1)}{2} = \frac{2}{2} = 1$$

Step 9. Compute $[N(N - 1)]/2$, where N refers to the total number of ranks in each column (10 in this example).

$$\frac{N(N - 1)}{2} = \frac{10(10 - 1)}{2} = \frac{10(9)}{2} = \frac{90}{2} = 45$$

Step 10. Subtract the result of Step 7 from the result of Step 9.

$$45 - 4 = 41$$

Step 11. Subtract the result of Step 8 from the result of Step 9.

$$45 - 1 = 44$$

Step 12. Multiply the result of Step 10 by the result of Step 11.

$$41 \times 44 = 1804$$

Then, take the square root of the product.

$$\sqrt{1804} = 42.474$$

Step 13. Divide the result of Step 6 by the result of Step 12. This yields the value of Kendall's *tau.*

$$tau = +\frac{28}{42.474} = +.66$$

Supplement

Testing the significance of *tau* (i.e., the hypothesis that *tau* = 0): When there are 10 or more ranks and when the number of tied ranks is not too great, the following significance test will yield approximate values.

Step 14. Compute $[N(N-1)(2N+5)]/18$, where N is the total number of ranks in each column of Step 1 (10 in this example).

$$\frac{10(10-1)[2(10)+5]}{18} = \frac{10(9)(20+5)}{18} = \frac{90(25)}{18} = \frac{2250}{18} = 125$$

Step 15. Take the square root of the result of Step 14.

$$\sqrt{125} = 11.18$$

Step 16. Divide the result of Step 6 by the answer of Step 15. This yields a *z* statistic.

$$z = +\frac{28}{11.18} = +2.50$$

If the result of Step 16 is larger than ± 1.96 (see Appendix A), then *tau* is significant at the .05 level using a two-tailed test.

SECTION 4.4
Point-Biserial Correlation

The point-biserial correlation is used when a coefficient of relationship is desired between one measure that is continuous and another that is dichotomous.

Example

Assume that an experimenter wishes to determine the relationship between scores on a certain college entrance examination and whether the student obtained a grade point average of C or better during the freshman year. Twelve students are picked at random, and their entrance-examination scores and their course grades are recorded.

The continuous variable in this example is the array of scores on the college entrance examination. The dichotomous variable is the grade point category, which is determined by whether the student got (1) a C or better or (2) lower than a C. The number 1 represents the grade of each person who obtained a C or better. A cipher, 0, represents the grade of those who got lower than a C.

The formula for the point-biserial correlation is

$$r_{pb} = \frac{\overline{Y}_1 - \overline{Y}_0}{s_y} \sqrt{\frac{N_1 N_0}{N(N-1)}}$$

where $\overline{Y}_1$ = the mean of the values of the continuous variable for persons in dichotomous category 1

$\overline{Y}_0$ = the mean of the values of the continuous variable for persons in dichotomous category 0

N_1 = the number of persons in dichotomous category 1

N_0 = the number of persons in dichotomous category 0

s_y = The estimated standard deviation of the population of continuous-variable values from which the sample was taken

N = the total number of persons in the sample $(N_1 + N_0)$

The formula for the estimated standard deviation of the population is

$$s_y = \sqrt{\frac{\Sigma Y^2}{N-1} - \frac{(\Sigma Y)^2}{N(N-1)}}$$

Step 1. Table the data as follows.

Student	College-Entrance-Examination Score (Y)	GPA Category
A	104	1
B	82	0
C	92	1
D	76	0
E	101	1
F	111	1
G	98	1
H	85	0
I	88	1
J	67	0
K	106	1
L	95	0

Step 2. Make a list of all those examination scores paired with the GPA category 1 and a list of all those scores paired with the GPA category 0.

Y_1 Scores (Paired with GPA 1)	Y_0 Scores (Paired with GPA 0)
104	82
92	76
101	85
111	67
98	95
88	
106	

Step 3. Compute the means of the two groups of scores.

$$\bar{Y}_1 = \frac{104 + 92 + \cdots + 106}{7} = \frac{700}{7} = 100$$

$$\bar{Y}_0 = \frac{82 + 76 + \cdots + 95}{5} = \frac{405}{5} = 81$$

Step 4. Perform the subtraction $\bar{Y}_1 - \bar{Y}_0$, noting and recording the sign of the result.

$$100 - 81 = +19$$

Step 5. Obtain $\sqrt{(N_1 N_0)/[N(N-1)]}$. (In our example, $N_1 = 7$, $N_0 = 5$, and $N = N_1 + N_0 = 12$.)

$$\sqrt{\frac{(7)(5)}{12(11)}} = \sqrt{\frac{35}{132}} = \sqrt{.265} = .515$$

Step 6. Square all the examination scores of Step 1, and add the squares. (*Note:* Steps 6 and 8 can be done at the same time if you are using a calculating machine.)

$$104^2 + 82^2 + \cdots + 95^2 = 103,625$$

Step 7. Multiply the value obtained in Step 6 by N (12 in our example).

$$103,625 \times 12 = 1,243,500$$

Step 8. Add all the examination scores.

$$104 + 82 + \cdots + 95 = 1105$$

Step 9. Square the value obtained in Step 8.

$$1105^2 = \mathbf{1,221,025}$$

Step 10. Subtract the result of Step 9 from the result of Step 7.

$$1,243,500 - 1,221,025 = \mathbf{22,475}$$

Step 11. Divide the result of Step 10 by $N(N - 1)$.

$$\frac{22,475}{N(N - 1)} = \frac{22,475}{12(11)} = \frac{22,475}{132} = \mathbf{170.265}$$

Step 12. Take the square root of the result of Step 11.

$$\sqrt{170.265} = \mathbf{13.05}$$

Step 13. Divide the result of Step 4 by the result of Step 12.

$$+\frac{19}{13.05} = \mathbf{+1.456}$$

Step 14. Multiply the result of Step 13 by the result of Step 5. This yields the value of r_{pb}. (We see from Step 4 that the value will be positive when $\overline{Y}_1$ is larger than $\overline{Y}_0$.)

$$r_{pb} = +1.456 \times .515 = \mathbf{+.75}$$

Supplement

Testing the significance of r_{pb}: The hypothesis that $r_{pb} = 0$ can be tested by the *t*-test

$$t = r_{pb} \sqrt{\frac{N - 2}{1 - r_{pb}^2}}$$

which has $N - 2$ degrees of freedom. In our example, $r_{pb} = +.75$ and $N = 12$. Thus,

$$t = (+.75) \sqrt{\frac{12 - 2}{1 - (+.75)^2}} = (+.75) \sqrt{\frac{10}{1 - .562}}$$

$$= (+.75) \sqrt{\frac{10}{.438}} = (+.75) \sqrt{22.831} = (+.75) (4.78) = +3.58$$

With $N - 2 = 12 - 2 = 10$ degrees of freedom, a *t* value larger than ± 2.23 is significant at the .05 level using a two-tailed test (see Appendix B).

SECTION 4.5

The Correlation Ratio (*eta*)

As a measure of relationship between two variables, *eta* can be useful in two different situations: (1) when the relationship is sufficiently non-linear to make you hesitant about using the Pearson correlation or (2) when the experiment involves independent and dependent measures and you wish to know the degree of their relationship.

Step 1. *Eta* is most clearly defined as

$$eta = \sqrt{\frac{SS_{\text{between groups}}}{SS_{\text{total}}}}$$

where SS_{between} and SS_{total} are sums of squares that are computed in a typical analysis of variance. For the purposes of illustration, we will use the SS_b and SS_t from the last step of Section 2.1.

$$SS_b = 567 \qquad SS_t = 1132$$

Step 2. From these values we compute

$$eta = \sqrt{\frac{567}{1132}} = \sqrt{.50} = .71$$

This value of *eta* gives a rough indication of the correlation between the independent and dependent variables in the experiment discussed in Section 2.1.

Step 3. In order to compute *eta* from raw data, turn to Section 2.1 and follow all the steps for the completely randomized design. To test the *eta* for significance, complete the *F*-test for significance of the group means as given in Section 2.1. The results of that test are also the results of a significance test of *eta*. For the example above, we see that *eta* is judged to be different from zero.

$$F = \frac{ms_{\text{between}}}{ms_{\text{within}}} = \frac{189}{12.84} = 14.71 \qquad df = 3/44$$

Regardless of which use of *eta* (as presented at the beginning of this section) you wish to make, compute the sums of squares according to the steps presented in Section 2.1; then follow the directions in the present section.

Supplement

It is sometimes important that one know the *eta* for research reported in a publication. The relationship between a particular independent variable and the dependent measure used as a criterion can be found if the *F* ratio and the degrees of freedom are reported. The procedure for computing *eta* from a given *F* value is as follows. We will use the data presented in Section 2.7 for the purposes of illustration.

Step 1. Obtain the *F* of interest (here, the *F* for trials) and the appropriate degrees of freedom.

$$F = 52.46 \qquad df = 2/60$$

Step 2. Multiply the *F* value of Step 1 by the numerator *df*.

$$52.46 \times 2 = 104.92$$

Step 3. Add the result of Step 2 to the denominator *df* from Step 1.

$$104.92 + 60 = 164.92$$

Step 4. Divide the result of Step 2 by the result of Step 3. This yields the square of *eta.*

$$\frac{104.92}{164.92} = .64$$

Step 5. Take the square root of the final value of Step 4.

$$eta = \sqrt{.64} = .80$$

This value of *eta* indicates the degree of relationship between the trials variable and the dependent variable, which is the number of correct responses per trial in the verbal-learning situation.

SECTION 4.6

Partial Correlation: Three Variables

Partial correlation is used when you have three sets of measures that are related and you wish to find the relationship between any two when the relationship effect of the third has been taken out of both variables.

Example

Suppose you have measures of (a) incomes of a group of fathers, (b) incomes of their eldest sons, and (c) number of years of formal education of the sons. A correlation of $+.60$ between incomes of fathers and sons might be interpreted as being indicative that money-making characteristics run in families. However, an observer might argue that the most probable contributor to the similarity in incomes of fathers and sons is level of education. A partial correlation between incomes of fathers and sons with the effect of education "partialed" out would help solve the controversy.

The notation for partial correlation is $r_{ab.c}$, which is read as "the partial correlation between variables a and b with the relational effects of c taken out." The formula for partial correlation is

$$r_{ab.c} = \frac{r_{ab} - r_{ac}r_{bc}}{\sqrt{1 - r_{ac}^2}\,\sqrt{1 - r_{bc}^2}}$$

Step 1. Compute the simple correlation between each of the three possible pairs of measures—r_{ab}, r_{ac}, and r_{bc}. To do this, follow the steps provided in Section 4.1 for each of the three correlation coefficients needed. Let us assume that the three correlations for this example turned out to be

$$r_{ab} = +.60 \qquad r_{ac} = +.70 \qquad r_{bc} = +.80$$

Step 2. Compute the partial correlation between a and b.

$$r_{ab.c} = \frac{.60 - (.70)\,(.80)}{\sqrt{1 - .70^2}\,\sqrt{1 - .80^2}} = \frac{.60 - .56}{\sqrt{1 - .49}\,\sqrt{1 - .64}}$$

$$= \frac{+.04}{\sqrt{.51}\,\sqrt{.36}} = \frac{+.04}{(.714)\,(.60)} = \frac{+.04}{.428} = +.09$$

Thus, the correlation $r_{ab} = +.60$ is reduced to $r_{ab.c} = +.09$ when the mutual relationships of the variables with c are partialed out. These data suggest that the hypothesis that money-making ability is hereditary is false and that the reason fathers and sons tend to have similar incomes is that they tend to have similar levels of education.

Step 3. The other two partial correlations could be computed in a similar manner if desired.

$$r_{ac.b} = \frac{r_{ac} - r_{ab}r_{cb}}{\sqrt{1 - r_{ab}^2}\,\sqrt{1 - r_{cb}^2}} = \frac{.70 - (.60)\,(.80)}{\sqrt{1 - .60^2}\,\sqrt{1 - .80^2}} = \frac{+.22}{.48} = +.46$$

$$r_{bc.a} = \frac{r_{bc} - r_{ba}r_{ca}}{\sqrt{1 - r_{ba}^2}\,\sqrt{1 - r_{ca}^2}} = \frac{.80 - (.70)\,(.60)}{\sqrt{1 - .70^2}\,\sqrt{1 - .60^2}} = \frac{+.38}{.571} = +.67$$

SECTION 4.7
Partial Rank-Order Correlation (Using Kendall's *tau*)

The *tau* statistics of Section 4.3 may be extended to the case of partial correlation in either of two cases: (1) when there are no tied ranks or (2) when there are tied ranks but the rank-order correlations are expected to yield only rough indications of true values. No significance test for partial *tau* can be presented since the sampling distribution is unknown.

Example

Suppose that a judge has rank ordered fifteen individuals on the basis of IQ, personality, and poise. In this case, the experimenter wishes to determine the correlation between each possible pair of rankings when the effects of the third have been taken out. The data might appear as follows.

Subject	Ranking 1 (IQ)	Ranking 2 (Personality)	Ranking 3 (Poise)
S_1	1	3	1
S_2	2	5	4
$\vdots$	$\vdots$	$\vdots$	$\vdots$
S_{15}	15	13	12

Step 1. First, go to Section 4.3 and compute the *tau* value for each of the three possible pairs of rankings—1 with 2 (tau_{12}), 1 with 3 (tau_{13}), and 2 with 3 (tau_{23}). For this example, suppose

$$tau_{12} = +.50 \qquad tau_{13} = +.60 \qquad tau_{23} = +.80$$

Step 2. Compute the partial correlation of 1 with 2, partialing out the relationships with 3.

$$tau_{12.3} = \frac{tau_{12} - tau_{13}\, tau_{23}}{\sqrt{1 - tau_{12}^2}\ \sqrt{1 - tau_{23}^2}} = \frac{.50 - (.60)\,(.80)}{\sqrt{1 - .60^2}\ \sqrt{1 - .80^2}}$$

$$= \frac{.50 - .48}{\sqrt{1 - .36}\ \sqrt{1 - .64}} = \frac{+.02}{\sqrt{.64}\ \sqrt{.36}} = \frac{+.02}{(.80)\,(.60)}$$

$$= \frac{+.02}{.48} = +.04$$

Step 3. In a similar manner, compute the partial correlation of 2 with 3, partialing out the relationships with 1.

$$tau_{23.1} = \frac{tau_{23} - tau_{21}\, tau_{31}}{\sqrt{1 - tau_{21}{}^2}\ \sqrt{1 - tau_{31}{}^2}} = \frac{.80 - (.50)\,(.60)}{\sqrt{1 - .50^2}\ \sqrt{1 - .60^2}}$$

$$= \frac{.80 - .30}{\sqrt{1 - .25}\ \sqrt{1 - .36}} = \frac{+.50}{\sqrt{.75}\ \sqrt{.64}} = \frac{+.50}{(.866)\,(.80)}$$

$$= \frac{+50}{.693} = +.72$$

Step 4. Finally, a compute the partial correlation of 1 with 3, partialing out the relationships with 2.

$$tau_{13.2} = \frac{tau_{13} = tau_{12}\, tau_{23}}{\sqrt{1 - tau_{12}{}^2}\ \sqrt{1 - tau_{23}{}^2}} = \frac{.60 - (.50)\,(.80)}{\sqrt{1 - .50^2}\ \sqrt{1 - .80^2}}$$

$$= \frac{.60 - .40}{\sqrt{1 - .25}\ \sqrt{1 - .64}} = \frac{+.20}{\sqrt{.75}\ \sqrt{.36}} = \frac{+.20}{(.866)\,(.60)}$$

$$= \frac{+.20}{.520} = +.38$$

SECTION 4.8
Multiple Correlation: Three Variables

Multiple correlation is used when you wish to determine the relationship of one set of numbers (measures on a variable) with two other sets of numbers.

Example

In a typical multiple-correlation situation, the first set of numbers represents measures on a criterion variable—in this example, freshman grade point average—and the other two sets of numbers are measures on predictors—here, high-school grades and college-entrance-examination scores. The multiple-correlation coefficient between the criterion variable and the two predictor variables will give an indication of the degree to which the predictors, taken together, actually predict.

Step 1. If you have measures on two predictor variables and one criterion variable, go to Section 4.1 and compute the three simple correlation coefficients according to the instructions there. Let's abbreviate the three variables as:

1 = the criterion variable: freshman grade point average (GPA)
2 = the first predictor: high-school grades
3 = the second predictor: entrance-examination scores

The first correlation coefficient, r_{12}, is between GPA and high-school grades; r_{13} is between GPA and examination scores; and r_{23} is between high-school grades and examination scores. For the purposes of illustration, assume that the simple correlations in this case turned out to be

$$r_{12} = +.60 \qquad r_{13} = +.50 \qquad r_{23} = +.70$$

Step 2. The multiple correlation of 1 with 2 and 3 is given by

$$R_{1.23} = \sqrt{\frac{r_{12}{}^2 + r_{13}{}^2 - 2r_{12}r_{13}r_{23}}{1 - r_{23}{}^2}}$$

Step 3. Simply plug the values of the simple correlations in at the right spots, and carry out the operations.

$$R_{1.23} = \sqrt{\frac{.60^2 + .50^2 - 2(.60)\,(.50)\,(.70)}{1 - .70^2}}$$

$$= \sqrt{\frac{.36 + .25 - 1.20(.50)\,(.70)}{1 - .49}} = \sqrt{\frac{.61 - 1.20(.35)}{.51}}$$

$$= \sqrt{\frac{.19}{.51}} = \sqrt{.373} = .61$$

This is the value of the multiple correlation between the criterion variable and the two predictor variables. The correlation between high-school grades and freshman GPA was .60, and the correlation between the entrance-examination scores and freshman GPA was .50. By combining the high-school grades with the entrance-examination scores, the multiple correlation with freshman GPA was raised to .61. Thus, the combination of predictors produced a higher correlation with the criterion variable than either predictor taken separately.

Concerning the size of the correlation between the predictor variables and their joint correlation with the criterion: If the simple correlations between each of the predictors and the criterion remain unchanged, the value of R, the multiple correlation, will *increase* as the correlation between the predictors becomes smaller than .65, and the value of R will *decrease* as the correlation between predictors increases above .65. In other words, the more independent of each other the predictors are, the greater their value in jointly predicting the criterion, provided the simple correlations between each predictor and the criterion remain constant.

Computation of the multiple-correlation coefficient becomes much more complicated if there are more than two predictor variables. A procedure such as the Doolittle solution, as explained by McNemar,* can be used or, probably better, a competent computer programmer can find a ready-made program that will do the job quickly and accurately.

SECTION 4.9
Simple Regression: *X* Variable to Predict *Y*

Simple regression is used when an experimenter wishes to determine the weight for the *X* variable which will fit the straight line which minimizes the discrepancies between the observed and the predicted values of the criterion variable (i.e., minimizes the discrepancies between *Y* and *Y'*).

Example

A college football coach wishes to determine the best prediction equation for the grade-point averages of potential freshmen recruits. ACT (*X*) test scores for the current group of recruits as well as their grade-point averages ($Y_{observed}$) are available. Based on the available *X* (ACT) and *Y* (GPA) scores for this year's class, the prediction equation for next year's class can be calculated.

Step 1. After the data have been collected, table them as follows. (*Note:* Since sums, sums of squares, and sums of cross products are required for computation of the regression equation, these can be calculated most easily by setting up the following table.)

Subjects	X (ACT)	Y (GPA)	XY (Cross Products)
S_1	20	2.8	56.0
S_2	14	1.8	25.2
S_3	18	2.2	39.6
S_4	22	2.9	63.8
S_5	23	3.5	80.5

*Q. McNemar. *Psychological Statistics*, 4th ed. New York: John Wiley & Sons, Inc., 1969. Chapter 11, pp. 199–202.

Subjects	X (ACT)	Y (GPA)	XY (Cross Products)
S_6	27	3.4	91.8
S_7	28	3.3	92.4
S_8	21	3.6	75.6
S_9	22	2.8	61.6
S_{10}	19	2.7	51.3
S_{11}	18	2.7	48.6
S_{12}	16	2.5	40.0
S_{13}	20	2.6	52.0
S_{14}	20	2.9	58.0
S_{15}	15	2.0	30.0
Sums:	303	41.7	866.4
Sums of Sqs.:	6337	119.83	

Step 2. Note: The computational steps for prediction of grade-point average from ACT scores will be demonstrated in the present example. A similar set of computations, but with different results, would be undertaken if prediction of ACT scores from GPA were desired (X and Y would be reversed).

The formula for calculation of the regression line is

$$Y_{\text{predicted}} = b_{yx}X + a$$

Where b_{yx} = the slope of the line which minimizes the discrepancies between the observed and predicted values of Y (observed and predicted GPAs)

a = the Y intercept, or the value of Y when $X = 0$.

Step 4. The computational formula for b_{yx} is

$$b_{yx} = \frac{N\Sigma XY - (\Sigma X)(\Sigma Y)}{N\Sigma X^2 - (\Sigma X)^2}$$

Using the data calculated in Step 2, the following results are obtained:

$$b_{yx} = \frac{15(866.4) - (303)(41.7)}{15(6337) - (303)^2}$$

$$= \frac{12,996.0 - 12,635.1}{95,055 - 91,809}$$

$$= \frac{360.9}{3246.0}$$

$$= .111$$

Step 5. Computation of a: The a coefficient is found by use of the following formula:

$$a = \overline{Y} - b_{yx}\overline{X}$$

Using the values calculated in Step 2, we have:

$$\overline{Y} = 41.7/15 = \textbf{2.78}$$
$$\overline{X} = 303/15 = \textbf{20.2}$$

Insertion of these values into the formula gives the following:

$$a = 2.78 - (.111)\,(20.2)$$
$$= 2.78 - 2.24$$
$$= \textbf{.54}$$

Step 6. The needed values have been computed. Using the present example:

$$Y_{\text{predicted}} = .111X + .54$$

Thus, if a recruit had an ACT score of 22, the recruit's predicted GPA would be 2.982, i.e.,

$$Y_{\text{predicted}} = (.111)\,(22) + .54$$
$$= 2.442 + .54$$
$$= \textbf{2.98}$$

Step 7. It is rare that the predicted Y score will be identical with the obtained score. The estimate of the deviation is called the *standard error of estimate* and can be calculated using the raw score formula below.* The necessary sums, squares, and cross products were computed and recorded in Step 2.

$$S_{yx} = \sqrt{\frac{\left[\Sigma Y^2 - \frac{(\Sigma Y)^2}{N}\right] - \left[\Sigma XY - \frac{(\Sigma X)(\Sigma Y)}{N}\right]^2 \Big/ \left[\Sigma X^2 - \frac{(\Sigma X)^2}{N}\right]}{N - 2}}$$

Using the values calculated in Step 2, we have

$$S_{yx} = \sqrt{\frac{\left[119.83 - \frac{(41.7)^2}{15}\right] - \frac{\left[866.4 - \frac{(303)(41.7)}{15}\right]^2}{\left[6337 - \frac{(303)^2}{15}\right]}}{15 - 2}}$$

---

*Equivalent formulas often reported in measurement textbooks include:

$$S_{yx} = \sqrt{\frac{\Sigma y^2 - \frac{(\Sigma xy)^2}{\Sigma x^2}}{N - 2}} \qquad \text{and,} \qquad S_{yx} = s_y \sqrt{1 - r_{xy}^2} \cdot \sqrt{N/N - 2}$$

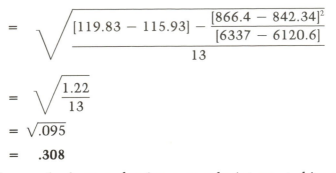

$$= \sqrt{\frac{[119.83 - 115.93] - \frac{[866.4 - 842.34]^2}{[6337 - 6120.6]}}{13}}$$

$$= \sqrt{\frac{1.22}{13}}$$

$$= \sqrt{.095}$$

$$= \quad .308$$

The standard error of estimate may be interpreted in much the same way as a standard deviation. Both are measures of variation—the standard deviation around the mean, the standard error of estimate around the regression line values. In the present example, use of the standard error of estimate would lead to the expectancy that for a given value of X (ACT score), the grade-point averages of approximately 68 percent of the new recruits will fall between $\pm.308$ of their predicted value. Extending the range to two standard deviations, you can expect approximately 95 percent of the grade point averages to fall between $\pm.616$ of their predicted values.

Using the example calculated in Step 6, it would be predicted that 68 percent of new football recruits who score 22 on their ACT test will achieve a GPA between 2.67 and 3.29 and that 95 percent will achieve a grade point between 2.36 and 3.60.

SECTION 4.10
Multiple Regression: Two X Variables

Multiple regression is used when an experimenter wishes to determine the weights for each of the X variables which will give the highest possible correlation between the predicted and the observed values of the criterion variable (Y and Y').

Example

An experimenter wishes to determine the best prediction equation for a group of disadvantaged college students. The ACT score (X_1) and the score on a reading comprehension test (X_2) are available as well as the grade point average earned by each student during his or her first year in college ($Y_{observed}$). Based on the X_1 and X_2 scores, the prediction equation for $Y_{predicted}$ can be obtained.

Step 1. After the data have been collected, table them as follows:

Subjects	X_1	X_2	Y
S_1	12	18	2.2
S_2	7	13	1.4
S_3	9	9	1.8
S_4	11	18	1.6
S_5	13	17	2.5
S_6	10	12	1.9
S_7	14	12	2.0
S_8	12	16	2.4
S_9	8	17	2.6
S_{10}	6	11	1.8
S_{11}	13	18	1.7
S_{12}	15	15	2.0
S_{13}	11	10	1.7
S_{14}	14	14	2.0
S_{15}	15	18	2.3

Step 2. Several sums, sums of squares, and sums of cross-products are required for the computation of the regression equation. These are done most easily by setting up the following table.

Subjects	X_1	X_2	Y	X_1Y	X_2Y	X_1X_2
S_1	12	18	2.2	26.4	39.6	216
S_2	7	13	1.4	9.8	18.2	91
S_3	9	9	1.8	16.2	16.2	81
S_4	11	18	1.6	17.6	28.8	198
S_5	13	17	2.5	32.5	42.5	221
S_6	10	12	1.9	19.0	22.8	120
S_7	14	12	2.0	28.0	24.0	168
S_8	12	16	2.4	28.8	38.4	192
S_9	8	17	2.6	20.8	44.2	136
S_{10}	6	11	1.8	10.8	19.8	66
S_{11}	13	18	1.7	22.1	30.6	234
S_{12}	15	15	2.0	30.0	30.0	225
S_{13}	11	10	1.7	18.7	17.0	110
S_{14}	14	14	2.0	28.0	28.0	196
S_{15}	15	18	2.3	34.5	41.4	270
Sums:	170	218	29.9	343.2	441.5	2524
Sum X^2:	2040	3310	61.29			

Step 3. Three Pearson product-moment correlations must be calculated. The basic computational formula for the Pearson product-moment correlation is

$$r = \frac{N\Sigma XY - (\Sigma X)(\Sigma Y)}{\sqrt{[N\Sigma X^2 - (\Sigma X)^2][N\Sigma Y^2 - (\Sigma Y)^2]}}$$

Values calculated in Step 2 can easily be inserted in appropriate positions in the above formula. The reader who so desires may turn to Section 4.1 for a step-by-step set of instructions for calculating r.

$$r_{x_1y} = \frac{15\,(343.2) - (170)\,(29.9)}{\sqrt{[15\,(2040) - (170)^2]\,[15\,(61.29) - (29.9)^2]}} = \frac{65.0}{207.5} = .31$$

$$r_{x_2y} = \frac{15\,(4415) - (218)\,(29.9)}{\sqrt{[15\,(3310) - (218)^2]\,[15\,(61.29 - (29.9)^2]}} = \frac{104.3}{2321} = .45$$

$$r_{x_1x_2} = \frac{15\,(2524) - (170)\,(218)}{\sqrt{[15\,(2040) - (170)^2]\,[15\,(3310) - (218)^2]}} = \frac{800}{1901} = .42$$

Step 4. The standard deviation of the three variables must also be calculated. The basic formula for the standard deviation is

$$s = \sqrt{\frac{\Sigma X^2 - \dfrac{(\Sigma X)^2}{N}}{N}}$$

Again, values calculated in Step 2 can be inserted into the above formula. If the reader desires, a step by step summary of the computations can be found in Section 1.2

$$s_{x_1} = \sqrt{\frac{2040 - \dfrac{(170)^2}{15}}{15}} = 2.75$$

$$s_{x_2} = \sqrt{\frac{3310 - \dfrac{(218)^2}{15}}{15}} = 3.07$$

$$s_y = \sqrt{\frac{61.29 - \dfrac{(29.9)^2}{15}}{15}} = .336$$

Step 5. All preliminary calculations have been completed. The general formula for the regression equation is:

$$Y_{\text{predicted}} = b_1X_1 + b_2X_2 + \cdots + b_xX_x + a$$

Since there are two variables in the present example, the formula would be

$$Y_{\text{predicted}} = b_1X_1 + b_2X_2 + a$$

Step 6. First, the regression coefficients, b_1 and b_2, must be obtained.

$$b_1 = \frac{s_y \left(r_{x_1y} - r_{x_2y} r_{x_1x_2} \right)}{s_{x_2} \left(1 - r^2_{x_1x_2} \right)}$$

$$= \frac{.336[.31 - (.45)(.42)]}{2.75(1 - .42^2)} = \frac{.0407}{2.265} = .018$$

and,

$$b_2 = \frac{s_y \left(r_{x_2y} - r_{x_1y} r_{x_1x_2} \right)}{s_{x_2} \left(1 - r^2_{x_1x_2} \right)}$$

$$= \frac{.336[.45 - (.31)(.42)]}{3.07(1 - .42^2)} = \frac{.1075}{2.528} = .0425$$

Step 7. Next, the constant, or adjustment, term, a, must be calculated. To calculate this term, the mean of X_1, X_2, and $Y_{observed}$ must be obtained. From the tabled data in Step 2 we find:

$$\overline{X}_1 = \frac{170}{15} = 11.33$$

$$\overline{X}_2 = \frac{218}{15} = 14.53$$

$$\overline{X}_{observed} = \frac{29.9}{15} = 1.99$$

Then, substituting into the formula below, we have

$$a = \overline{Y}_{observed} - b_1 \overline{X} - b_2 \overline{X}_2$$
$$= 1.99 - (.018)(11.33) - (.0425)(14.53)$$
$$1.99 - 2.04 - .618$$
$$= 1.168$$

Step 8. All calculations needed to create the multiple-regression equation needed to predict a Y score based on observed X scores have now been completed. Substituting into the formula presented in Step 3, we have

$$Y_{predicted} = b_1 \overline{X}_1 + b_2 \overline{X}_2 + a$$
$$= .018X_1 + .0425X_2 + 1.168$$

Continuing the initial example, if a student scored 15 on Test 1 (X_1) and 17 on Test 2 (X_2), the $Y_{predicted}$ would be

$$Y_{predicted} = (.018)(15) + (.0425)(17) + 1.168$$
$$= .270 + .7225 + 1.168$$
$$= 2.1605$$

Step 9. If, in addition to the regression equation, a question is raised regarding the multiple correlation between $Y_{observed}$ and $Y_{predicted}$, the multiple correlation can be computed according to the following formula:

$$R_{yy'}^2 = R_{y \cdot x_1 x_2}^2 = \frac{r_{yx_1}^2 + r_{x_2}^2 - 2 r_{yx_1} r_{yx_2} r_{x_1 x_2}}{1 - r_{x_1 x_2}^2}$$

$$= \frac{(.31)^2 + (.45)^2 - 2 (.31) (.45) (.42)}{1 - .1764}$$

$$= \frac{.1814}{.8236}$$

$$= .22$$

and

$$R_{yy'} = \sqrt{.22} = .469$$

Step 10. Test of significance of $R_{yy'}$. If both b_1 and b_2 are equal to zero, then $R_{yy'}$ will be equal to zero. A test of the null hypothesis that the population values β_1 and β_2 are equal to zero will be given by:

$$F = \frac{R_{yy'}^2 / k}{1 - R_{yy'}^2 / (n - k - 1)}$$

$$= \frac{.22/2}{(1 - .22)/(15 - 2 - 1)}$$

$$= \frac{.11}{.065} = 1.69 \text{ with } df = k, n - k - 1 = 2, 12$$

Looking in the F tables under 2 and 12 *df*, we find that an F ratio of 3.88 is required for significance. Since the computed F was less than 3.88, we conclude that the R^2 is nonsignificant.

SECTION 4.11
Simple Analysis of Covariance: One Treatment Variable

When experimental control of an important variable—such as age or IQ—is impossible or impractical, statistical control is possible by use of a covariance analysis. This section presents the procedures for such an analysis when there is only one treatment variable.

Example

In order to make a decision concerning future equipment needs, an elementary-school principal designed the following experiment to evaluate three different methods of teaching spelling. Eighteen first-grade children were given a vocabulary test (the control measure) to determine their verbal ability before the program was begun. The X values below represent these test scores. The three different teaching methods were used for three months, and then a standard spelling test was given. The Y values represent these criterion spelling scores.

Step 1. Table the data as follows.

Teaching Method 1			Teaching Method 2			Teaching Method 3		
Student	X	Y	Student	X	Y	Student	X	Y
A	32	62	G	21	72	M	38	95
B	46	66	H	24	85	N	45	88
C	27	64	I	18	61	O	52	104
D	35	48	J	32	87	P	48	86
E	31	51	K	35	69	Q	41	72
F	40	74	L	26	74	R	37	63

Step 2. Add the scores in each column.

	Method 1		Method 2		Method 3	
	X	Y	X	Y	X	Y
	32	62	21	72	38	95
	46	66	24	85	45	88
	⋮	⋮	⋮	⋮	⋮	⋮
Sums:	211	365	156	448	261	508

Step 3. Square all the numbers in the X columns (Step 1), and add the squared values.

$$32^2 + 46^2 + \cdots + 40^2 + 21^2 + 24^2$$
$$+ \cdots + 26^2 + 38^2 + 45^2 + \cdots + 37^2 = \mathbf{23,448}$$

Step 4. Add the sums of the X columns (Step 2).

$$211 + 156 + 261 = \mathbf{628}$$

Step 5. Square the result of Step 4, and divide by the number of X values (18 in this example).

$$\frac{628^2}{18} = \frac{394,384}{18} = \mathbf{21,910.222}$$

Step 6. Subtract the result of Step 5 from the result of Step 3.

$$23,448 - 21,910.222 = \mathbf{1537.778}$$

Step 7. Square the three sums of the X columns (Step 2) and add the squares; then divide by the number of scores composing each sum (6 in this example).

$$\frac{211^2 + 156^2 + 261^2}{6} = \frac{136,978}{6} = \mathbf{22,829.667}$$

Step 8. Subtract the result of Step 5 from the result of Step 7.

$$22,829.667 - 21,910.222 = \mathbf{919.445}$$

Step 9. Subtract the value of Step 8 from that of Step 6.

$$1537.778 - 919.445 = \mathbf{618.333}$$

Step 10. Square all the numbers in the Y columns (Step 1), and add the squares.

$$62^2 + 66^2 + \cdots + 74^2 + 72^2 + 85^2 + \cdots$$
$$+ 74^2 + 95^2 + 88^2 + \cdots + 63^2 = \mathbf{100,747}$$

Step 11. Add the sums of the Y columns (Step 2).

$$365 + 448 + 508 = \mathbf{1321}$$

Step 12. Square the result of Step 11, and divide by the number of Y values (18 in this example).

$$\frac{1321^2}{18} = \frac{1,745,041}{18} = \mathbf{96,946.722}$$

Step 13. Subtract the result of Step 12 from the result of Step 10.

$$100,747 - 96,946.722 = \mathbf{3800.278}$$

Step 14. Square the three sums of the Y columns (Step 2), and add the squares; then divide by the number of scores composing each sum (6 in this example).

$$\frac{365^2 + 448^2 + 508^2}{6} = \mathbf{98,665.500}$$

Step 15. Subtract the result of Step 12 from the result of Step 14.

$$98,665.500 - 96,946.722 = \mathbf{1718.778}$$

Step 16. Subtract the final value of Step 15 from that of Step 13.

$$3800.278 - 1718.778 = \mathbf{2081.500}$$

Step 17. Multiply each X score by its paired Y score (see Step 1), and add all the products.

$$(32 \times 62) + (46 \times 66) + \cdots + (37 \times 63) = \mathbf{47{,}131}$$

Step 18. Multiply the result of Step 4 by the result of Step 11, and divide by the number of XY pairs (18 in this example).

$$\frac{628 \times 1321}{18} = \frac{829{,}588}{18} = \mathbf{46{,}088.222}$$

Step 19. Subtract the final value of Step 18 from that of Step 17.

$$47{,}131 - 46{,}088.222 = \mathbf{1042.778}$$

Step 20. Multiply each X sum in Step 2 by its paired Y sum, and add the three products; then divide by the number of pairs under each method (6 in this example).

$$\frac{(211 \times 365) + (156 \times 448) + (261 \times 508)}{6} = \frac{279{,}491}{6} = \mathbf{46{,}581.833}$$

Step 21. Subtract the result of Step 18 from the result of Step 20.

$$46{,}581.833 - 46{,}088.222 = \mathbf{493.611}$$

Step 22. Subtract the value of Step 21 from that of Step 19.

$$1042.778 - 493.611 = \mathbf{549.167}$$

Step 23. Square the result of Step 19, and divide by the result of Step 6.

$$\frac{1042.778^2}{1537.778} = \frac{1{,}087{,}385.957}{1537.778} = \mathbf{707.115}$$

Step 24. Subtract the result of Step 23 from the result of Step 13.

$$3800.278 - 707.115 = \mathbf{3093.163}$$

Step 25. Square the result of Step 22, and divide by the result of Step 9.

$$\frac{549.167^2}{618.333} = \frac{301{,}584.394}{618.333} = \mathbf{487.738}$$

Step 26. Subtract the result of Step 25 from the result of Step 16.

$$2081.500 - 487.738 = \mathbf{1593.762}$$

Step 27. Divide the value of Step 26 by the number of degrees of freedom for the adjusted within-groups measures. The degrees of freedom will always be $N - a - 1$, where N is the total number of XY pairs (18 in our example) and a is the number of different experimental groups (3 in our example, corresponding to the three methods of teaching).

$$\frac{1593.762}{N - a - 1} = \frac{1593.762}{18 - 3 - 1} = \frac{1593.762}{14} = \mathbf{113.840}$$

Step 28. Subtract the value of Step 26 from that of Step 24.

$$3093.163 - 1593.762 = \mathbf{1499.401}$$

Step 29. Divide the result of Step 28 by the number of degrees of freedom for the adjusted between-groups measures. This degrees-of-freedom value will always be $a - 1$, where a is the number of different experimental groups (3 in our example).

$$\frac{1499.401}{a - 1} = \frac{1499.401}{3 - 1} = \frac{1499.401}{2} = \mathbf{749.700}$$

Step 30. Compute the F ratio by dividing the result of Step 29 by the result of Step 27.

$$F = \frac{749.700}{113.840} = \mathbf{6.59}$$

This F has $df = (a - 1)/(N - a - 1)$. (This equals 2/14 in our example.) From Appendix E it is seen that with $df = 2/14$, an F ratio larger than 6.51 would be expected by chance alone less than one time in a hundred. Therefore, the F value is said to be significant at the .01 level. We conclude, then, that when the covariance analysis was used to equate statistically the groups according to vocabulary-test score (this equating was necessary since the groups differed at the beginning of training), the methods of teaching provided significantly different results on the spelling test.

Supplement

In order to determine whether the individual group regressions are homogeneous with the overall regression, first compute the sum of squares (SS_x) and the sum of the cross products (SP) for each group of data.

Step 1. Using the data from Step 1 of the original analysis, compute SS_X for each group. The basic formula for each computation is $\Sigma X^2 - (\Sigma X)^2/N_{(\text{per gp.})}$.

$$\text{Group 1 } SS_X = 7655 - \frac{211^2}{6} = \quad 234.8$$

$$\text{Group 2 } SS_X = 4266 - \frac{156^2}{6} = \quad 210.0$$

$$\text{Group 3 } SS_X = 11{,}527 - \frac{261^2}{6} = 173.5$$

Step 2. The SP values are computed in comparable fashion using the formula $\Sigma XY - \Sigma X \, \Sigma Y/N_{(\text{pairs of scores})}$.

$$\text{Group 1 SP} = 12{,}969 - \frac{211 \times 365}{6} = 133.2$$

$$\text{Group 2 SP} = 11{,}773 - \frac{156 \times 448}{6} = 125.0$$

$$\text{Group 3 SP} = 22{,}389 - \frac{261 \times 508}{6} = 291.0$$

Step 3. Now compute SP^2/SS_X for each group.

$$\text{Group 1} = \frac{133.2^2}{234.8} = 75.56$$

$$\text{Group 2} = \frac{125.0^2}{210.0} = 74.40$$

$$\text{Group 3} = \frac{291.0^2}{173.5} = 488.10$$

Step 4. Add the results of Step 3.

$$75.56 + 74.4 + 488.1 = \textbf{638}$$

Step 5. Get the error sum of squares for Y from Section 4.9, Step 16, and subtract the result of Step 4 from it.

$$2081.5 - 638 = \textbf{1443.5}$$

Step 6. Divide the result of Step 5 by $df = N - 2a = 18 - 6 = 12$

$$\frac{1443.5}{12} = \textbf{120.3}$$

Step 7. Get the adjusted error sum of squares for Y from Section 4.11, Step 26, and subtract the result of Step 5 from it.

$$1593.762 - 1443.5 = \mathbf{150.26}$$

Step 8. Divide the result of Step 7 by $df = a - 1 = 3 - 1 = \mathbf{2}$

$$\frac{150.26}{2} = \mathbf{75.13}$$

Step 9. Compute the F ratio.

$$\frac{75.13}{120.3} = \mathbf{.62}$$

Since this F is very small we accept the hypothesis that the group regressions are homogeneous.

SECTION 4.12
Factorial Analysis of Covariance: Two Treatment Variables

The following example presents the computational steps to be used when there are two treatment variables and it is necessary to have statistical control of some other variable to draw meaningful conclusions.

Example

A certain university was engaged in teaching Peace Corps volunteers the foreign languages they would need during their tour of duty. The following experiment was carried out to determine the effectiveness of the different teaching methods that were used. All students were first given a language-aptitude test, which provided the control measures (the X values presented in Step 1 below). It is obvious from these X values that the students' language-aptitude scores differed widely. Since all the students were not able to participate in the experiment at the same time, they were placed in experimental groups on the basis of when they participated rather than on the basis of their aptitude scores. The average aptitude scores of the resulting experimental groups differed considerably. It is for this reason that a covariance analysis was needed. Using statistical procedures, the covariance analysis equated the groups on aptitude scores, so that any differences found after the experiment could be interpreted as results of the experimental manipulations rather than of the original differences in aptitudes.

The experiment that was carried out was to evaluate two methods of teaching the foreign languages and to determine the value of language laboratory sessions. The two teaching methods were (1) formal classroom meetings with lectures and (2) no formal classroom meetings—only conversation periods held in a congenial atmosphere. In addition, half of the students being taught by each teaching method spent three hours a day in the language laboratory using the tape-recording equipment. The other half of the students in each method group never entered the language lab.

Two years later, when the volunteers returned from overseas, they were asked to evaluate the degree to which their language training prepared them for their work. The ratings were on a 10-point scale ranging from 1 to 10. These ratings served as the criterion measures (the Y scores below).

Step 1. Table the data as follows.

Classroom Group 1 (3 Hrs./Day in Lab)			Conversation Group 1 (3 Hrs./Day in Lab)		
Subject	X	Y	Subject	X	Y
S_1	62	5	S_{11}	46	5
S_2	75	7	S_{12}	53	4
S_3	41	3	S_{13}	57	3
S_4	88	8	S_{14}	49	7
S_5	72	7	S_{15}	62	6

Classroom Group 2 (No Time in Lab)			Conversation Group 2 (No Time in Lab)		
Subject	X	Y	Subject	X	Y
S_6	84	2	S_{16}	58	9
S_7	91	3	S_{17}	72	10
S_8	68	1	S_{18}	61	8
S_9	77	1	S_{19}	65	8
S_{10}	85	3	S_{20}	59	10

Step 2. Add the scores in each column, and record the sums.

	Class 1/Lab		Converse 1/ Lab	
	X	Y	X	Y
	62	5	46	5
	75	7	53	4
	⋮	⋮	⋮	⋮
Sums:	338	30	267	25

	Class 2/No Lab		Converse 2/ No Lab	
	X	Y	X	Y
	84	2	58	9
	91	3	72	10
	⋮	⋮	⋮	⋮
Sums:	**405**	**10**	**315**	**45**

The Control Variable (Steps 3–15)

Step 3. Square all the numbers in the X columns (Step 1), and add the squares.

$$62^2 + \cdots + 72^2 + 84^2 + \cdots + 85^2 + 46^2$$
$$+ \cdots + 62^2 + 58^2 + \cdots + 59^2 = \mathbf{91{,}587}$$

Step 4. Add the sums for all four X columns (Step 2).

$$338 + 405 + 267 + 315 = \mathbf{1325}$$

Step 5. Square the result of Step 4, and divide by the total number of X values (20 in this example).

$$\frac{1325^2}{20} = \frac{1{,}755{,}625}{20} = \mathbf{87{,}781.250}$$

Step 6. Subtract the result of Step 5 from the result of Step 3.

$$91{,}587 - 87{,}781.25 = \mathbf{3805.750}$$

Step 7. Add the sums of the X columns in the two classroom groups (Step 2), square that number, and divide it by the number of cases on which the sum is based (10 in our example).

$$\frac{(338 + 405)^2}{10} = \frac{743^2}{10} = \frac{552{,}049}{10} = \mathbf{55{,}204.900}$$

Step 8. Add the sums of the X columns in the two conversation groups (Step 2), square that number, and divide it by the number of cases on which the sum is based (10 in our example).

$$\frac{(267 + 315)^2}{10} = \frac{582^2}{10} = \frac{338{,}724}{10} = \mathbf{33{,}872.400}$$

Step 9. Add the result of Step 7 to the result of Step 8, and subtract the value of Step 5.

$$55,204.9 + 33,872.4 - 87,781.250 = \textbf{1296.050}$$

Step 10. Add the sums of the X columns in the two lab groups (Step 2), square that number, and divide it by the number of cases on which the sum is based (10 in our example).

$$\frac{(338 + 267)^2}{10} = \frac{605^2}{10} = \frac{366,025}{10} = \textbf{36,602.500}$$

Step 11. Add the sums of the X columns in the two no-lab groups (Step 2), square that number, and divide it by the number of cases on which the sum is based (10 in our example).

$$\frac{(405 + 315)^2}{10} = \frac{720^2}{10} = \frac{518,400}{10} = \textbf{51,840}$$

Step 12. Add the result of Step 10 to the result of Step 11; then subtract the value of Step 5.

$$36,602.500 + 51,840 - 87,781.250 = \textbf{661.250}$$

Step 13. Square each of the sums under the four X columns (Step 2), divide each square by the number of cases on which each sum is based (5 in our example), and then add the four quotients.

$$\frac{338^2}{5} + \frac{405^2}{5} + \frac{267^2}{5} + \frac{315^2}{5} = \frac{448,783}{5} = \textbf{89,756.6}$$

Step 14. Subtract the results of Steps 5, 9, and 12 from the value of Step 13.

$$89,756.6 - 87,781.250 - 1296.050 - 661.250 = \textbf{18.05}$$

Step 15. Subtract the result of Step 13 from that of Step 3.

$$91,587 - 89,756.6 = \textbf{1830.4}$$

The Criterion Variable (Steps 16–28)

Step 16. Square all the numbers in the Y columns (Step 1), and add the squares.

$$5^2 + \cdots + 7^2 + 2^2 + \cdots + 3^2 + 5^2$$
$$+ \cdots + 6^2 + 9^2 + \cdots + 10^2 = \textbf{764}$$

Step 17. Add the sums for all the Y columns (Step 2).

$$30 + 10 + 25 + 45 = \mathbf{110}$$

Step 18. Square the result of Step 17, and divide by the total number of Y values (20 in this example).

$$\frac{110^2}{20} = \frac{12,100}{20} = \mathbf{605}$$

Step 19. Subtract the result of Step 18 from the result of Step 16.

$$764 - 605 = \mathbf{159}$$

Step 20. Add the sums of the Y columns in the two classroom groups (Step 2), square that number, and divide it by the number of cases on which the sum is based (10 in our example).

$$\frac{(30 + 10)^2}{10} = \frac{40^2}{10} = \frac{1600}{10} = \mathbf{160}$$

Step 21. Add the sums under the Y columns in the two conversation groups (Step 2), square that number, and divide it by the number of cases on which the sum is based (10 in our example).

$$\frac{(25 + 45)^2}{10} = \frac{70^2}{10} = \frac{4900}{10} = \mathbf{490}$$

Step 22. Add the result of Step 20 to the result of Step 21, and subtract the result of Step 18.

$$160 + 490 - 605 = \mathbf{45}$$

Step 23. Add the sums under the Y columns in the two lab groups (Step 2), square that number, and divide it by the number of cases on which the sum is based (10 in our example).

$$\frac{(30 + 25)^2}{10} = \frac{55^2}{10} = \frac{3025}{10} = \mathbf{302.5}$$

Step 24. Add the sums under the Y columns in the two no-lab groups (Step 2), square that number, and divide it by the number of cases on which the sum is based (10 in our example).

$$\frac{(10 + 45)^2}{10} = \frac{55^2}{10} = \frac{3025}{10} = \mathbf{302.5}$$

Step 25. Add the result of Step 23 to the result of Step 24, and subtract the result of Step 18.

$$302.5 + 302.5 - 605 = 0$$

Step 26. Square each of the sums under the four Y columns (Step 2), divide each square by the number of cases on which each sum is based (5 in our example), and then add the quotients.

$$\frac{30^2}{5} + \frac{10^2}{5} + \frac{25^2}{5} + \frac{45^2}{5} = \frac{3650}{5} = 730$$

Step 27. Subtract the results of Steps 18, 22, and 25 from the result of Step 26.

$$730 - 605 - 45 - 0 = 80$$

Step 28. Subtract the value of Step 26 from that of Step 16.

$$764 - 730 = 34$$

Products of the Two Variables (Steps 29–53)

Step 29. Multiply each of the X scores in Step 1 by its paired Y score, and add the products.

$$(62 \times 5) + \cdots + (72 \times 7) + (84 \times 2) + \cdots + (85 \times 3) +$$
$$(46 \times 5) + \cdots + (62 \times 6) + (58 \times 9) + \cdots + (59 \times 10) = 7175$$

Step 30. Multiply the result of Step 4 by the value of Step 17; then divide by the number of XY pairs (20 in our example).

$$\frac{1325 \times 110}{20} = \frac{145,750}{20} = 7287.500$$

Step 31. Subtract the result of Step 30 from the result of Step 29.

$$7175 - 7287.5 = -112.5$$

Step 32. From Step 2, obtain the X sum and the Y sum for the two classroom groups.

$$338 + 405 = 743 = X \text{ sum for classroom groups}$$
$$30 + 10 = 40 = Y \text{ sum for classroom groups}$$

Multiply the X sum by the Y sum, and divide by the number of scores in each sum (10 in our example).

$$\frac{743 \times 40}{10} = \frac{29,720}{10} = \textbf{2972}$$

Step 33. From Step 2, obtain the X sum and the Y sum for the two conversation groups.

$$267 + 315 = \textbf{582} = X \text{ sum for conversation groups}$$
$$25 + 45 = \ \ \textbf{70} = Y \text{ sum for conversation groups}$$

Multiply the X sum by the Y sum, and divide by the number of scores in each sum (10 in our example).

$$\frac{582 \times 70}{10} = \frac{40,740}{10} = \textbf{4074}$$

Step 34. Add the result of Step 32 to the result of Step 33, and subtract the value of Step 30.

$$2972 + 4074 - 7287.5 = \textbf{-241.5}$$

Step 35. From Step 2, obtain the X sum and the Y sum for the two lab groups.

$$338 + 267 = \textbf{605} = X \text{ sum for lab groups}$$
$$30 + 25 = \ \ \textbf{55} = Y \text{ sum for lab groups}$$

Multiply the X sum by the Y sum, and divide by the number of scores in each sum (10 in our example).

$$\frac{605 \times 55}{10} = \frac{33,275}{10} = \textbf{3327.5}$$

Step 36. From Step 2, obtain the X sum and the Y sum for the two no-lab groups.

$$405 + 315 = \textbf{720} = X \text{ sum for no-lab groups}$$
$$10 + 45 = \ \ \textbf{55} = Y \text{ sum for no-lab groups}$$

Multiply the X sum by the Y sum, and divide by the number of scores in each sum (10 in our example).

$$\frac{720 \times 55}{10} = \frac{39,600}{10} = \textbf{3960}$$

Step 37. Add the result of Step 35 to the result of Step 36, and subtract the result of Step 30.

$$3327.5 + 3960 - 7287.5 = \textbf{0}$$

Step 38. Multiply each X sum in Step 2 by its paired Y sum, divide each product by the number of scores on which each sum is based (5 in our example), and then add the quotients.

$$\frac{338 \times 30}{5} + \frac{405 \times 10}{5} + \frac{267 \times 25}{5} + \frac{315 \times 45}{5} = \frac{35,040}{5} = \textbf{7008}$$

Step 39. Subtract the results of Steps 30, 34, and 37 from the result of Step 38.

$$7008 - 7287.5 - (-241.5) - 0 = \textbf{-38}$$

Step 40. Subtract the value of Step 38 from the value of Step 29.

$$7175 - 7008 = \textbf{167}$$

Step 41. Square the result of Step 40, and divide by the value of Step 15.

$$\frac{167^2}{1830.4} = \frac{27,889}{1830.4} = \textbf{15.237}$$

Then, subtract that quotient from the result of Step 28.

$$34 - 15.237 = \textbf{18.763}$$

Step 42. Add the result of Step 9 to the result of Step 15.

$$1296.050 + 1830.4 = \textbf{3126.45}$$

Step 43. Add the value of Step 22 to that of Step 28.

$$45 + 34 = \textbf{79}$$

Step 44. Add the result of Step 34 to the result of Step 40.

$$(-241.5) + 167 = \textbf{-74.5}$$

Step 45. Square the result of Step 44, and divide by the result of Step 42.

$$\frac{(-74.5)^2}{3126.5} = \frac{5550.25}{3126.5} = \textbf{1.775}$$

Then, subtract that quotient and the result of Step 41 from the result of Step 43.

$$79 - 1.775 - 18.763 = \textbf{58.462}$$

Step 46. Add the final value of Step 12 to that of Step 15.

$$661.250 + 1830.4 = \textbf{2491.65}$$

Step 47. Add the result of Step 25 to that of Step 28.

$$0 + 34 = 34$$

Step 48. Add the result of Step 37 to that of Step 40.

$$0 + 167 = 167$$

Step 49. Square the result of Step 48, and divide by the result of Step 46.

$$\frac{167^2}{2491.65} = \frac{27,889}{2491.65} = 11.193$$

Then subtract that quotient and the result of Step 41 from the result of Step 47.

$$34 - 11.193 - 18.763 = 4.044$$

Step 50. Add the value of Step 14 to that of Step 15.

$$18.05 + 1830.4 = 1848.450$$

Step 51. Add the value of Step 27 to that of Step 28.

$$80 + 34 = 114$$

Step 52. Add the value of Step 39 to that of Step 40.

$$(-38) + 167 = 129$$

Step 53. Square the result of Step 52, and divide by the result of Step 50.

$$\frac{129^2}{1848.45} = \frac{16,641}{1848.45} = 9.003$$

Then, subtract that quotient and the result of Step 41 from the result of Step 51.

$$114 - 9.003 - 18.763 = 86.234$$

Mean Squares and F Ratios (Steps 54–60)

Step 54. Divide the result of Step 41, which is the adjusted-error sum of squares, by the degrees of freedom for adjusted error. The adjusted-error *df* are always $N - CL - 1$, where N is the total number of XY pairs (20 in this example), C is the number of experimental conditions on the first variable (2 in this example, for classroom versus conversation), and L is the number of experimental conditions on the other variable (2 in this example, for lab versus no lab). Thus, $N - CL - 1 = 20 - (2 \times 2) - 1 = 20 - 5 = 15$.

$$\frac{18.763}{N - CL - 1} = \frac{18.763}{15} = \mathbf{1.251}$$

This final value is the error mean square, which will make up the denominator for the three F-tests to follow.

Step 55. Divide the result of Step 45, which is the adjusted sum of squares for teaching methods, by the appropriate degrees of freedom. The *df* are $C - 1$, where C is the number of experimental conditions in this variable (2 in this example, for classroom versus conversation).

$$\frac{58.462}{C - 1} = \frac{58.462}{1} = \mathbf{58.462}$$

Step 56. Divide the value of Step 55 by that of Step 54. This will yield the F for teaching methods. The *df* are $(C - 1)/(N - CL - 1) = 1/15$.

$$F = \frac{58.462}{1.251} = \mathbf{46.73}$$

From Appendix E it is seen that with $df = 1/15$, an F ratio larger than 16.59 would be expected by chance alone less than one time in a thousand. Therefore, the F is said to be significant at the .001 level. This means that the different teaching methods were rated as being significantly different in their effectiveness.

Step 57. Divide the result of Step 49, which is the adjusted sum of squares for language-lab usage, by the appropriate degrees of freedom. The *df* are $L - 1$, where L is the number of experimental conditions in this variable (2 in this example, for lab versus no lab).

$$\frac{4.004}{L - 1} = \frac{4.044}{1} = \mathbf{4.044}$$

Step 58. Divide the value of Step 57 by that of Step 54. This will yield the F for language-lab usage. The *df* are $(L - 1)/(N - CL - 1) = (2 - 1)/(20 - 5) = 1/15$.

$$F = \frac{4.044}{1.251} = \mathbf{3.23}$$

Appendix E shows that the probability of getting a value larger than this F is between five and ten in one thousand. Therefore, we conclude that there is no difference in language effectiveness depending upon whether one spends three hours in the lab or no time in the lab.

Step 59. Divide the result of Step 53, which is the adjusted sum of squares for the methods-by-lab interaction, by the appropriate degrees of freedom. The *df* are $(C - 1)(L - 1)$, or the product of the *df* from Step 55 and the *df* from Step 57.

$$\frac{86.234}{(C - 1)(L - 1)} = \frac{86.234}{1 \times 1} = \mathbf{86.234}$$

Step 60. Divide the value of Step 59 by that of Step 54. This will yield the *F* for the methods-by-lab interaction.

$$F = \frac{86.234}{1.251} = \mathbf{68.93}$$

The *df* for this *F* are

$$\frac{(C - 1)(L - 1)}{N - CL - 1} = \frac{1}{15}$$

Appendix E shows that this *F* is significant at the .001 level. Since we have found that the teaching methods affect (interact with) the lab experience, we would wish to seek out the source of the interaction. The statistical procedures to use for such an undertaking are presented in Section 3.10. If you find a significant interaction in your data, go to Section 3.10 and proceed.

Supplement

The test for homogeneity of regression when a factorial analysis of covariance is used is computed as follows:

Step 1. Using the data from Step 1 of the original analysis, compute the sum of squares (SS_X) and the sum of the cross products (SP). The formula for SS_X is $\Sigma X^2 - (\Sigma X)^2/N_{(per\ gp.)}$ and for SP is $\Sigma XY - \Sigma X \Sigma Y/N_{(pairs)}$ for each group.

$$\text{Group 1 } SS_X = 24078 - \frac{338^2}{5} = \mathbf{1229.2}$$

$$\text{Group 2 } SS_X = 14419 - \frac{267^2}{5} = \mathbf{161.2}$$

$$\text{Group 3 } SS_X = 33115 - \frac{405^2}{5} = \mathbf{310}$$

$$\text{Group 4 } SS_X = 19975 - \frac{315^2}{5} = \mathbf{130}$$

$$\text{Group 1 SP} = 2166 - \frac{338 \times 30}{5} = \textbf{138}$$

$$\text{Group 2 SP} = 1328 - \frac{267 \times 25}{5} = \textbf{-7}$$

$$\text{Group 3 SP} = 841 - \frac{405 \times 10}{5} = \textbf{31}$$

$$\text{Group 4 SP} = 2840 - \frac{315 \times 45}{5} = \textbf{5}$$

Step 2. Compute SP^2/SS_X values for each group.

$$\text{Group 1} = \frac{138^2}{12229.2} = \textbf{15.5}$$

$$\text{Group 2} = \frac{(-7)^2}{161.2} = \textbf{.3}$$

$$\text{Group 3} = \frac{31^2}{310} = \textbf{3.1}$$

$$\text{Group 4} = \frac{5^2}{130} = \textbf{.2}$$

Step 3. Add the results of Step 2.

$$15.5 + .3 + 3.1 + .2 = \textbf{19.1}$$

Step 4. Get the error sum of squares for Y from Section 4.10, Step 28, and subtract the result of Step 3, above, from it.

$$34 - 19.1 = \textbf{14.9}$$

This SS represents the within-group regressions.

Step 5. Get the adjusted error sum of squares for Y from Section 4.12, Step 41, and subtract the result of Step 4, above, from it.

$$18.763 - 14.9 = \textbf{3.86}$$

This SS represents the between-group regression.

Step 6. Divide the result of Step 4 by its degrees of freedom.

$$df = n - 2a = 20 - 8 = \textbf{12}$$

$$\frac{14.9}{12} = \textbf{12.4}$$

Step 7. Divide the result of Step 5 by its degrees of freedom.

$$df = a - 1 = 4 - 1 = 3$$

$$\frac{3.86}{3} = 1.29$$

Step 8. Form the F ratio by dividing the result of Step 7 by the result of Step 6.

$$\frac{1.29}{1.24} = 1.04$$

Since this F is small, we accept the hypothesis that the regressions are homogeneous.

SECTION 4.13
Reliability of Measurement: The Test as a Whole (Test-Retest, Parallel Forms, and Split Halves)

The three coefficients of reliability presented here all start with the computation of a Pearson product-moment correlation coefficient.

1. The *test-retest* reliability measure is computed when you have pairs of scores obtained from two different administrations of the same test to the same people. If you have such pairs of scores, label the scores obtained on the first administration as X and those obtained on the second administration as Y. Then, go to Section 4.1 and follow the instructions. The computed value will be the test-retest reliability coefficient (sometimes called the *coefficient of stability*).
2. The *parallel-forms* reliability measure is computed when you have pairs of test scores from two tests composed of "equated" items. If you have such pairs of scores, label them X and Y, go to Section 4.1, and follow the instructions. The computed value will be the parallel-forms reliability coefficient (sometimes called the *measure of equivalence and stability*).
3 The *split-halves* reliability measure is computed when you have given a test to several individuals and then formed pairs of scores by dividing the test items into two equal groups. (The usual method is to put even-numbered items in one group and odd-numbered items in the other group.) The pairs of scores are then used to compute the correlation. Carry out the following steps in order to compute the split-halves reliability coefficient.

Step 1. Go to Section 4.1 and follow the instructions. When you have the correlation coefficient, return to this section and go on to the next step.

Step 2. In order to compute the correlation, you divide the test in half. This means that the real or whole test is twice as long as either half used to compute the reliability coefficient. When a test is lengthened, the reliability increases. You must therefore "correct" the value obtained in Step 1 so that it effectively pertains to the whole test rather than to just half of it.

In order to correct the value of Step 1, first multiply the answer of Step 1 by the number 2. For example, suppose the correlation value from Step 1 is .83.

$$.83 \times 2 = 1.66$$

Step 3. Now add the number 1 to the result of Step 1.

$$1 + .83 = 1.83$$

Step 4. Divide the result of Step 2 by the result of Step 3.

$$\frac{1.66}{1.83} = .91$$

This value is the split-halves reliability coefficient corrected so that it applies to the whole test. A high reliability value (.70 or higher) shows that the test is reliably (accurately) measuring the characteristic it was designed to measure.

SECTION 4.14

Reliability of Measurement: The Individual Items (Kuder-Richardson and Hoyt)

When your main concern is whether the items in a test are fairly homogeneous in terms of how the individuals responded to the items, the measure of reliability to be used is the Kuder-Richardson coefficient. The computational method presented here was actually developed by Hoyt. It uses the analysis of variance and is somewhat simpler than the original formulation. Hoyt's basic formula for reliability is

$$r_{tt} = \frac{V_e - V_r}{V_e}$$

where V_r = variance for remainder sum of squares

V_e = variance for examinees

Example

Suppose that an experimenter wishes to test the reliability of a certain ten-item test administered to eight subjects. Then, for the purposes of determining reliability, the experimenter records for each subject on each test item whether the question was answered correctly (indicated by the number 1) or incorrectly (indicated by the cipher, 0).

Step 1. Table the data as follows.

Subject	Test Items									
	1	2	3	4	5	6	7	8	9	10
S_1	1	1	1	1	1	1	1	1	1	1
S_2	0	0	0	1	1	0	1	1	1	1
S_3	0	0	1	1	1	0	0	0	0	0
S_4	0	0	1	1	1	1	1	0	1	1
S_5	0	1	1	1	1	0	1	0	1	1
S_6	1	1	0	1	0	0	1	1	1	0
S_7	0	0	1	1	0	0	1	0	0	1
S_8	1	1	0	1	1	1	1	0	1	1

Step 2. Count the number of items that each subject answered correctly (in this example, 10 for the first subject, 6 for the second subject, etc.). List the totals for each subject.

Subject	No. of Correct Answers
S_1	10
S_2	6
S_3	3
S_4	7
S_5	7
S_6	6
S_7	4
S_8	8

Step 3. Add the number of correct answers (Step 2), and record the sum.

$$10 + 6 + \cdots + 8 = 51$$

Step 4. Square each number of correct answers in Step 2; then add the squares and divide that sum by the number of items in the test (10 in this example).

$$\frac{10^2 + 6^2 + \cdots + 8^2}{10} = \frac{359}{10} = \textbf{35.9}$$

Step 5. Square the result of Step 3, and divide by the product of the number of people by the number of items ($8 \times 10 = 80$ in this example).

$$\frac{51^2}{80} = \frac{2601}{80} = \textbf{32.512}$$

Step 6. Subtract the result of Step 5 from the result of Step 3.

$$51 - 32.512 = \textbf{18.488}$$

Step 7. Subtract the result of Step 5 from the result of Step 4.

$$35.9 - 32.512 = \textbf{3.388}$$

Step 8. Count the number of subjects who correctly answered each item. List the totals for each item.

Item	No. of Persons Correct
1	3
2	4
3	5
4	8
5	6
6	3
7	7
8	3
9	6
10	6

Step 9. Square each number of persons correct in Step 8; then add the squares and divide that sum by the number of people who took the test (8 in this example).

$$\frac{3^2 + 4^2 + \cdots + 6^2}{8} = \frac{259}{8} = \textbf{36.125}$$

Step 10. Subtract the result of Step 5 from the result of Step 9.

$$36.125 - 35.512 = \textbf{3.613}$$

Step 11. Subtract the result of Step 7 and the result of Step 10 from the result of Step 6.

$$18.488 - 3.388 - 3.613 = \mathbf{11.487}$$

Step 12. Divide the result of Step 7 by $N - 1$, where N is the number of subjects who took the test (8 in this example).

$$\frac{3.388}{N - 1} = \frac{3.388}{8 - 1} = \frac{3.388}{7} = \mathbf{.484}$$

Step 13. Divide the result of Step 11 by $(N - 1)(I - 1)$, where N is the number of subjects who took the test (8 in our example) and I is the number of items in the test (10 in our example).

$$\frac{11.487}{(N - 1)(I - 1)} = \frac{11.487}{7 \times 9} = \frac{11.487}{63} = \mathbf{.182}$$

Step 14. Subtract the result of Step 13 from the result of Step 12.

$$.484 - .182 = \mathbf{.302}$$

Step 15. Divide the result of Step 14 by the result of Step 12. This yields the value of the Kuder-Richardson (or Hoyt) reliability coefficient.

$$\frac{.302}{.484} = \mathbf{.62}$$

A high reliability coefficient (.70 or higher) would mean that the test was accurately measuring some characteristic of the people taking it. Further, it would mean that the individual items on the test were producing similar patterns of responding in different people. Therefore, a high value would mean that the test items were homogeneous and reliable.

SECTION 4.15
Test for Difference Between Independent Correlations

If you have two correlations computed from data that were gathered from two different groups of individuals, the correlation coefficients will be experimentally independent. In such a case, you may use the following procedure to test for significance of the difference between the correlations.

Example

Suppose we have a correlation coefficient of +.68 that was computed between grades in an English class and IQ scores for thirty-eight people. Suppose further, that we have a correlation coefficient of +.36 between grades in a similar English class and IQ scores for a different group of seventy-three people. We wish to know whether these coefficients are different.

Step 1. First, change the two correlations into Fisher z scores. This can be done by means of any table of such transformations (see Appendix F):

Correlation of .68 = z of **.829**

Correlation of .36 = z of **.377**

Step 2. Subtract either z score of Step 1 from the other.

$$.829 - .377 = .452$$

Step 3. Subtract 3 from the number of people in the group for which the first correlation was computed (38 in this example). (*Note:* The number 3 is always used.)

$$38 - 3 = 35$$

Step 4. Divide the result of Step 3 into the number 1 (i.e., take the reciprocal of 35). Carry the answer to four decimal places.

$$\frac{1}{35} = .0286$$

Step 5. Subtract 3 from the number of people in the group for which the second correlation was computed (73 in this example).

$$73 - 3 = 70$$

Step 6. Divide the result of Step 5 into the number 1 (i.e., take the reciprocal of 70). Carry the answer to four decimal places.

$$\frac{1}{70} = .0143$$

Step 7. Add the result of Step 4 to the result of Step 6.

$$.0286 + .0143 = .0429$$

Then take the square root of the sum.

$$\sqrt{.0429} = .207$$

Step 8. Divide the result of Step 2 by the result of Step 7. This yields a z statistic.

$$z = \frac{.452}{.207} = 2.18$$

A z larger than 1.96 is significant at the .05 level using a two-tailed test (see Appendix A). A significant z tells us that the two correlation values are very likely really different.

SECTION 4.16
Test for Difference Between Dependent Correlations

The following procedure is used to determine the significance of the difference between experimentally dependent correlations—i.e., correlations based on data taken from the same group of people.

Example

Suppose it is known that the correlation between grades in a statistics course and overall grade point average (GPA) for sixty-three students is +.70. Suppose it is also known that the correlation between grades in an introductory psychology course and overall GPA for those same sixty-three students is +.40. If you wish to test for the significance of the difference between these two correlations, you must first be aware of the fact that they are related or dependent. Then, you must find the remaining correlation between statistics grades and introductory-psychology grades for the sixty-three students. Suppose that correlation is +.30.

Step 1. You have the following three correlations:

Statistics grade with GPA = +.70
Introductory psychology grade with GPA = +.40
Statistics grade with psychology grade = +.30

Compute the difference between the two correlations of interest (in this example, the first two).

$$.70 - .40 = .30$$

Step 2. Subtract 3 from the number of individuals involved in the correlations (63 in this example). (*Note:* The number 3 is always used.)

$$63 - 3 = 60$$

Step 3. Add 1 to the third correlation in Step 1—i.e., the correlation that you are not presently interested in (+.30 in this example). (*Note:* The number 1 is always used.)

$$.30 + 1 = 1.30$$

Step 4. Multiply the result of Step 2 by the result of Step 3.

$$60 \times 1.30 = 78$$

Then, take the square root of the product.

$$\sqrt{78} = 8.832$$

Step 5. Multiply the result of Step 1 by the result of Step 4.

$$.30 \times 8.832 = 2.65$$

Step 6. Square each of the three correlation values from Step 1, and add the squares.

$$.70^2 + .40^2 + .30^2 = .49 + .16 + .09 = .74$$

Step 7. Multiply the three correlation values from Step 1.

$$.70 \times .40 \times .30 = .084$$

Step 8. Multiply the result of Step 7 by 2, and then add 1 to the product. (*Note:* The numbers 2 and 1 are always used.)

$$(2 \times .084) + 1 = .168 + 1 = 1.168$$

Step 9. Subtract the result of Step 6 from the result of Step 8.

$$1.168 - .74 = .428$$

Step 10. Multiply the result of Step 9 by 2.

$$2 \times .428 = .856$$

Then, take the square root of the product.

$$\sqrt{.856} = .925$$

Step 11. Divide the result of Step 5 by the result of Step 10. This yields a *t* statistic.

$$t = \frac{2.65}{.925} = 2.86$$

The appropriate degrees of freedom are given as the result of Step 2, i.e., 60. A *t* larger than 2.00, with 60 *df*, is significant at the .05 level using a two-tailed test (see Appendix B). A significant *t* tells us that the two correlation values are very likely really different.

Part 5

Multivariate Analyses

The six sections presented in this part represent an extension of the univariate analyses to the multivariate situation. In the case of the univariate analyses presented in Part 2, the analyses are conducted to determine whether the sample means differ significantly from the hypothesized population means or from each other. In multivariate analyses, the tests of significance refer to the differences between the coordinates of the means of several variables, or centroids.

In almost all instances, multivariate analyses require the use of a computer. For example, sample size should be fairly large for MANOVA experiments. While appropriate sample size is debatable, a rough rule of thumb is that the total size of the sample should be at least 20 times the number of dependent variables times the number of experimental groups compared. Thus, if three groups were utilized with two dependent variables recorded, a sample of 120 should be used, i.e., 20 $\times$ 3 $\times$ 2 = 120. However, for simple problems and for understanding of the concepts, hand calculations of problems are rather easily accomplished with significant enhancement of the understanding of the relationship among univariate, correlational, and multivariate techniques.

The six analyses presented in Part 5 include the following:

Multivariate—One-Sample Tests (Sections 5.1 and 5.2).

The first example, presented in Section 5.1, involves an assumption that the population values of the centroid are known. In fact, this rarely occurs, so the actual utility of this presentation is limited. The second example, presented in Section 5.2, demonstrates the use of difference scores and is the multivariate counterpart of the *t*-test for matched pairs or related measures described in Section 1.7.

Multivariate—Two-Group Tests (Sections 5.3 and 5.4).

The multivariate tests appropriate for two-sample situations represent the counterpart of the univariate t-test between two independent means presented in Section 1.6. The first example, presented in Section 5.3, represents a direct extension of the univariate t-test. The second example, presented in Section 5.4, also compares two groups but makes use of difference scores [(pre test)-(post test)] on each of the variables.

Multivariate—Multiple-Group Tests (Sections 5.5 and 5.6).

The analyses presented in Sections 5.5 and 5.6 represent the multivariate counterparts of the F-test in one-way analysis of variance. Section 5.5 represents the multivariate extension of the completely randomized design presented in Section 2.1, while Section 5.6 represents the multivariate counterpart of the univariate factorial analysis presented in Section 2.2.

In this part, computational formulas will be presented in those instances where a standard set of symbols and terms are used in nearly all textbooks. A minimal understanding of matrix notation is required but this should pose no problem for anyone with even a limited knowledge of algebra since only bivariate cases will be presented.

SECTION 5.1

Multivariate Analysis to Test for a Difference Between a Sample Centroid and the Population Values

This analysis, using Hotelling's T^2 is the logical multivariate extension of the univariate t-test presented in Section 1.5. Use of T^2 assumes that the population values of the centroid are known. If this information is known, the experimenter can then determine whether the sample centroid is significantly different from the population values.

Example

Assume that an experimenter gives a mathematical aptitude test (Test 1) and an intellectual curiosity test (Test 2) to a group of twelve high school seniors. She knows that the population means (μ) for these tests are 12 and 26, respectively. Use of the following test can determine whether the scores made by her subjects are significantly different from the population centroid.

Step 1. Table the data as follows. Care must be taken to ensure that the scores for each subject are recorded in the proper rows and columns.

Subjects	Test 1 (Mathematics)	Test 2 (Curiosity)
S_1	11	26
S_2	13	29
S_3	18	38
S_4	14	27
S_5	16	30
S_6	15	26
S_7	19	37
S_8	10	22
S_9	10	23
S_{10}	21	39
S_{11}	19	36
S_{12}	20	35

Step 2. Calculate the sum of the scores on Test 1 and Test 2.

Subjects	Test 1	Test 2
S_1	11	26
S_2	13	29
⋮	⋮	⋮
S_{12}	20	35
	186	368

Step 3. Calculate the sum of squared scores for Test 1.

$$11 + 13 + 18 + \cdots + 20 = \textbf{3,054}$$

Step 4. Calculate the sum of squared scores for Test 2.

$$26 + 29 + 38 + \cdots + 35 = \textbf{11,690}$$

Step 5. Calculate the sum of the cross products.

$$(11 \times 26) + (13 \times 29) + \cdots + (20 \times 35) = \textbf{5,951}$$

Step 6. Calculate the sample centroid (from Step 2).

$$\overline{X}_1 = \frac{186}{12} = \textbf{15.5}$$

$$\overline{X}_2 = \frac{368}{12} = \textbf{30.67}$$

Step 7. Calculate the difference between the population mean (μ) and the sample mean ($\overline{X}$). For this example, it is assumed that the population values for Test 1 and Test 2 are known to be 12 and 26 respectively. Thus,

$$\begin{aligned} \overline{X}_1 - \mu_1 &= 15.5 - 12 = \\ \overline{X}_2 - \mu_2 &= 30.67 - 26 = \end{aligned} \quad \begin{bmatrix} 3.5 \\ 4.67 \end{bmatrix}$$

Step 8. Calculation of the values required to set up the sample dispersion matrix: For a problem with two variables, the sum of squares for the first variable, the sum of squares for the second variable, and the sum of squares of the cross products must be calculated.

First, calculation for the sum of squares for Test 1: The formula for this computation is

$$\Sigma X^2 - \frac{(\Sigma X)^2}{N}$$

Sum of X^2 (Step 3) = **3054**
Sum of X (Step 2) = **186**

and

$$3054 - \frac{(186)^2}{12} = 3054 - 2883 = 171$$

Next, calculation of the sum of squares for Test 2:

Sum of X^2 (Step 3) = **11,690**
Sum of X (Step 2) = **368**

and

$$11,690 - \frac{(368)^2}{12} = 11,690 - 11,285.3 = 404.7$$

Finally, calculation of the sum of cross products:

$$\Sigma X_1 X_2 - \frac{(\Sigma X_1)(\Sigma X_2)}{N}$$

Sum of cross products (Step 4) = **5951**

and

$$\Sigma X_1 \times \Sigma X_2 \text{ (Step 2)} = 186 \times 368 = \mathbf{68,448}$$

thus,

$$5,951 - \frac{68,448}{12} = 5,951 - 5704 = 247$$

Step 9. Calculation of the sample dispersion matrix: The values will be inserted into a symmetric dispersion matrix with the elements arranged like this:

$$\begin{bmatrix} a & b \\ c & d \end{bmatrix}$$

The formula for the dispersion matrix is

$$\mathbf{C}_d = \frac{1}{N-1} \begin{bmatrix} SS_{x_1} & SSCP \\ SSCP & SS_{x_2} \end{bmatrix}$$

Using the calculated values

$$\mathbf{C}_d = \frac{1}{11} \begin{bmatrix} 171 & 247 \\ 247 & 404.7 \end{bmatrix}$$

$$\mathbf{C}_d = \begin{bmatrix} (a) & (b) \\ 15.54 & 22.45 \\ (c) & (d) \\ 22.45 & 36.79 \end{bmatrix}$$

Step 10. Calculation of the inverse matrix, $\mathbf{C}_d^{-1}$: The computational formula is

$$\mathbf{C}_d^{-1} = \frac{1}{ad-bc} \begin{bmatrix} d & -b \\ -c & a \end{bmatrix}$$

or, in terms of the current problem (from Step 9),

$$\mathbf{C}_d^{-1} = \frac{1}{(15.54)(36.79) - (-22.45)(-22.45)} \begin{bmatrix} 36.79 & -22.45 \\ -22.45 & 15.54 \end{bmatrix}$$

$$= \frac{1}{67} \begin{bmatrix} 36.79 & -22.45 \\ -22.45 & 15.54 \end{bmatrix}$$

$$= \begin{bmatrix} .55 & -.33 \\ -.33 & .23 \end{bmatrix}$$

Note: For subsequent calculations, the above elements of the inverse matrix will be identified in terms of standard matrix notation, i.e.,

$$\mathbf{C}_d^{-1} = \begin{bmatrix} (a) & (b) \\ .55 & -.33 \\ (c) & (d) \\ -.33 & .23 \end{bmatrix}$$

Step 11. All necessary preparatory calculations have been completed. The statistical test to be applied to determine whether the sample centroid differs significantly from the population value is Hotelling's T^2.

The formula for calculation of T^2 (values from Steps 7 and 10) is

$$T^2 = N\bar{\mathbf{x}}'\mathbf{C}_d^{-1}\bar{\mathbf{x}}$$

$$= 12\ [3.5 \quad 4.67] \begin{bmatrix} .55 & -.33 \\ -.33 & .23 \end{bmatrix} \begin{bmatrix} 3.5 \\ 4.67 \end{bmatrix}$$

In a symmetric matrix such as that presented above, where element $b = c$, the computational formula reduces to

$$T^2 = N[a(\bar{X}_{T1} - \mu_{T1})^2 + 2b(\bar{X}_{T1} - \mu_{T1})(\bar{X}_{T2} - \mu_{T2}) + d(\bar{X}_{T2} - \mu_2)^2]$$
$$T^2 = 12[(.55)(3.5)^2 + (2)(-.33)(3.5)(4.67) + (.23)(4.67)^2]$$
$$= 12[(6.737) + (-10.788) + (5.016)]$$
$$= 12(.965)$$
$$= 11.586$$

Step 12. Calculation of the F ratio: The relationship between T^2 and F is described by the following formula:

$$F = \frac{N - p}{p(N - 1)}\ T^2$$

Where N = the number of subjects

p = the number of dependent variables measured

Inserting the appropriate values, we have

$$F = \frac{12 - 2}{2(12 - 1)} \times 11.586$$

$$= \frac{10}{22} \times 11.586$$

$$= 5.27$$

Step 13. Computation of the degrees of freedom: The *df* for this F test are equal to the number of test measures, p, and the number of subjects minus the number of test measures, or $N - p$. Since two tests, mathematics and curiosity, were given, the *df* for this example would be 2 and $12 - 2$.

Step 14. Test of significance: Looking in the F tables under 2 and 10 degrees of freedom, an F ratio of 3.88 is needed for significance at the .05 level. Since the calculated F ratio exceeds 3.88, it is concluded that the sample centroid differs significantly from the population value.

SECTION 5.2

Multivariate Analysis for Single-Sample Matched Pairs (Difference Scores)

This analysis also makes use of Hotelling's T^2 as the statistical test and represents the multivariate extension of the univariate t-test for related measures presented in Section 1.7.

Points to Consider When Analyzing Single-Sample Difference Scores

1. Usually, two measures are recorded on each variable (pre and post), and the change or difference between the first and second measurements are the scores actually analyzed.
2. Occasionally, as in the t for related measures, matched pairs of subjects will be employed rather than measuring the same subject's performance twice (see univariate example for the t for related measures in Section 1.7).

Example

Assume that an experimenter is interested in determining whether subjects change their attitudes toward education for the poor (Test 1) and taxation (Test 2) after watching a short film on the plight of migrant workers. Attitudes are measured by the use of two tests, each administered before the film is shown and then again after. Scores recorded are the number of statements each subject "agrees" with before and after seeing the film. The difference scores represent the change in positive responses.

Step 1. After the experiment is completed, table the data as follows:

Subjects	Test 1 (Education)		Test 2 (Taxation)	
	Pre	Post	Pre	Post
S_1	12	19	9	12
S_2	16	19	11	11
S_3	11	10	17	20
S_4	15	14	9	11
S_5	18	25	9	16
S_6	20	22	14	15
S_7	10	11	8	9
S_8	20	23	10	12
S_9	13	21	14	17
S_{10}	7	9	6	11
S_{11}	21	22	14	18
S_{12}	19	20	8	19

Step 2. Compute the difference between each pair of pre- and post-test scores.

Subjects	Test 1 difference	Test 2 difference
S_1	7	3
S_2	3	0
S_3	-1	3
$\vdots$	$\vdots$	$\vdots$
S_{12}	1	11

Step 3. Sum the difference scores. Calculate the algebraic sum of the difference scores for Test 1 and Test 2.

$$\text{For Test 1: } 7 + 3 + (-1) + \cdots + 1 = \mathbf{33}$$
$$\text{For Test 2: } 3 + 0 + 3 + \cdots + 11 = \mathbf{42}$$

Step 4. Square the difference scores and sum the squared values.

$$\text{For Test 1: } 7^2 + 3^2 + (-1)^2 + \cdots + 1^2 = \mathbf{193}$$
$$\text{For Test 2: } 3^2 + 0^2 + 3^2 + \cdots + 11^2 = \mathbf{248}$$

Step 5. Calculation of the sum of the cross products: Multiply the Test 1 difference scores times the Test 2 difference scores. Then add these cross products.

$$(7 \times 3) + (3 \times 0) + (-1 \times 3) + \cdots + (1 \times 11) = \mathbf{123}$$

Step 6. Calculation of the sample centroid (from Step 3):

$$\begin{aligned} 33/12 &= \\ 42/12 &= \end{aligned} \begin{bmatrix} 2.75 \\ 3.50 \end{bmatrix}$$

Step 7. All of the necessary preliminary calculations have been completed. Now the sample dispersion matrix must be calculated. The computational formulas are

$$\text{For the Sum of Squares Test 1: } \Sigma d_1^2 - \frac{(\Sigma d_1)^2}{N}$$

$$\text{For the Sum of Squares Test 2: } \Sigma d_2^2 - \frac{(\Sigma d_2)^2}{N}$$

$$\text{For the Sum of Cross Products: } \Sigma d_1 d_2 - \frac{(\Sigma d_1)(\Sigma d_2)}{N \text{ (pairs)}}$$

These values will now be inserted into a symmetric dispersion matrix with the elements arranged as follows:

$$\begin{bmatrix} a & b \\ c & d \end{bmatrix}$$

The formula for the dispersion matrix is

$$\mathbf{C}_d = \frac{1}{N-1} \begin{bmatrix} SS_{x_1} & SSCP \\ SSCP & SS_{x_2} \end{bmatrix}$$

Step 8. The calculations are as follows (values from Steps 3 and 4):

$$\mathbf{C}_d = \frac{1}{11} \begin{bmatrix} 193 - \dfrac{33^2}{12} & 123 - \dfrac{(33)(42)}{12} \\ 123 - \dfrac{(33)(42)}{12} & 248 - \dfrac{42^2}{12} \end{bmatrix}$$

$$= \frac{1}{11} \begin{bmatrix} 102.25 & 7.50 \\ 7.50 & 101.00 \end{bmatrix}$$

$$= \begin{bmatrix} (a) & (b) \\ 9.30 & .68 \\ (c) & (d) \\ .68 & 9.18 \end{bmatrix}$$

Step 9. Next the inverse of $\mathbf{C}_d$ must be found. The formula for C_d^{-1} is

$$\mathbf{C}_d^{-1} = \frac{1}{ad - bc} \begin{bmatrix} d & -b \\ -c & a \end{bmatrix}$$

$$= \frac{1}{(9.30 \times 9.18) - (-.68 \times -.68)} \begin{bmatrix} 9.18 & -.68 \\ -.68 & 9.30 \end{bmatrix}$$

$$= \frac{1}{84.91} \begin{bmatrix} 9.18 & -.68 \\ -.68 & 9.38 \end{bmatrix}$$

$$= \begin{bmatrix} .108 & -.008 \\ -.008 & .110 \end{bmatrix}$$

Note: For subsequent calculations, the above elements in the inverse matrix will be referred to in relation to common matrix positioning, i.e.

$$\begin{bmatrix} (a) & (b) \\ .108 & -.008 \\ (c) & (d) \\ -.008 & .110 \end{bmatrix}$$

Step 10. Computation of T^2. The general formula for T^2 is

$$T^2 = N\bar{\mathbf{x}}'\mathbf{C}_d^{-1}\bar{\mathbf{x}}$$

In this example,

$$T^2 = 12 \begin{bmatrix} 2.75 & 3.50 \end{bmatrix} \begin{bmatrix} .108 & -.008 \\ -.008 & .110 \end{bmatrix} \begin{bmatrix} 2.75 \\ 3.50 \end{bmatrix}$$

Note: The computational formula for a symmetric matrix where the off-diagonal terms (elements b and c) are equal reduces to the following:

$$T^2 = N[a(\overline{X}diff_{T_1})^2 + 2(b)(\overline{X}diff_{T_1})(\overline{X}diff_{T_2}) + d(\overline{X}diff_{T_2})^2]$$

Therefore (values from Steps 6 and 9),

$$T^2 = 12[(.108)(2.75^2) + (2)(-.008)(2.75)(3.50) + (.110)(3.50^2)]$$
$$= 12 \times 2.011$$
$$= 24.132$$

Step 11. Computation of the F statistic:

$$F = \left(\frac{N - p}{p(N - 1)}\right) T^2$$

Where N = the number of subjects

p = the number of dependent variables measured

For the bivariate case in this example, $N = 12$ and $p = 2$, thus,

$$F = \left(\frac{12 - 2}{2(12 - 1)}\right) 24.132$$
$$= \frac{241.32}{22}$$
$$= 10.97$$

Step 12. Computation of the *df*: The *df* for the F statistic are

$$F_{p, N-p}$$

In this example, the *df* would be 2 and 10.

Step 13. Test of significance: Since the computed F of 10.97 exceeds the tabled value of 4.10 for 2 and 10 *df*, the conclusion is drawn that the experimental treatment (seeing the movie) did result in a change in attitude toward education of the poor (Test 1) and taxation (Test 2).

SECTION 5.3
Multivariate Analysis to Test for a Difference Between Two Independent Groups

This analysis represents the multivariate extension of the *t*-test presented in Section 1.6. As in the case of the two-group univariate experiment, the two groups can be formed by random assignment, or the groups already may be constituted (males versus females). The major difference from the univariate experiment is, of course, that two or more responses (dependent variables) are measured and analyzed in combination.

Points to Consider When Using T^2 to Test for Differences Between Independent Groups

1. For each subject, there will be two or more different measures taken and recorded.
2. It is not necessary to have an equal number of subjects in each group, but it makes comparisons easier if there are equal numbers.

Example

Assume that fifteen psychology and twelve history majors are randomly selected to participate in an experiment to determine whether choice of major is related to particular intellectual skills. Two tests, Reading Comprehension (Test 1) and Mathematical Concepts (Test 2) are given to each subject. The number of correct answers on each of the tests is recorded for each subject.

Step 1. After the experiment is completed, table the data as follows:

Psychology Students			History Students		
Subjects	Test 1	Test 2	Subjects	Test 1	Test 2
S_1	20	21	S_{16}	26	28
S_2	27	27	S_{17}	20	24
S_3	20	23	S_{18}	19	23
S_4	23	22	S_{19}	17	19
S_5	30	31	S_{20}	14	20
S_6	21	21	S_{21}	26	27
S_7	26	28	S_{22}	28	29
S_8	18	18	S_{23}	14	19
S_9	19	23	S_{24}	19	19
S_{10}	27	30	S_{25}	16	21
S_{11}	14	19	S_{26}	14	17
S_{12}	15	14	S_{27}	28	27
S_{13}	17	19			
S_{14}	19	22			
S_{15}	16	21			

Step 2. Obtain the sum of the two test scores for both groups of subjects.

Psychology, Test 1 Sum = **312**

Psychology, Test 2 Sum = **339**
History, Test 1 Sum = **241**
History, Test 2 Sum = **273**

Step 3. Obtain the sum of squared scores of the two test scores for both groups of subjects.

Psychology, Sum of Test 1 scores squared
$$20^2 + 27^2 + \cdots + 16^2 = \textbf{6816}$$
Psychology, Sum of Test 2 scores squared
$$21^2 + 27^2 + \cdots + 21^2 = \textbf{7965}$$
History, Sum of Test 1 scores squared
$$26^2 + 20^2 + \cdots + 28^2 = \textbf{5175}$$
History, Sum of Test 2 scores squared
$$28^2 + 24^2 + \cdots + 27^2 = \textbf{6401}$$

Step 4. Obtain the sum of the cross products for each group.

Psychology, Sum of cross products for Test 1 and Test 2
$$(20 \times 21) + (27 \times 27) + \cdots + (16 \times 21) = \textbf{7338}$$
History, Sum of cross products for Test 1 and Test 2
$$(26 \times 28) + (20 \times 24) \cdots + (28 \times 27) = \textbf{5719}$$

Step 5. Obtain the centroids for each of the experimental groups.

Centroid for psychology group (from Step 2)
$$\begin{matrix} 312/15 = \\ 339/15 = \end{matrix} \begin{bmatrix} 20.8 \\ 22.6 \end{bmatrix}$$
Centroid for history group (from Step 2)
$$\begin{matrix} 241/12 = \\ 273/12 = \end{matrix} \begin{bmatrix} 20.08 \\ 22.75 \end{bmatrix}$$

Step 6. Calculate the sum of squares for Group 1 (Psychology; data from Steps 2 and 3). The general computational formula is

$$\Sigma X^2 - \frac{(\Sigma X)^2}{N}$$

For Test 1: $SS_{Gp_1T_1} = 6816 - \dfrac{(312)^2}{15} = \textbf{326.4}$

For Test 2: $SS_{Gp_1T_2} = 7965 - \dfrac{(339)^2}{15} = \textbf{303.6}$

Step 7. Calculate the sum of cross products for Group 1 (Psychology; data from Steps 2 and 4). The general computational formula is

$$\Sigma X_1 X_2 - \frac{(\Sigma X_1)(\Sigma X_2)}{N(\text{pairs})}$$

Sum of cross products: $SSCP_{Gp1} = 7338 - \dfrac{(312)(339)}{15} = \textbf{286.8}$

Step 8. Calculate the sum of squares for Group 2 (History).

For Test 1: $SS_{Gp2T_1} = 5175 - \dfrac{(241)^2}{12} = \textbf{334.92}$

For Test 2: $SS_{Gp2T_2} = 6401 - \dfrac{(273)^2}{12} = \textbf{190.25}$

Step 9. Calculate the sum of cross products for Group 2.

Sum of cross products: $SSCP_{Gp2} = 5719 - \dfrac{(241)(273)}{12} = \textbf{236.25}$

Step 10. All the preliminary calculations have been completed. The next step requires the calculation of the within-groups dispersion matrix, $\mathbf{C}_w$. These values will be inserted into a symmetric matrix with the elements arranged as follows:

$$\begin{bmatrix} a & b \\ c & d \end{bmatrix}$$

The diagonal elements of this matrix (a and d) are the within-groups variance estimates, the off-diagonal elements (b and c) are the within-groups covariance estimates.

The matrix elements to be calculated are as follows:

$$\mathbf{C}_w = \begin{bmatrix} SS_{wT_1} & SSCP_w \\ SSCP_w & SS_{wT_2} \end{bmatrix}$$

The computational formulas for each of the elements are

$$\text{Psychology} \qquad \text{History}$$

Test 1: $SS_{wT_1} = \dfrac{SS_{Gp1T_1} + SS_{Gp2T_1}}{n_{Gp1} + n_{Gp2} - 2}$

Test 2: $SS_{wT_2} = \dfrac{SS_{Gp1T_2} + SS_{Gp2T_2}}{n_{Gp1} + n_{Gp2} - 2}$

Cross Products: $SSCP_w = \dfrac{SSCP_{Gp1} + SSCP_{Gp2}}{n_1(\text{pairs}) + n_2(\text{pairs}) - 2}$

For Test 1, the SS_{wT_1} is equal to (from Steps 6 and 8):

$$\frac{326.4 + 334.92}{25} = \frac{661.32}{25} = \textbf{26.45}$$

For Test 2, the SS_{wT_2} is equal to (from Steps 6 and 8):

$$\frac{303.6 + 190.25}{25} = \frac{493.85}{25} = \textbf{19.75}$$

The $SSCP_w$ elements are equal to (from Steps 7 and 9):

$$\frac{286.8 + 236.25}{25} = \frac{523.05}{25} = \textbf{20.92}$$

Step 11. Computation of the within-groups dispersion matrix:

$$C_w = \begin{bmatrix} (a) & (b) \\ 26.45 & 20.92 \\ (c) & (d) \\ 20.92 & 19.75 \end{bmatrix}$$

Step 12. The inverse matrix must now be computed. The formula is

$$C_w^{-1} = \frac{1}{ad - bc} \begin{bmatrix} d & -b \\ -c & a \end{bmatrix}$$

$$= \frac{1}{(26.45)(19.75) - (-20.92)(-20.92)} \begin{bmatrix} 19.75 & -20.92 \\ -20.92 & 26.45 \end{bmatrix}$$

$$= \frac{1}{84.74} \begin{bmatrix} 19.75 & -20.92 \\ -20.92 & 26.45 \end{bmatrix}$$

$$= \begin{bmatrix} .2331 & -.2469 \\ -.2469 & .3121 \end{bmatrix}$$

Note: For subsequent calculations, the above elements of the inverse matrix will be referred to in relation to accepted matrix positioning, i.e.

$$\begin{bmatrix} (a) & (b) \\ .2331 & -.2469 \\ (c) & (d) \\ -.2469 & .3121 \end{bmatrix}$$

Step 13. Computation of the difference between the centroids of the two groups (from step 5):

$$\begin{bmatrix} \overline{X}_{Gp_1T_1} \\ \overline{X}_{Gp_1T_2} \end{bmatrix} - \begin{bmatrix} \overline{X}_{Gp_2T_1} \\ \overline{X}_{Gp_2T_2} \end{bmatrix} = \begin{bmatrix} 20.8 \\ 22.6 \end{bmatrix} - \begin{bmatrix} 20.08 \\ 22.75 \end{bmatrix} = \begin{bmatrix} .72 \\ -.15 \end{bmatrix}$$

Step 14. Compuation of T^2: The general computational formula for T^2 in matrix form is

$$T^2 = N\bar{\mathbf{x}}'\mathbf{C}_w^{-1}\bar{\mathbf{x}}$$

For the two-group example, the formula is

$$T^2 = \left(\frac{n_1 n_2}{n_1 + n_2}\right) \bar{\mathbf{x}}'\mathbf{C}_w^{-1}\bar{\mathbf{x}}$$

$$= \left(\frac{15 \times 12}{15 + 12}\right) [.72 - .15] \begin{bmatrix} .2331 & -.2469 \\ -.2469 & .3121 \end{bmatrix} \begin{bmatrix} .72 \\ -.15 \end{bmatrix}$$

The computational formula for a symmetric matrix where the off-diagonal terms are equal $(b = c)$ reduces to the following:

$$T^2 = \left(\frac{n_1 n_2}{n_1 + n_2}\right) (a)(\bar{X}_{\text{diffT}_1})^2 + 2(b)(\bar{X}_{\text{diffT}_1})(\bar{X}_{\text{diffT}_2}) + (d)(\bar{X}_{\text{diffT}_2})^2$$

$$= 6.67(.2331)(.72^2) + (2)(-.2469)(.72)(-.15) + (.3121)(-.15^2)$$

$$= 6.67 \times .1812$$

$$= \mathbf{1.208}$$

Step 15. Transforming T^2 into F: The formula for transforming T^2 into F when there are two independent groups is

$$F = \left(\frac{N - p - 1}{p(N - 2)}\right) T^2$$

Where N = the number of subjects

$\quad\quad p$ = the number of dependent variables measured

$$= \left(\frac{27 - 2 - 1}{2(27 - 2)}\right) 1.208$$

$$= \mathbf{.58}$$

Step 16. Calculating the *df* for this F ratio:

The *df* for $F = p$ and $N - p - 1$
In this example, *df* = 2 and $27 - 2 - 1$, or **24**.

Step 17. Test of significance: Looking in the F tables for 2 and 24 degrees of freedom, it is seen that an F ratio of 3.40 is needed for significance at the .05 level. Since the computed F in the present example was only .58, it can be concluded that the Psychology and History subjects did not respond differentially to the tests.

SECTION 5.4

Multivariate Analysis to Test for a Difference in the Change Scores of Two Groups

This analysis is an extension of the multivariate example presented in Section 5.2. In this example, the performance of two groups will be compared to determine whether the changes (pre- versus post-experiment measurement) in one group are larger than those in the other.

Points to Consider When Using T^2 to Test for Differences in Change Scores

1. For each subject, there will be two or more response measures (dependent variables) recorded. In addition, each response measure will be recorded twice (usually before experimental manipulation and then again, after).
2. It is not necessary to have an equal number of subjects in each group, but comparisons and understanding of results are easier with equal numbers.

Example

For this example, assume that a special curriculum enrichment program is initiated in a school system. The school psychologist wishes to determine whether such a program benefits talented and gifted students more than average students. Two tests, one of art appreciation (Test 1) and the other of abstract thinking (Test 2), will be administered to the students. Since it is likely that the talented and gifted students would score higher initially due to superior intellectual ability, the experimenter decides to give the tests prior to the initiation of the program and then administer alternative forms of the tests after the curriculum has been in place for one year. The measures to be analyzed will be the change between the pre and post scores of the two groups.

Step 1. Table the data as follows:

Talented and Gifted Group

Subjects	Test 1 (Art Appreciation)		Test 2 (Abstract Thinking)	
	Pre	Post	Pre	Post
S_1	10	10	21	27
S_2	12	13	19	25

Talented and Gifted Group

	Test 1 (Art Appreciation)		Test 2 (Abstract Thinking)	
Subjects	Pre	Post	Pre	Post
S_3	9	9	16	22
S_4	15	17	22	27
S_5	16	18	31	36
S_6	14	19	27	31
S_7	13	16	36	38
S_8	10	11	32	38
S_9	10	12	30	36
S_{10}	13	16	17	19

Average Group

S_{11}	11	13	21	21
S_{12}	14	15	25	25
S_{13}	13	18	27	28
S_{14}	19	17	29	29
S_{15}	17	17	31	32
S_{16}	16	17	31	32
S_{17}	14	18	34	35
S_{18}	15	16	27	30
S_{19}	18	17	29	29
S_{20}	16	22	22	23

Step 2. Calculate the difference score between the pre- and post-special curriculum experience scores for each subject.

Talented and Gifted

	Test 1 Difference	Test 2 Difference
S_1	0	6
S_2	1	6
$\vdots$	$\vdots$	$\vdots$
S_{10}	3	2

Average

S_{11}	2	0
$\vdots$	$\vdots$	$\vdots$
S_{20}	6	1

Step 3. Calculate the sum of the difference scores (Σd) for each condition.

For Test 1, Talented and Gifted: $0 + 1 + \cdots + 3 = \mathbf{19}$

For Test 2, Talented and Gifted: $6 + 6 + \cdots + 2 = \mathbf{48}$

For Test 1, Average Group: $2 + 1 + \cdots + 6 = 17$
For Test 2, Average Group: $0 + 0 + \cdots + 1 = 8$

Step 4. Calculate the sum of the difference scores squared (d^2)

For Test 1, Talented and Gifted: $0^2 + 1^2 + \cdots 3^2 = 57$
For Test 2, Talented and Gifted: $6^2 + 6^2 + \cdots 2^2 = 254$
For Test 1, Average Group: $2^2 + 1^2 + \cdots + 6^2 = 89$
For Test 2, Average Group: $0^2 + 0^2 + \cdots + 1^2 = 14$

Step 5. Calculate the sum of the cross products of the difference scores on Test 1 and Test 2 for each group. $\Sigma(d_{test_1} \times d_{test_2})$

For Talented and Gifted: $(0 \times 6) + (1 \times 6) + \cdots + (3 \times 2) = 76$
For the Average Group: $(2 \times 0) + (1 \times 0) + \cdots + (6 \times 1) = 19$

Step 6. Calculate the centroid for each group (from Step 3).

Centroid for Talented and Gifted: $\begin{aligned} 19/10 &= 1.9 \\ 48/10 &= 4.8 \end{aligned}$

Centroid for Average Group: $\begin{aligned} 17/10 &= 1.7 \\ 8/10 &= .8 \end{aligned}$

Step 7. Calculation of the sum of squares of the difference scores for Group 1 (Talented and Gifted): The necessary values were computed in Steps 3 and 4. The basic computational formula is

$$SS_d = \Sigma d^2 - \frac{(\Sigma d)^2}{n_{\text{per group}}}$$

For Test 1: $SS_{Gp1T_1} = 57 - \frac{(19)^2}{10} = 20.9$

For Test 2: $SS_{Gp1T_2} = 254 - \frac{(48)^2}{10} = 23.6$

Step 8. Calculation of the sum of cross products of the difference scores for Group 1 (Talented and Gifted). The basic computational formula is

$$SSCP = \Sigma d_1 d_2 - \frac{(\Sigma d_1)(\Sigma d_2)}{n(\text{pairs})}$$

The necessary values for the calculations were obtained in Steps 3 and 5.

$$SSCP_{Gp1} = 76 - \frac{(19)(48)}{10} = -15.2$$

Step 9. Calculation of the sum of squares for Group 2 (Average): Repeat the calculations of Step 7 using the data from the average group.

$$\text{For Test 1: } SS_{Gp2T_1} = 89 - \frac{(17)^2}{10} = 60.1$$

$$\text{For Test 2: } SS_{Gp2T_2} = 14 - \frac{(8)^2}{10} = 7.6$$

Step 10. Calculation of the sum of cross products for Group 2: Again, repeat the calculations of Step 8 using the data from the Average group.

$$SSCP_{Gp2} = 19 - \frac{(17)(8)}{10} = 5.4$$

Step 11. These values will now be arranged into a symmetric dispersion matrix with the elements set up as follows:

$$\begin{bmatrix} a & b \\ c & d \end{bmatrix}$$

The elements of the dispersion matrix (C_w) will now be obtained. The diagonal elements of this matrix are the within-groups variance estimates; the off-diagonal elements are the within-groups covariance estimates.

The elements of the matrix are calculated as follows:

$$\mathbf{C}_w = \begin{bmatrix} SS_{wT_1} & SSCP_w \\ SSCP_w & SS_{wT_2} \end{bmatrix}$$

The computational formulas for the elements are

Talented and Gifted   Average

$$\text{Test 1:} \qquad SS_{wT_1} = \frac{SS_{Gp1T_1} + SS_{Gp2T_1}}{n_1 + n_2 - 2}$$

$$\text{Test 2:} \qquad SS_{wT_2} = \frac{SS_{Gp1T_2} + SS_{Gp2T_2}}{n_1 + n_2 - 2}$$

$$\text{Cross products: } SSCP_w = \frac{SSCP_{Gp1} + SSCP_{Gp2}}{n_1(\text{pairs}) + n_2(\text{pairs}) - 2}$$

For Test 1, the SS_{wT_1} element is equal to (from Steps 7 and 9):

$$SS_{wT_1} = \frac{20.9 + 60.1}{18} = 4.50$$

For Test 2, the SS_{wT_2} element is equal to (also from Steps 7 and 9):

$$SS_{wT_2} = \frac{23.6 + 7.6}{18} = \mathbf{1.733}$$

And the cross products elements $(SSCP_w)$ are equal to (from Steps 8 and 10):

$$SSCP_w = \frac{-15.2 + 5.4}{18} = \mathbf{-.544}$$

Thus, the $\mathbf{C}_w$ matrix $= \begin{bmatrix} 4.50 & -.544 \\ -.544 & 1.733 \end{bmatrix}$

Step 12. The inverse matrix must now be computed. The formula for this matrix is

$$\mathbf{C}_w^{-1} = \frac{1}{ad - bc} \begin{bmatrix} d & -b \\ -c & a \end{bmatrix}$$

$$= \frac{1}{(4.50)(1.733) - (-.544)(-.544)} \begin{bmatrix} 1.733 & .544 \\ .544 & 4.50 \end{bmatrix}$$

$$= \frac{1}{7.502} \begin{bmatrix} 1.733 & .544 \\ .544 & 4.50 \end{bmatrix}$$

$$= \begin{bmatrix} .231 & .072 \\ .072 & .600 \end{bmatrix}$$

Note: For subsequent calculations, the above elements of the inverse matrix will be referred to in relation to accepted matrix positioning, i.e.,

$$\mathbf{C}_w^{-1} = \begin{bmatrix} (a) & (b) \\ .231 & .072 \\ (c) & (d) \\ .072 & .600 \end{bmatrix}$$

Step 13. Computation of the difference between the change score centroids:

$$\begin{bmatrix} 1.9 \\ 4.8 \end{bmatrix} - \begin{bmatrix} 1.7 \\ .8 \end{bmatrix} = \begin{bmatrix} .2 \\ 4.0 \end{bmatrix}$$

Step 14. Computation of T^2: The general computational formula for T^2 is

$$T^2 = \left(\frac{n_1 n_2}{n_1 + n_2} \right) \mathbf{x}' \mathbf{C}_w^{-1} \mathbf{x}$$

$$= \left(\frac{10 \times 10}{10 + 10}\right) [.2 \quad 4.0] \begin{bmatrix} .231 & 0.72 \\ .072 & .600 \end{bmatrix} \begin{bmatrix} .2 \\ 4.0 \end{bmatrix}$$

In the special case where C_w^{-1} is symmetric so that element $b = c$, the above expression simplifies to:

$$T^2 = \left(\frac{n_1 n_2}{n_1 + n_2}\right) [(a)(\overline{X}_{\text{diff}Gp1T1 - Gp2T1})^2 + 2(b)(\overline{X}_{\text{diff}G1T1 - G2T1})(\overline{X}_{\text{diff}G1T2 - G2T2})$$
$$+ (d)(\overline{X}_{\text{diff}G1T2 - G2T2})^2]$$
$$= 5[(.231)(.2)^2 + (2)(.072)(4.0)(.2) + (.6)(4.0)^2]$$
$$= 5 \times 9.724 = \textbf{48.622}$$

Step 15. Tranforming T^2 to F: The relationship between T^2 and F is as follows:

$$F = \left(\frac{N - p - 1}{p(N - 2)}\right) T^2$$

Where N = the number of subjects

p = the number of dependent variables measured

$$= \left(\frac{20 - 2 - 1}{2(20 - 2)}\right) 48.622$$
$$= \left(\frac{17}{36}\right) 48.622$$
$$= \textbf{22.96}$$

Step 16. Calculation of the *df* for the *F* ratio: The *df* are equal to p and $N - p - 1$. In the present example, $p = 2$ and $N = 20$. Thus, the *df* = 2 and 17.

Step 17. Test of Significance: Looking in the *F* tables for 2 and 17 *df*, we find that a *F* ratio larger than 3.59 is needed for significance beyond the .05 level. Since the computed *F* was 22.96, it is concluded that the change score centroids differ significantly.

SECTION 5.5
Multivariate Analysis of Variance: More Than Two Groups

This multivariate example is the counterpart of the univariate completely randomized design presented in Section 2.1. As pointed out in the preceding sections, more than one response measure will be recorded for each subject. While the *F*-test is the standard test employed in univariate analyses, several tests are available in the multiple group–multivariate case. The test most widely used is Wilks' *lambda* criterion. This test, which can be related directly to the *F*-test, will be used in the following example.

Points to Consider When Designing an Experiment for More Than Two Groups

1. For each subject, there will be two or more response measures (dependent variables) recorded.
2. More than two groups of subjects will be employed. The number of groups to be compared is arbitrary, but it is rare that more than four or five groups are compared in one experiment.
3. Although not necessary, it is advantageous to have an equal number of subjects in each of the experimental groups.

Example

For this example, assume that three groups of subjects are selected and classified on the basis of their self-reported stress levels. Seven subjects are classified as low stress, nine as medium stress, and eight as high stress.

All groups undergo the same fitness and exercise program. After six months in the program, all subjects are measured in terms of self-confidence (T_1) and weight loss (T_2).

Step 1. Following completion of the experiment, table the data as follows:

Low Stress			Medium Stress			High Stress		
Subjects	T_1	T_2	Subjects	T_1	T_2	Subjects	T_1	T_2
S_1	10	14	S_8	11	14	S_{17}	5	6
S_2	14	15	S_9	15	19	S_{18}	7	8
S_3	13	13	S_{10}	16	16	S_{19}	9	9
S_4	17	19	S_{11}	14	19	S_{20}	11	9
S_5	19	22	S_{12}	11	19	S_{21}	9	8
S_6	14	16	S_{13}	9	14	S_{22}	13	13
S_7	9	8	S_{14}	9	13	S_{23}	9	7
			S_{15}	8	15	S_{24}	7	9
			S_{16}	14	19			

Step 2. Obtain the sum for Test 1 and Test 2 for each of the three groups.

$$\text{Low Group} \quad \begin{array}{ll} \text{Test 1:} & 10 + 14 + \cdots + 9 = \mathbf{96} \\ \text{Test 2:} & 14 + 15 + \cdots + 8 = \mathbf{107} \end{array}$$

$$\text{Med Group} \quad \begin{array}{ll} \text{Test 1:} & 11 + 15 + \cdots + 14 = 107 \\ \text{Test 2:} & 14 + 19 + \cdots + 19 = 148 \end{array}$$

$$\text{High Group} \quad \begin{array}{ll} \text{Test 1:} & 5 + 7 + \cdots + 7 = 70 \\ \text{Test 2:} & 6 + 8 + \cdots + 9 = 69 \end{array}$$

Step 3. Obtain the sum of the scores squared for each test and group.

$$\text{Low Group} \quad \begin{array}{ll} \text{Test 1:} & 10^2 + 14^2 + \cdots + 9^2 = 1392 \\ \text{Test 2:} & 14^2 + 15^2 + \cdots + 8^2 = 1755 \end{array}$$

$$\text{Med Group} \quad \begin{array}{ll} \text{Test 1:} & 11^2 + 15^2 + \cdots + 14^2 = 1341 \\ \text{Test 2:} & 14^2 + 19^2 + \cdots + 19^2 = 2486 \end{array}$$

$$\text{High Group} \quad \begin{array}{ll} \text{Test 1:} & 5^2 + 7^2 + \cdots + 7^2 = 656 \\ \text{Test 2:} & 6^2 + 8^2 + \cdots + 9^2 = 625 \end{array}$$

Step 4. Obtain the sum of the cross products for each group.

Low Group: $(10 \times 14) + (14 \times 15) + \cdots + (9 \times 8) = 1556$
Med Group: $(11 \times 14) + (15 \times 18) + \cdots + (14 \times 19) = 1799$
High Group: $(5 \times 6) + (7 \times 8) + \cdots + (7 \times 9) = 633$

Step 5. The necessary preliminary calculations have been completed. The test will determine whether the mean vectors represented by the experimental groups are equal. In the univariate ANOVA, the test of significance is the F ratio represented by the following general formula:

$$F = \frac{MS_{\text{treatments}}}{MS_{\text{error}}}$$

In the case of the MANOVA, the test to be employed will be Wilks' criterion, which is defined as:

$$lambda = \Lambda = \frac{|\mathbf{W}|}{|\mathbf{T}|}$$

in which $|\mathbf{W}|$ is the determinant of the within-groups sums of squares and cross-products matrix and $|\mathbf{T}|$ is the determinant of the total sums of squares and cross-products matrix.

Step 6. Computation of the elements of the $\mathbf{W}$ matrix: The elements will be arranged in a symmetric matrix as follows:

$$\begin{bmatrix} a & b \\ c & d \end{bmatrix}$$

Note: The determinant of the above matrix is required for later calculations. It is computed as $ad - bc$.

The diagonal elements of the $\mathbf{W}$ matrix (elements a and d) consist of the sum of the SS_w components for Test 1 (self-confidence) and the sum of the SS_w components for Test 2 (weight loss). The basic computational formula for each component is

$$\Sigma X^2 - \frac{(\Sigma X)^2}{N}$$

The components for each group will then be combined to obtain the a and d elements (data from Steps 2 and 3).

	Low—Test 1	Med—Test 1	Hi—Test 1

$$\text{Element } a = 1392 - \frac{(96)^2}{7} + 1341 - \frac{(107)^2}{9} + 656 - \frac{(70)^2}{8}$$

$$= 75.43 \qquad + 68.89 \qquad + 43.50$$

$$= \mathbf{187.82}$$

	Low—Test 2	Med—Test 2	Hi—Test 2

$$\text{Element } d = 1755 - \frac{(107)^2}{7} + 2486 - \frac{(148)^2}{9} + 625 - \frac{(69)^2}{8}$$

$$= \mathbf{201.53}$$

Step 7. Computation of the *SSCP* elements (elements b and c): In this case, the *SSCP* consists of the sums of the cross-products for each of the three experimental groups. Since two test measures were recorded ($p = 2$), elements b and c will be identical. (Data are from Steps 2 and 4.)

	Low	Medium

$$\text{Elements } b \text{ and } c = 1556 - \frac{(96)(107)}{7} + 1799 - \frac{(107)(148)}{9}$$

High

$$+ 633 - \frac{(70)(69)}{8} = \mathbf{157.26}$$

Step 8. Collecting the elements:

$$\mathbf{W} = \begin{bmatrix} (a) & (b) \\ 187.82 & 157.26 \\ (c) & (d) \\ 157.26 & 201.53 \end{bmatrix}$$

Step 9. Computation of the total sum of squares and total sum of cross products required to set up matrix $\mathbf{T}$: The necessary scores must first be totaled so the computations can be completed.

Sum of scores on Test 1 (ΣT_1) (from Step 2): $96 + 107 + 70 = \mathbf{273}$

Sum of scores on Test 2 (ΣT_2) (from Step 2): $107 + 148 + 69 = \mathbf{324}$

Sum of T_1 scores squared (ΣT_1^2) (from Step 3): $1392 + 1341 + 656$
$$= 3389$$

Sum of T_2 scores squared (ΣT_2^2) (from Step 3): $1755 + 2486 + 625$
$$= 4866$$

Sum of cross products $(\Sigma T_1 T_2)$ (from Step 4): $1556 + 1799 + 633$
$$= 3988$$

Step 10. Computation of the elements of the **T** matrix:

Element a: (self-confidence) $= 3389 - \dfrac{(273)^2}{24} = $ **283.6**

Element d: (weight loss) $= 4866 - \dfrac{(324)^2}{24} = $ **492**

Elements b and c: (cross products) $= 3988 - \dfrac{(273)(324)}{24} = $ **302.5**

Step 11. Collecting the elements:

$$
\mathbf{T} =
\begin{bmatrix}
(a) & (b) \\
283.6 & 302.5 \\
(c) & (d) \\
302.5 & 492.0
\end{bmatrix}
$$

Step 12. The final calculation for the necessary components of the lambda ratio requires computation of the determinants. In both cases, the calculation represents subtraction of elements $b \times c$ from $a \times d$.

$|\mathbf{W}| = ad - bc = (187.82)(201.53) - (157.26)(157.26) = $ **13,120.66**
$|\mathbf{T}| = ad - bc = (283.6)(492) - (302.5)(302.5) = $ **48,024.95**

Step 13. Calculation of

$$
\Lambda = \frac{|\mathbf{W}|}{|\mathbf{T}|} = \frac{13,120.66}{48,024.95} = .2732
$$

Step 14. The distribution of *lambda* for particular values of G (number of groups) and p (number of dependent variables) in a one-way MANOVA is as follows:

G	P	Distribution as F	Degrees of Freedom
Any number of groups	2	$\dfrac{1 - \sqrt{\Lambda}}{\sqrt{\Lambda}} \left(\dfrac{N - G - 1}{G - 1} \right)$	$2(G - 1), 2(N - G - 1)$
2	Any number of measures	$\dfrac{1 - \Lambda}{\Lambda} \left(\dfrac{N - p - 1}{p - 1} \right)$	$p, N - p - 1$

G	P	Distribution as F	Degrees of Freedom
3	Any number of measures	$\dfrac{1 - \sqrt{\Lambda}}{\sqrt{\Lambda}}\left(\dfrac{N - p - 2}{2}\right)$	$2p,\ 2(N - p - 2)$

Thus, for the special case where the number of variables measured is two ($p = 2$), the relationship between lambda and F is described as follows:

$$F = \frac{1 - \sqrt{\Lambda}}{\sqrt{\Lambda}}\left(\frac{N - G - 1}{G - 1}\right)$$

Step 15. The degrees of freedom are described by the following:

$$df = 2(G - 1) \text{ and } 2(N - G - 1)$$

In the present example,

$$
\begin{aligned}
df &= 2(3 - 1) \text{ and } 2(24 - 3 - 1) \\
&= 2(2) \text{ and } 2(20) \\
&= \mathbf{4} \text{ and } \mathbf{40}
\end{aligned}
$$

Step 16. Test of significance:

$$F = \frac{1 - \sqrt{.2732}}{.\sqrt{2732}}\left(\frac{24 - 3 - 1}{3 - 1}\right) = .913(10) = \mathbf{9.13}, \text{ with } df = 4, 40$$

Step 17. Looking in the F tables with 4 and 40 df, we see that an F larger than 2.61 is required for significance at the .05 level. Since the computed F is 9.13, it is concluded that the centroids of the three groups are significantly different.

Step 18. Summary Table: After the analysis has been completed, all the calculations are usually collected into a single summary table.

Summary Table for One-Way MANOVA

Source	Matrix	lambda	df of F	F
Groups	not needed			
Within Groups	$\begin{bmatrix} 187.82 & 157.26 \\ 157.26 & 201.53 \end{bmatrix}$	$\dfrac{13{,}120.66}{48{,}024.95}$		
Total	$\begin{bmatrix} 283.6 & 302.5 \\ 302.5 & 492.0 \end{bmatrix}$	$= .2732$	4, 40	9.13

Additional Analyses

In the above example, the three groups of subjects (Low, Medium, and High Stress) clearly responded differently when the criterion measures, self-confidence (T_1) and weight loss (T_2), where considered concurrently. The next likely question would be whether the subjects in the three groups responded differently in terms of the response measures when the measures were considered separately.

To answer this question, the usual practice is to carry out two separate univariate ANOVAs, one on T_1 and the other on T_2. To compute these analyses, the user should turn to Section 2.1 and follow Steps 2 through 13, using the data recorded for T_1 and then repeating that analysis using the data recorded for T_2.

If still further univariate analyses are needed to determine which specific groups differ from each other, the experimenter should turn to Part 3 and undertake one of the appropriate supplemental computations presented in Section 3.4 through 3.9.

SECTION 5.6
Multivariate Analysis: Two-Factor Comparisons

The following example represents the multivariate counterpart of the factorial analysis of variance presented in Section 2.2. As in the preceding section, more than one response measure (dependent variable) is recorded for each subject. Wilks' *lambda* criterion will be the test of significance employed. As in the case of the univariate factorial design, the multivariate two-factor design permits investigation of factors in combination with each other.

Points to Consider When Using the Two-Factor Multivariate Design

1. For each subject there will be two or more response measures (dependent variables) recorded. *Note:* If more than two response measures are recorded, the analysis almost certainly will require use of a computer.
2. It is usually best to have an equal number of subjects in each of the experimental groups and conditions. If equality cannot be achieved, the groups must be proportional in size.
3. The number of treatment groups within each factor is determined by the nature of the problem. However, it is rare that more than four or five groups are included in either of the two factors.

Example

For this example, assume that an experimenter wishes to determine whether there are differences between kindergarten students, first graders, and second graders in their intellectual curiosity. He also wishes to determine whether boys and girls show the same changes. To investigate these questions, the experimenter sets up a factorial design involving three grade levels (kindergarten, first, and second) and two sexes (boys and girls). Two response measures (dependent variables) are recorded. The first (T_1) is a standardized test of creativity; the second (T_2) is based on teachers' judgments of each child's behavior in a variety of standardized problem situations.

Step 1. After the experiment is completed, table the data as follows:

<table>
<tr><th colspan="9">GIRLS</th></tr>
<tr><th colspan="3">Kindergarten</th><th colspan="3">First Grade</th><th colspan="3">Second Grade</th></tr>
<tr><th>S_s</th><th>T_1</th><th>T_2</th><th>S_s</th><th>T_1</th><th>T_2</th><th>S_s</th><th>T_1</th><th>T_2</th></tr>
<tr><td>S_1</td><td>10</td><td>14</td><td>S_{11}</td><td>15</td><td>30</td><td>S_{21}</td><td>17</td><td>30</td></tr>
<tr><td>S_2</td><td>11</td><td>18</td><td>S_{12}</td><td>17</td><td>26</td><td>S_{22}</td><td>19</td><td>35</td></tr>
<tr><td>S_3</td><td>19</td><td>19</td><td>S_{13}</td><td>18</td><td>25</td><td>S_{23}</td><td>22</td><td>36</td></tr>
<tr><td>S_4</td><td>16</td><td>7</td><td>S_{14}</td><td>15</td><td>17</td><td>S_{24}</td><td>22</td><td>35</td></tr>
<tr><td>S_5</td><td>18</td><td>16</td><td>S_{15}</td><td>15</td><td>26</td><td>S_{25}</td><td>19</td><td>31</td></tr>
<tr><td>S_6</td><td>17</td><td>14</td><td>S_{16}</td><td>12</td><td>25</td><td>S_{26}</td><td>23</td><td>29</td></tr>
<tr><td>S_7</td><td>9</td><td>13</td><td>S_{17}</td><td>14</td><td>21</td><td>S_{27}</td><td>26</td><td>31</td></tr>
<tr><td>S_8</td><td>12</td><td>15</td><td>S_{18}</td><td>17</td><td>21</td><td>S_{28}</td><td>23</td><td>29</td></tr>
<tr><td>S_9</td><td>12</td><td>15</td><td>S_{19}</td><td>19</td><td>23</td><td>S_{29}</td><td>21</td><td>30</td></tr>
<tr><td>S_{10}</td><td>16</td><td>15</td><td>S_{20}</td><td>21</td><td>26</td><td>S_{30}</td><td>21</td><td>25</td></tr>
<tr><th colspan="9">BOYS</th></tr>
<tr><td>S_{31}</td><td>31</td><td>31</td><td>S_{41}</td><td>15</td><td>30</td><td>S_{51}</td><td>10</td><td>12</td></tr>
<tr><td>S_{32}</td><td>30</td><td>40</td><td>S_{42}</td><td>19</td><td>26</td><td>S_{52}</td><td>12</td><td>13</td></tr>
<tr><td>S_{33}</td><td>26</td><td>43</td><td>S_{43}</td><td>17</td><td>22</td><td>S_{53}</td><td>14</td><td>15</td></tr>
<tr><td>S_{34}</td><td>26</td><td>36</td><td>S_{44}</td><td>16</td><td>21</td><td>S_{54}</td><td>19</td><td>18</td></tr>
<tr><td>S_{35}</td><td>25</td><td>32</td><td>S_{45}</td><td>18</td><td>25</td><td>S_{55}</td><td>14</td><td>14</td></tr>
<tr><td>S_{36}</td><td>28</td><td>31</td><td>S_{46}</td><td>16</td><td>21</td><td>S_{56}</td><td>18</td><td>15</td></tr>
<tr><td>S_{37}</td><td>21</td><td>36</td><td>S_{47}</td><td>14</td><td>21</td><td>S_{57}</td><td>16</td><td>17</td></tr>
<tr><td>S_{38}</td><td>21</td><td>36</td><td>S_{48}</td><td>12</td><td>19</td><td>S_{58}</td><td>12</td><td>14</td></tr>
<tr><td>S_{39}</td><td>32</td><td>32</td><td>S_{49}</td><td>14</td><td>28</td><td>S_{59}</td><td>10</td><td>14</td></tr>
<tr><td>S_{40}</td><td>27</td><td>34</td><td>S_{50}</td><td>16</td><td>25</td><td>S_{60}</td><td>19</td><td>14</td></tr>
</table>

Step 2. Calculate the sum of each group.

<table>
<tr><th colspan="9">GIRLS</th></tr>
<tr><td>S_1</td><td>10</td><td>14</td><td>S_{11}</td><td>15</td><td>30</td><td>S_{21}</td><td>17</td><td>30</td></tr>
<tr><td>S_2</td><td>11</td><td>18</td><td>S_{12}</td><td>17</td><td>26</td><td>S_{22}</td><td>19</td><td>35</td></tr>
<tr><td></td><td>+</td><td>+</td><td></td><td>+</td><td>+</td><td></td><td>+</td><td>+</td></tr>
<tr><td></td><td>⋮</td><td>⋮</td><td></td><td>⋮</td><td>⋮</td><td></td><td>⋮</td><td>⋮</td></tr>
<tr><td>S_{10}</td><td>16</td><td>15</td><td>S_{20}</td><td>21</td><td>26</td><td>S_{30}</td><td>21</td><td>25</td></tr>
<tr><td>Sum =</td><td>140</td><td>146</td><td></td><td>163</td><td>240</td><td></td><td>213</td><td>311</td></tr>
<tr><th colspan="9">BOYS</th></tr>
<tr><td>S_{31}</td><td>31</td><td>31</td><td>S_{41}</td><td>15</td><td>30</td><td>S_{51}</td><td>10</td><td>12</td></tr>
<tr><td>S_{32}</td><td>30</td><td>40</td><td>S_{42}</td><td>19</td><td>26</td><td>S_{52}</td><td>12</td><td>13</td></tr>
</table>

	BOYS							
	+ ⋮	+ ⋮		+ ⋮	+ ⋮		+ ⋮	+ ⋮
S_{40}	27	34	S_{50}	16	25	S_{60}	19	14
Sum =	267	351		157	238		144	146

Step 3. Calculate the sum of the squared scores for each group.

	GIRLS							
S_1	10^2 +	14^2 +	S_{11}	15^2 +	30^2 +	S_{21}	17^2 +	30^2 +
S_{10}	16^2	15^2	S_{20}	21^2	26^2	S_{30}	21^2	25^2
Sum =	2076	2226		2719	5878		4595	9775

	BOYS							
S_{31}	31^2 +	31^2 +	S_{41}	15^2 +	30^2 +	S_{51}	10^2 +	12^2 +
S_{40}	27^2	34^2	S_{50}	16^2	25^2	S_{60}	9^2	14^2
Sum =	7257	12,463		2503	5778		2182	2160

Step 4. Calculate the sum of the cross products.

	GIRLS				
S_1	$10 \times 14 = 140$ +	S_{11}	$15 \times 30 = 450$ +	S_{21}	$17 \times 30 = 510$ +
S_{10}	$6 \times 15 = 90$	S_{20}	$21 \times 26 = 546$	S_{30}	$21 \times 25 = 525$
Sum =	2054		3921		6621

	BOYS				
S_{31}	$31 \times 31 = 961$ +	S_{41}	$15 \times 30 = 450$ +	S_{51}	$10 \times 12 = 120$ +
S_{40}	$27 \times 34 = 918$	S_{50}	$16 \times 25 = 400$	S_{60}	$9 \times 14 = 126$
Sum =	9337		3754		2140

Step 5. Calculate the correction term for T_1 (creativity is this example). First, add the sums of all of the T_1 sub-totals (Step 2) to obtain the grand sum of the T_1 scores for the entire table.

$$140 + 163 + \cdots + 144 = 1084$$

Then square this value and divide by the total number of T_1 measures recorded. (See Step 1.) This yields the T_1 correction term.

$$\frac{1084^2}{60} = \frac{1,175,056}{60} = \textbf{19,584.267}$$

Step 6. Calculation of the correction term for T_2 (teacher's judgment in this example). Repeat the above steps for T_2.

$$146 + 240 + \cdots + 146 = \textbf{1432}$$

and

$$\frac{1432^2}{60} = \frac{2,050,624}{60} = \textbf{34,177.067}$$

Step 7. Calculate the cross products correction term. First, multiply the sum of the T_1 scores (Step 5) times the sum of the T_2 scores (Step 6).

$$1084 \times 1432 = \textbf{1,552,288}$$

Then divide by the number of measures on which each sum was based.

$$\frac{1,552,288}{60} = \textbf{25,871.467}$$

Step 8. Calculate the total sum of squares for T_1. First, obtain the sum of the T_1 scores squared (from Step 3).

$$2076 + 2719 + 4595 + 7257 + 2503 + 2182 = \textbf{21,332}$$

Then, subtract the correction term obtained in Step 5.

$$21,332 - 19,584.267 = \textbf{1747.733}$$

Step 9. Calculate the total sum of squares for T_2. First, obtain the sum of the T_2 scores squared (from Step 3).

$$2226 + 5878 + 9775 + 12463 + 5778 + 2160 = \textbf{38,280}$$

Then, subtract the correction term obtained in Step 6.

$$38,280 - 34,177.067 = \textbf{4102.933}$$

Step 10. Calculate the total sum of squares for cross products. First, obtain the total sum of cross products (from Step 4).

$$2054 + 3921 + 6621 + 9337 + 3754 + 2140 = \textbf{27,827}$$

Then, subtract the correction term obtained in Step 7.

$$27,827 - 25,871.467 = \textbf{1955.533}$$

Step 11. Calculate the sum of squares for rows on T_1 (the overall effects of girls versus boys on the creativity test).

Sum of girls' scores on the creativity test: $140 + 163 + 213 = $ **516**
Sum of boys' scores on the creativity test: $267 + 157 + 144 = $ **568**

Then, square the above sums, divide by the number of measures on which each was based, and add the quotients.

$$\frac{516^2}{30} + \frac{568^2}{30} = \frac{588,880}{30} = \mathbf{19{,}629.333}$$

Finally, subtract the correction term for T_1 (Step 5) from this value.

$$19{,}629.333 - 19{,}584.267 = \mathbf{45.066}$$

Step 12. Calculation of the sum of squares for rows on T_2 (the overall difference between girls and boys based on teachers' judgments): These calculations repeat Step 11, using the scores recorded as T_2.

Sum of girls' scores (judgment): $146 + 240 + 311 = $ **697**
Sum of boys' scores (judgment): $351 + 238 + 146 = $ **735**

Then, squaring and dividing:

$$\frac{697^2}{30} + \frac{735^2}{30} = \frac{1{,}026{,}034}{30} = \mathbf{34{,}201.133}$$

Then, subtract the correction term (Step 6).

$$34{,}201.133 - 34{,}177.067 = \mathbf{24.066}$$

Step 13. Calculation of the sum of squares of cross products for rows (boys versus girls.)
First, obtain the overall sum of the T_1 and T_2 scores for the first (girls) condition (from Step 2).

$$140 + 163 + 213 = \mathbf{516}$$

and

$$146 + 240 + 311 = \mathbf{697}$$

Then, multiply the overall sum of the T_1 scores times the overall sum of the T_2 scores.

$$697 \times 516 = \mathbf{359{,}652}$$

Then, divide by the number of scores on which each sum was based.

$$\frac{359{,}652}{30} = \mathbf{11{,}988.4}$$

Now, repeat the above steps for the second (boys) condition.

Sum of the T_1 scores: $267 + 157 + 144 = $ **568**

Sum of the T_2 scores: $351 + 238 + 146 = $ **735**

and

$$\frac{568 \times 735}{30} = \frac{417,480}{30} = 13,916$$

Finally, add the sums of cross products for girls and boys computed above.

$$11,988.4 + 13,916 = 25,904.4$$

From that value, subtract the sum of cross products correction term (Step 7).

$$25,904.4 - 25,871.467 = 32.933$$

Step 14. Calculation of the sum of squares for columns on T_1 (the overall effects of grade level on creativity):

First, add the sums of the kindergarten, first, and second grade level on the creativity test (T_1) (from Step 2).

K on T_1: $140 + 267 = $ **407**

First on T_1: $163 + 157 = $ **320**

Second on T_1: $213 + 144 = $ **357**

Square the above sums, divide by the number of measures on which each of the sums was based, and add the quotients.

$$\frac{407^2}{20} + \frac{320^2}{20} + \frac{357^2}{20} = \frac{684,342}{20} = 19,774.9$$

Then, subtract the correction term for T_1 (Step 5).

$$19,774.9 - 19,584.267 = 190.633$$

Step 15. Computation of the sum of squares for columns on T_2 (the overall effects of grade level on judgments).

Again, add the sums of the teacher judgment scores recorded for the kindergarten, first and second graders.

K on T_2: $146 + 351 = $ **497**

First on T_2: $240 + 238 = $ **478**

Second on T_2: $311 + 146 = $ **457**

Square the sums, divide by the number of measures on which each was based, and add the quotients.

$$\frac{497^2}{20} + \frac{478^2}{20} + \frac{457^2}{20} = \frac{684,342}{20} = \mathbf{34,217.1}$$

Then, subtract the correction term for T_2 (Step 6).

$$34,217.1 - 34,177.067 = \mathbf{40.033}$$

Step 16. Calculation of the sum of squares for cross products for columns (grade levels):

First, obtain the overall sum of T_1 scores and the overall sum of T_2 scores for kindergarten group (from Step 2).

Creativity/kindergarten: $140 + 267 = \mathbf{407}$
Judgments/kindergarten: $146 + 351 = \mathbf{497}$

Then multiply these sums times each other and divide by the number of scores on which each sum was based.

$$\frac{407 \times 497}{20} = \frac{202,279}{20} = \mathbf{10,113.95}$$

Then, repeat the same set of computations for first and second graders.

Creativity/first graders: $163 + 157 = \mathbf{320}$
Judgments/first graders: $240 + 238 = \mathbf{478}$

and

$$\frac{320 \times 478}{20} = \frac{152,900}{20} = \mathbf{7,648}$$

Creativity/second graders: $213 + 144 = \mathbf{357}$
Judgments/second graders: $311 + 146 = \mathbf{457}$

and

$$\frac{357 \times 457}{20} = \frac{163,149}{20} = \mathbf{8,157.45}$$

Then, add the sums of cross products for the three groups (kindergarten, first and second graders) computed above.

$$10,113.95 + 7,520 + 8,157.45 = \mathbf{25,919.4}$$

Finally, from that value, subtract the sum of squares for cross-products correction term (Step 7).

$$25,919.4 - 25,871.467 = \mathbf{47.933}$$

Note: The sum of squares for cross products can be a negative value; the sum of squares value is always positive.

Step 17. Calculation of the sum of squares of rows by columns ($SS_{r \times c}$) for T_1 (the interactive effects of boys and girls by grade level for the creativity test):

First, square the sum of each of the group's scores on T_1, divide each by the number of scores added to obtain that sum, then sum these quotients (data from Step 2).

$$\frac{140^2}{10} + \frac{163^2}{10} + \frac{213^2}{10} + \frac{267^2}{10} + \frac{157^2}{10} + \frac{144^2}{10} = \frac{208,212}{10} = \mathbf{20,821.2}$$

Then, subtract the correction term for T_1 (Step 5), the SS for rows$_{T_1}$ (Step 11), and the SS for columns$_{T_1}$ (Step 14).

$$20,821.2 - 19,584.267 - 45.066 - 190.633 = \mathbf{1001.234}$$

Step 18. Calculation of the $SS_{r \times c}$ for T_2: Repeat the above calculations for the performance of boys and girls at each grade level for the comprehension test (data from Step 2). Square the group scores on T_2, and divide by the number of scores added to obtain the sum.

$$\frac{146^2}{10} \times \frac{240^2}{10} + \frac{311^2}{10} + \frac{351^2}{10} + \frac{238^2}{10} + \frac{146^2}{10} = \mathbf{37,679.8}$$

Then subtract the correction term for T_2 (Step 6), the SS for rows (Step 12), and the SS for columns (Step 15).

$$37,679.8 - 34,177.067 - 24.066 - 40.033 = \mathbf{3,438.634}$$

Step 19. Calculation of the sum of squares of cross products for the rows by columns ($SSCP_{r \times c}$):

First, multiply the sum of T_1 scores for each condition times the sum of the T_2 scores for each condition (from Step 2). Then divide by the number of pairs of scores.

$$\frac{140 \times 146}{10} + \frac{163 \times 240}{10} + \frac{213 \times 311}{10} + \frac{267 \times 351}{10}$$

$$+ \frac{157 \times 238}{10} + \frac{144 \times 146}{10} = \frac{277,910}{10} = \mathbf{27,791}$$

Then subtract the SSCP correction term (Step 7), the $SSCP_{T_1}$ for rows (Step 13), and the $SSCP_{T_2}$ for columns (Step 16).

$$27,791 - 25,871.467 - 32.933 - 47.933 = \mathbf{1838.667}$$

Step 20. Computation of the $SS_{\text{within groups}}$ for each of the components. These values are computed as residuals according to the following formula:

Total − Rows − Columns − Rows × Columns = Within Variability

For $SS_{W_{T_1}}$: 1747.733 (Step 8) − 45.066 (Step 11)

$\qquad$ − 190.633 (Step 14) − 1001.234 (Step 17) = **510.8**

For $SS_{W_{T_2}}$: 4102.933 (Step 9) − 24.066 (Step 12)

$\qquad$ − 40.033 (Step 15) − 3438.634 (Step 18) = **600.2**

For $SSCP_{W_{T_1T_2}}$: 1955.533 (Step 10) − 974.633 (Step 13)

$\qquad$ − 47.933 (Step 16) − 896.967 (Step 19) = **36**

Note: As a double check of the computations, each of the above terms can be computed directly. For example, for the cross products values $(SSCP_{T_1T_2})$:

$$2054 - \frac{140 \times 146}{10} + 3921 - \frac{163 \times 240}{10}$$

$$+ \cdots + 2140 - \frac{144 \times 146}{10} = \mathbf{36}$$

Step 21. The above computed values will now be inserted into a series of symmetric matricies with the elements identified as follows:

$$\begin{bmatrix} a & b \\ c & d \end{bmatrix}$$

The above symmetric matricies will serve as the bases for all of the calculations needed to set up the necessary tests of significance. The general formula for Wilks' *lambda* is

$$lambda = \Lambda = \frac{|\mathbf{W}|}{|\mathbf{T}|}$$

where $|\mathbf{W}|$ = the *error* sums of squares and cross products determinant

and $|\mathbf{T}|$ = the determinant of the *error* sums of squares and cross products added to the determinant of the sums of squares and cross products of the effects being measured.

The ratio calculation will involve the determinants for each of several matrices. Each determinant will be calculated as:

$$(a \times d) - (b \times c)$$

Step 22. Summary of computational steps as they relate to the several matrix elements.

The several tests of significance will be based on the following matrices. In each, SS_{T_1} will form element a, SS_{T_2} will form element d, and $SSCP_{T_1T_2}$ will form elements b and c.

$$\text{Total} = \begin{bmatrix} 1747.733 & 1955.533 \\ 1955.533 & 4102.933 \end{bmatrix} \quad \text{(computed in Steps 8, 9, and 10)}$$

$$\text{Rows} = \begin{bmatrix} 45.066 & 32.933 \\ 32.933 & 24.066 \end{bmatrix} \quad \text{(computed in Steps 11, 12, and 13)}$$

$$\text{Columns} = \begin{bmatrix} 190.633 & 47.933 \\ 47.933 & 40.033 \end{bmatrix} \quad \text{(computed in Steps 14, 15, and 16)}$$

$$\text{Rows} \times \text{Cols} = \begin{bmatrix} 1001.234 & 1838.667 \\ 1838.667 & 3438.634 \end{bmatrix} \quad \text{(computed in Steps 17, 18, and 19)}$$

$$\text{Within} = \begin{bmatrix} 510.8 & 36.0 \\ 36.0 & 600.2 \end{bmatrix} \quad \text{(computed in Step 20)}$$

Step 23. Computation of the necessary ratios: As stated above, the general relationship is

$$\Lambda = \frac{|\mathbf{W}|}{|\mathbf{T}|}$$

and the actual ratios are equal to:

$$\frac{|\mathbf{W}|}{|\mathbf{T}|} = \frac{\text{Determinant}_{\text{within}}}{\text{Determinant}_{\text{effects}} + \text{Determinant}_{\text{within}}}$$

Thus, the above ratios are calculated on the appropriate matrix addition followed by calculation of $ad - bc$.

Since the test ratios always involve the determinant of the within component, that value must be calculated first.

$$\text{Determinant}_{\text{within}} = \mathbf{D}_w = \overset{a}{(510.8)}\overset{d}{(600.2)} - \overset{b}{(36.0)}\overset{c}{(36.0)} = \mathbf{305{,}286.16}$$

$$\text{Determinant}_{\text{total}} = \mathbf{D}_t = \text{Not needed for significance test}$$

$$\text{Rows} = \mathbf{D}_r + \mathbf{D}_w$$

$$= \begin{bmatrix} 45.066 & 32.933 \\ 32.933 & 24.066 \end{bmatrix} + \begin{bmatrix} 510.8 & 36.0 \\ 36.0 & 600.2 \end{bmatrix} = \begin{bmatrix} 555.866 & 68.933 \\ 68.933 & 624.266 \end{bmatrix}$$

And, multiplication of $ad - bc = (555.866)(624.266)$
$$- (68.933)(68.933) = 347{,}008.244 - 4751.7578 = \mathbf{342{,}256.486}$$

$$\text{Columns} = \mathbf{D}_c + \mathbf{D}_w = \begin{bmatrix} 190.633 & 47.933 \\ 47.933 & 40.033 \end{bmatrix} + \begin{bmatrix} 510.8 & 36.0 \\ 36.0 & 600.2 \end{bmatrix}$$

$$= \begin{bmatrix} 701.433 & 83.933 \\ 83.933 & 640.233 \end{bmatrix}$$

$$= (701.433)(640.233) - (83.933)(83.933)$$

$$= 449{,}080.544 - 7{,}044.748 = \mathbf{442{,}035.806}$$

$$\text{Rows} \times \text{Cols} = \mathbf{D}_{/x_c} + \mathbf{D}_w = \begin{bmatrix} 1001.234 & 1838.667 \\ 1839.667 & 3438.634 \end{bmatrix} + \begin{bmatrix} 510.8 & 36.0 \\ 36.0 & 600.2 \end{bmatrix}$$

$$= \begin{bmatrix} 1512.034 & 1874.667 \\ 1874.667 & 4038.834 \end{bmatrix}$$

$$= (1512.034)(4038.834) - (1874.667)(1874.667)$$

$$= 6{,}106{,}854.328 - 3{,}514{,}372.444 = \mathbf{2{,}592{,}481.884}$$

Step 24. The relationship between **lambda** and F: The distribution of lambda for certain numbers of dependent variables (p), the df associated with the within-groups error (w), and the df for the particular effects (e) being analyzed are as follows:

Value of e	Value of p	Distribution of F	Degrees of Freedom of F
Any number	2	$\dfrac{1 - \sqrt{\Lambda_e}}{\sqrt{\Lambda_e}}\left(\dfrac{w-1}{e}\right)$	$2e,\ 2(w-1)$
1	Any number	$\dfrac{1 - \Lambda_e}{\Lambda_e}\left(\dfrac{w+e-p}{p}\right)$	$p,\ w+e-p$
2	Any number	$\dfrac{1 - \sqrt{\Lambda_e}}{\sqrt{\Lambda_e}}\left(\dfrac{w+e-p-1}{p}\right)$	$2p,\ 2(w+e-p-1)$

For the example just completed, there are two dependent variables ($p = 2$). Thus, the following formula is appropriate to describe the exact functional relationship between *lambda* and F.

$$F = \frac{1 - \sqrt{\Lambda_e}}{\sqrt{\Lambda_e}}\left(\frac{w-1}{e}\right)$$

Where $w = \text{rows} \times \text{columns} \times (n_{\text{per gp.}} - 1)$

and

$e = (\text{rows} - 1)$, or $(\text{columns} - 1)$, or $(\text{rows} - 1) \times (\text{columns} - 1)$

Step 25. Computation of the *lambda* ratios:

$$\Lambda_{\text{rows}} = \frac{305{,}286.160}{342{,}256.486} = \mathbf{.8920}$$

$$\Lambda_{\text{cols}} = \frac{305{,}286.160}{442{,}035.806} = \mathbf{.6906}$$

$$\Lambda_{\text{rows}\times\text{cols}} = \frac{305,286.160}{2,592,481.884} = .1178$$

Step 26. The degrees of freedom, *df*, for the F-ratio are equal to:

$$
\begin{aligned}
\text{Rows} &= 2(r - 1) \text{ and } 2[(r \times c)(n_{\text{per gp.}} - 1) - 1] \\
&= 2(1) \text{ and } 2[(2 \times 3)(10 - 1) - 1] \\
&= 2 \text{ and } 2[(6)(9) - 1] \\
&= \mathbf{2} \text{ and } \mathbf{106} \\
\text{Columns} &= 2(c - 1) \text{ and } 2[(r \times c)(n_{\text{per gp.}} - 1) - 1] \\
&= 2(2) \text{ and } 2[(6)(9) - 1] \\
&= \mathbf{4} \text{ and } \mathbf{106} \\
\text{Rows} \times \text{Columns} &= 2(r - 1)(c - 1) \text{ and } 2[(r \times c)(n_{\text{per gp.}} - 1) - 1] \\
&= \mathbf{4} \text{ and } \mathbf{106}
\end{aligned}
$$

Step 27. The tests of significance:

For Rows (boys versus girls): $F_{2,106} = \dfrac{1 - \sqrt{.8920}}{\sqrt{.8920}} \left(\dfrac{53}{1}\right) = \mathbf{3.1203}$

For Columns (grade level): $F_{4,106} = \dfrac{1 - \sqrt{.6906}}{\sqrt{.6906}} \left(\dfrac{53}{2}\right) = \mathbf{5.3893}$

For Rows by Columns: $F_{4,106} = \dfrac{1 - \sqrt{.1178}}{\sqrt{.1178}} \left(\dfrac{53}{2}\right) = \mathbf{50.7145}$

Step 28. First, looking in the F tables under the *df* for rows (2, 106) we find that an F ratio of 3.10 is needed for significance.

Since the computed F is 3.1203, we conclude that the centroid of T_1 and T_2 is significantly different for boys and girls.

Looking now in the F tables under the *df* for rows and rows × columns (4, 106), we find that an F ratio of 2.48 is needed for significance. Since the computed F of 5.3893 for columns exceeds that value, we conclude that the centroid of T_1 and T_2 is significantly different for the three grade levels. Similarly, the interaction ($F_{4,106} = 50.7145$) of boy/girls by grade level is also significant beyond the .05 level.

Step 29. Summary Table: Following completion of all the computations, the final results of the analysis are summarized as follows:

Source	df	Matrix		lambda_e	df of F	F
Rows (Boys/ Girls)	R − 1	555.866 68.933	68.933 624.266	305,286.160 342,256.486 = .8920	2, 106	3.1203
Columns (Grade level)	C − 1	701.433 83.933	83.933 640.233	305,286.160 442,035.806 = .6906	4, 106	5.3895

Source	df	Matrix	lambda$_e$	df of F	F
Rows × Cols (Interaction)	$(R - 1) \times (C - 1)$	$\begin{bmatrix} 1512.034 & 1874.667 \\ 1874.667 & 4038.834 \end{bmatrix}$	$\dfrac{305,286.160}{2,592,481.884}$ = .1178	4, 106	50.7145
Within Gps. (Error)	$RC(n - 1)$	$\begin{bmatrix} 510.800 & 36.000 \\ 36.000 & 600.200 \end{bmatrix}$			
Total	$nRC - 1$				

Further Analyses

In the above example, the boys and girls responded differently in terms of the standardized test scores of creativity and their teachers' judgments. Also, there were clear differences among the kindergarten, first, and second graders. Finally, there was a strong interaction between the sex of the subjects and grade level in terms of the response measures recorded.

The usual first step to analyze these differences in greater detail is to carry out two separate univariate, two-way ANOVAs. The first would be carried out using the T_1 scores, and the second carried out using the T_2 scores. To compute these analyses, the experimenter should turn to Section 2.2 and follow steps 2 through 13. If still further univariate analyses are felt necessary to determine which specific groups differ on each measure, the experimenter should turn to Part 3 and compute one or more of the appropriate supplemental computations presented in Sections 3.4 through 3.10.

Nonparametric Tests, Miscellaneous Tests of Significance, and Indexes of Relationship

In many cases where it is desirable to make comparisons of two proportions, or of two medians, or within a table of frequency counts, there are some very simple tests that can be used. Some of these simple tests are listed in the following six sections. Various names—such as *nonparametric tests* or *distribution-free tests*—have been applied to many such tests of significance. We will not attempt to argue the merits and deficiencies of these tests, but will merely point out that they are relatively easy to use and that they are often applicable when the various tests presented elsewhere (e.g., in Sections 1.5 through 2.13) are not.

The six sections in this part are discussed below in three sets of two each. These discussions should help you choose the correct analysis for your particular problem.

1. Significance tests for proportions
2. Significance tests for differences between two groups
3. Tests for (1) significance of differences or (2) relationships, when data are gathered into contingency tables

Tests for Proportions (Sections 6.1 and 6.2)

If you have a proportion that has been computed from experimental data (such as the proportion of all American men over forty who have been married at least two times) and you wish to find whether it differs from some specific value (such as a guess that one third, or .33, of all

American men over forty have been married at least two times), the material presented in Section 6.1 will help you solve the problem. If you wish to find whether two independently computed proportions are different, use the analysis presented in Section 6.2.

Tests for Differences Between Two Groups (Sections 6.3 and 6.4)

Section 1.6 (*t*-test for independent means) shows how to find whether two *mean* values are different. If you wish to find whether two independent *median* values are different, use the procedure in Section 6.3.

Section 1.7 (*t*-test for related measures) presents the method for finding whether two *means* are different when the samples are paired, matched, or otherwise related. If you wish to test whether two related groups differ on the basis of a simple sign test, use the analysis in Section 6.4.

Tests for Data in Contingency Tables (Sections 6.5 and 6.6)

When you have frequency data (counts of the people exhibiting the characteristic of interest to you) that can be put into the form of a contingency table, you probably wish to ask one side or the other of a two-sided question: (1) Do these variables (treatments) produce different results? or (2) Are these variables related, and if so, how highly related? Both questions can be answered by the tests presented in Sections 6.5 and 6.6. The first question is a chi-square question, and the first part of either Section 6.5 (if you have a 2×2 table of data) or Section 6.6 (if you have a large number of cells in the table) will help you answer it. The second question is a *phi*-coefficient question or a contingency-coefficient question depending on the number of cells in your table, and the last part of either Section 6.5 or 6.6 will help you answer it.

Tests for Trends, Runs, and Randomness (Section 6.7)

When scores are recorded which represent an event occurring over time, questions often arise whether the scores are patterned in some fashion or are random. The tests described in this section represent a few of the ways scores recorded to describe a recurring event can be tested to determine their degree of patterning.

In the following sections, computational formulas are again presented where they contribute to the understanding of the analysis. In those instances where the computation involves only a count or a ranking, the formula is omitted.

SECTION 6.1
Test for Significance of a Proportion

In a situation where a logically dichotomous variable, such as passing or failing a course, is considered, the use of proportions can sometimes facilitate the statistical analysis of significance. Before the test of significance can be executed, it is necessary to know the proportion to be expected. If you toss a coin, for example, you expect to get heads one half of the times (in the long run); thus, the expected proportion is .50. If you toss a die, you expect to get a one-spot one time in six; thus, the expected proportion is 1/6. When the dichotomy is on a variable such as passing or failing a course, you must either (1) know the expected proportion from some previous experience, (2) insist on some a priori proportion, such as "10 per cent will fail," or (3) in some other way determine or state the expected proportion. Then, when the actual proportion has been determined and the expected proportion has been decided upon, the test for significance can be carried out.

Example

Assume that a teacher wishes to determine whether a particular class of fifteen students is under par in its performance on an arithmetic achievement test. This can be done simply by determining if the proportions who passed and failed correspond to those proportions in classes taught in previous years.

Step 1. In order to determine the actual or observed proportion, the first step is to record and table the data on which the proportion will be based (in this case, the arithmetic scores).

Student	Score
A	64
B	35
C	81
D	77
E	65
F	48
G	56
H	31
I	64
J	72
K	43
L	44
M	63
N	59
O	38

Step 2. In order to compute the proportion of students who passed, the teacher must set a pass-fail cutoff point. Suppose it is decided that a score of 50 or more is passing and a score of 49 or less is failing. We can quickly see that Students B, F, H, K, L, and O failed. Six out of fifteen failed; thus, nine out of fifteen passed. The proportion (p) of students who passed is the number who passed divided by the total number of students.

$$p = \frac{9}{15} = .60$$

Step 3. Suppose the teacher's information from previous classes indicates that, on the average, 80 per cent (or .80) of the students got a passing grade on this examination. The .80 expected to pass is the *P* value. The significance test (z-test) for the obtained proportion is computed by use of the following formula.

$$z = \frac{p - P}{\sqrt{\dfrac{P(1 - P)}{N}}}$$

Thus, for this case,

$$z = \frac{.60 - .80}{\sqrt{\dfrac{.80(1 - .80)}{15}}} = \frac{-.20}{\sqrt{\dfrac{.80(.20)}{15}}} = \frac{-.20}{\sqrt{\dfrac{.16}{15}}}$$

$$= \frac{-.20}{\sqrt{.0107}} = \frac{-.20}{.1034} = -1.93$$

In order to have a significant difference, at the .05 level using a two-tailed test, the z would need to be at least ±1.96 (see Appendix A). A significant z would mean that the *p* value is significantly different from the *P* value. In this case, it is concluded that the scores of the fifteen students are not under par for the arithmetic achievement test.

SECTION 6.2
Test for Significance of Difference Between Two Proportions

When dealing with data on a logically dichotomous variable, it is easy to test the significance of a difference between the proportions.

Example

In an experiment on short-term memory for a single item, the following conditions were used. The item was presented to all subjects for 2 seconds. One group of twenty subjects was tested to determine if they recalled the item after 5 seconds; the other group of twenty subjects was tested after 10 seconds. If the item was correctly recalled, a "+" was recorded for that subject; if an incorrect response was made, a "0" was recorded.

The formula for the significance of the difference between two proportions is

$$z = \frac{P_1 - P_2}{\sqrt{\dfrac{p(1 - p)}{N_1} + \dfrac{p(1 - p)}{N_2}}}$$

where the value of p under the radical is computed as:

$$p = \frac{N_1 P_1 + N_2 P_2}{N_1 + N_2}$$

Step 1. In order to compute the two proportions P_1 and P_2, from Groups 1 and 2 respectively, first table the data as follows:

	Group 1		Group 2
Subject	Recall (5 sec.)	Subject	Recall (10 sec.)
S_1	+	S_{21}	0
S_2	+	S_{22}	0
S_3	+	S_{23}	+
S_4	+	S_{24}	0
S_5	+	S_{25}	+
S_6	+	S_{26}	+
S_7	+	S_{27}	0
S_8	+	S_{28}	+
S_9	+	S_{29}	0
S_{10}	0	S_{30}	0
S_{11}	+	S_{31}	+
S_{12}	0	S_{32}	0
S_{13}	0	S_{33}	0
S_{14}	+	S_{34}	0
S_{15}	+	S_{35}	0
S_{16}	0	S_{36}	+
S_{17}	+	S_{37}	0
S_{18}	+	S_{38}	0
S_{19}	+	S_{39}	+
S_{20}	+	S_{40}	+

Step 2. Subjects 10, 12, 13, and 16 in Group 1 missed the item; thus, a total of 16 subjects recalled correctly. The number correctly recalling, divided by the number of subjects in the group, gives the proportion who correctly recalled the item.

$$P_1 = \frac{16}{20} = .80$$

Step 3. Subjects 21, 22, 24, 27, 29, 30, 32, 33, 34, 35, 37, and 38 in Group 2 missed the item. Thus, 12 subjects missed, and only 8 correctly recalled the item. Again, the number who correctly recalled, divided by the number of subjects in the group, gives the proportion who correctly recalled the item.

$$P_2 = \frac{8}{20} = .40$$

Step 4. Multiply the result of Step 2 by the value of N_1 (the number of cases in the first sample).

$$.80 \times 20 = 16$$

Then multiply the result of Step 3 by the value of N_2 (the number of cases in the second sample).

$$.40 \times 20 = 8$$

Step 5. Add the two numbers computed in Step 4.

$$16 + 8 = 24$$

Then divide that answer by the total number of cases in both samples $(N_1 + N_2 = 40)$.

$$24/40 = .60 \quad \textit{This is the value of p.}$$

Step 6. Subtract the final result of Step 5 from the number 1.

$$1 - .60 = .40$$

Then multiply that number by the final result of Step 5.

$$.60 \times .40 = .24$$

Step 7. Divide the result of Step 6 by N_1 (the size of the first sample).

$$.24/20 = .012$$

Step 8. Divide the result of Step 6 by N_2 (the size of the second sample).

$$.24/20 = .012$$

Step 9. Add the result of Step 7 to the result of Step 8.

$$.012 + .012 = .024$$

Then take the square root of that sum.

$$\sqrt{.024} = .155$$

Step 10. Subtract the result of Step 3 from the result of Step 2.

$$.80 - .40 = .40$$

Step 11. Divide the result of Step 10 by the result of Step 9.

$$.40/.155 = \textbf{2.48}$$

A z having value greater than or equal to 1.96 or less than or equal to -1.96 is considered significant at the .05 level using a two-tailed test (see Appendix A). Now, a significant z tells us that the two proportions are significantly different. Thus, in the above problem, the two proportions significantly differ.

SECTION 6.3

The Mann-Whitney U-Test for Differences Between Independent Samples

Occasionally the data collected by an experimenter are badly skewed. For this reason, the Mann-Whitney U-test might be chosen rather than the t-test for independent groups.

The basic formula for the Mann-Whitney U-test is

$$U = n_1 n_2 + \frac{n_1(n_1 + 1)}{2} - R_1$$

or

$$U' = n_1 n_2 + \frac{n_2(n_2 + 1)}{2} - R_2$$

where n_1 = size of the smaller sample

n_2 = size of the larger sample

R_1 = sum of the ranks of the smaller sample

R_2 = sum of the ranks of the larger sample

When either U or U' has been computed, the other can be quickly found by

$$U = n_1 n_2 - U'$$

or

$$U' = n_1 n_2 - U$$

Therefore, it is necessary to compute only one; the other is then obtained by subtraction.

Example

Suppose that the following data have been collected representing weekly incomes for office managers who are members of two different ethnic groups. The experimenter wishes to determine whether there is a significant difference between these incomes.

| | Group 1 | | Group 2 |
| | Weekly Income | | Weekly Income |
Subject	(dollars)	Subject	(dollars)
S_1	387	S_{10}	431
S_2	372	S_{11}	394
S_3	365	S_{12}	377
S_4	354	S_{13}	388
S_5	367	S_{14}	416
S_6	376	S_{15}	390
S_7	373	S_{16}	387
S_8	382	S_{17}	376
S_9	404	S_{18}	395
		S_{19}	464
		S_{20}	427
		S_{21}	377

Step 1. Rank *all* the numbers in *both* groups taken together according to the size of the number, beginning with the smallest. (*Note:* When there are equal measures, the same rank is assigned to each. But notice that the rank assigned is the mean value of the tied ranks. For example, there are two incomes of 376 dollars, which are tied for the sixth and seventh ranking positions. Therefore, they are both assigned the rank of 6.5.)

| Group 1 | | Group 2 | |
Income	Rank	Income	Rank
387	11.5	431	20
372	4	394	15
365	2	377	8.5
354	1	388	13
367	3	416	18
376	6.5	390	14
373	5	387	11.5
382	10	376	6.5
404	17	395	16
		464	21
		427	19
		377	8.5

Step 2. Count the number of measures in each group, and then multiply these numbers. (Here, there are 9 measures in Group 1 and 12 in Group 2.)

$$9 \times 12 = 108$$

Step 3. Add 1 to the number of measures in Group 1; then multiply that sum by the original number (9 in this example), and divide the product by 2.

$$\frac{(9 + 1)9}{2} = \frac{(10)9}{2} = \frac{90}{2} = 45$$

Step 4. Add the ranks for Group 1.

$$11.5 + 4 + \cdots + 17 = 60$$

Step 5. Add the result of Step 2 to the result of Step 3, and subtract the sum of Step 4. This yields the value of U.

$$U = 108 + 45 - 60 = 93$$

Step 6. Subtract the result of Step 5 from the result of Step 2. This yields the value of U'.

$$U' = 108 - 93 = 15$$

If the number of measures in the larger group is greater than 8, and the number of measures in either group is less than 21, take whichever answer, from Steps 5 and 6, is smaller, and look in tables of U values to check for significance (see Appendix H). In the present example, n_1 = 9 and n_2 = 12 so both requirements are satisfied. Thus, go to Appendix H, look in the table for the appropriate alpha level, and see whether the value is smaller than would be expected by chance. For this example, using a two-tailed test, if alpha is .05, we see that if the value is 26 or less, the difference is significant.

If the number of measures in *either* group is greater than 20, carry out the following steps:

Step 7. Divide the result of Step 2 by 2.

$$\frac{108}{2} = 54$$

Then, subtract that quotient from the result of Step 5.

$$93 - 54 = 39$$

Step 8. Add together the number of measures in each group; then add the number 1.

$$9 + 12 + 1 = 22$$

Step 9. Multiply the result of Step 2 by the sum of Step 8.

$$108 \times 22 = 2376$$

Step 10. Divide the result of Step 9 by the number 12. (*Note:* The number 12 is always used.)

$$\frac{2376}{12} = 198$$

Then, take the square root of that quotient.

$$\sqrt{198} = 14.071$$

Step 11. Divide the result of Step 7 by the result of Step 10.

$$\frac{39}{14.071} = 2.77$$

This is a z, or critical-ratio, test. A value greater than ± 1.96 is significant at the .05 level using a two-tailed test (see Appendix A). The hypothesis tested by the Mann-Whitney analysis is that the *medians* of the two groups are equal. A z value that is large enough so that the hypothesis is rejected tells us that the chance of the medians being the same is very small. It is concluded, then, that there is a significant difference between the median incomes of office managers in the two ethnic groups.

SECTION 6.4
A Sign Test (Wilcoxon) for Differences Between Related Samples

The Mann-Whitney U-test presented in Section 6.3 is a nonparametric test for *independent* samples. The Wilcoxon signed-ranks test is applicable where the samples are related (i.e., not independent).

Example

Suppose we have collected the reading-rate scores for a group of ten people before and after they enrolled in a course in speed reading. We wish to determine if the course made any significant difference in reading rate.

Step 1. Table the data as follows. Since these are related measures, they must be tabled in appropriate pairs.

Subject	Rate Before	Rate After
A	150	145
B	135	138
C	102	121
D	96	115
E	127	134
F	118	132
G	132	138
H	124	145
I	115	126
J	103	94

Step 2. Subtract each number in the first column from its paired number in the second column, and record the differences.

Subject	After	Before	Difference
A	145	150	−5
B	138	135	3
C	121	102	19
D	115	96	19
E	134	127	7
F	132	118	14
G	138	132	6
H	145	124	21
I	126	115	11
J	94	103	−9

Step 3. Rank the difference values (Step 2) according to the size of the number, beginning with the smallest. Ignore the sign (i.e., use absolute values).

Difference	Rank
−5	2
3	1
19	8.5
19	8.5
7	4
14	7
6	3
21	10
11	6
−9	5

Step 4. Add the ranks for all the difference values that are negative (see Step 3). (In this example, the two negative differences, −5 and −9, are ranked 2 and 5, respectively.)

$$2 + 5 = 7$$

Note: Zero differences are not used in the computation. The N which is used in consulting the table (Appendix M) must be reduced by the number of zero difference cases.

Step 5. Add the ranks for all the difference values that are positive (see Step 3).

$$1 + 8.5 + 8.5 + 4 + 7 + 3 + 10 + 6 = 48$$

Step 6. The results of Steps 4 and 5 are the signed-rank values. Take the result that is smaller (7 in this example) and look in a table of Wilcoxon's signed-rank probabilities (see Appendix M) to check for significance. When the number of pairs is 10, as it is in this example, a value of 8 or less is significant at the .05 level using a two-tailed test. Since the result of Step 4 (i.e., 7) is smaller than 8, we have a significant difference. We conclude that the reading rate *after* training was significantly faster than *before* training.

SECTION 6.5
Simple Chi-Square and the *Phi* Coefficient

When you have frequency data comparing the effects of two variables and there are two groups on both variables, the *phi* coefficient can be quickly computed to show the degree of the relationship between the two variables. Then, the simple chi-square can be easily computed from the *phi* coefficient. The chi-square test will establish whether the variables are related. A significant chi-square is interpreted as showing a relationship between the two variables. The *phi* coefficient gives a numerical value, ranging from 0 to $+1$, for that relationship.

Example

Consider data collected on the two variables (1) gum-chewing activity and (2) educational background. The experimenter simply sampled two groups with differing educational backgrounds. Within each group, the subjects were classified according to their gum-chewing behavior.

The basic computational formula for chi-square is

$$\chi^2 = \frac{N(AD - BC)^2}{(A + B)(C + D)(A + C)(B + D)}$$

where the numbers represented by the letters A, B, C, and D come from the contingency table.

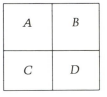

A	B
C	D

and the formula for *phi* (ϕ) is

$$\phi = \sqrt{\frac{\chi^2}{N}}$$

where N = the total frequency in the entire contingency table.

Step 1. Organize the data into a contingency table as follows. The values in the table represent the number of persons in each category.

		Gum Behavior	
		Chews	Eschews
Educational History	High-School Dropout	90	50
	College Graduate	20	30

Step 2. Add all the numbers in the table.

$$90 + 50 + 20 + 30 = \mathbf{190}$$

Step 3. Add the numbers in each row to get the row sums.

$$90 + 50 = \mathbf{140} = \text{sum for Row 1}$$
$$20 + 30 = \mathbf{50} = \text{sum for Row 2}$$

Then, add the numbers in each column to get the column sums.

	Column 1	Column 2
	90	50
	+20	+30
Sums:	**110**	**80**

Step 4. Multiply together the four sums from Step 3.

$$140 \times 50 \times 110 \times 80 = \mathbf{61{,}600{,}000}$$

Step 5. Multiply the number from the upper-left square by the number from the lower-right square (Step 1).

$$90 \times 30 = \textbf{2700}$$

Step 6. Multiply the number from the upper-right square by the number from the lower-left square (Step 1).

$$50 \times 20 = \textbf{1000}$$

Step 7. Get the results of Steps 5 and 6, and subtract the smaller from the larger.

$$2700 - 1000 = \textbf{1700}$$

Step 8. Square the results of Step 7, and then multiply by the total number of scores in the contingency table. In this example $N = 190$.

$$1700^2 = 2{,}890{,}000$$

and

$$2{,}890{,}000 \times 190 = \textbf{549{,}100{,}000}$$

Step 9. Divide the result of Step 8 by the result of Step 4. The answer is the chi-square.

$$\chi^2 = \frac{549{,}100{,}000}{61{,}600{,}000} = \textbf{8.91}$$

The number of degrees of freedom for this simple 2×2 chi-square will always be 1. A chi-square value larger than 3.8, with one *df* is significant at the .05 level (see Appendix D). It is concluded, therefore, that educational background and gum-chewing behavior are significantly related.

Step 10. Computation of *phi.* Divide the chi-square value obtained in Step 9 by N, and then take the square root of the quotient.

$$\frac{8.91}{190} = \textbf{.0468}$$

and

$$\phi = \sqrt{.0468} = \textbf{.217}$$

The value of *phi* is an indicator of the degree of relationship between the two variables in this example; it is similar in meaning to a correlation coefficient.

Supplement

In a 2 × 2 contingency table, if the frequency of any cell is less than 10, it is often recommended that Yates' correction for continuity be applied to account for the fact that an χ^2 computed with very small ns will overestimate the true χ^2 value. Yates' correction is applied to the basic χ^2 formula as follows (note the absolute value signs).

$$\chi^2 = \frac{N[\ |AD - BC| - N/2]^2}{(A + B)(C + D)(A + C)(B + D)}$$

In the data presented (even though it is not needed since the N in every cell is greater than 10), the computations would be

$$\chi^2 = \frac{190[\ |2700 - 1000| - 190/2]^2}{140 \times 50 \times 110 \times 80}$$

$$= \frac{190(1700 - 95)^2}{140 \times 50 \times 110 \times 80}$$

$$= \frac{489{,}444{,}750}{61{,}600{,}000}$$

$$= 7.94$$

SECTION 6.6
Complex Chi-Square and the Contingency Coefficient (*C*)

When you have frequency data comparing the effects of two variables and there are more than two groups on either of the two variables, the complex chi-square can be used to test the hypothesis of no relationship between the variables. If the chi-square test shows that there is most likely a relationship between the variables, then the contingency coefficient can be computed to give an indication of the degree of the relationship.

Example

Suppose we gather income information for three groups of people: (1) those under 26 years of age, (2) those between 26 and 60 years of age, (3) those 61 years of age and older. The incomes are categorized according to three-thousand-dollar ranges, and a frequency count is made for each category.

The basic formula for chi-square to be used here is

$$\chi^2 = \sum \frac{(O - E)^2}{E}$$

where O = the *observed* frequency for a particular cell of the contingency table (see Step 1)

E = the *expected* frequency for a cell, based upon marginal totals (see Steps 2 through 6 for a complete definition)

The relationship between the contingency coefficient (C) and chi-square is

$$C = \sqrt{\frac{\chi^2}{\chi^2 + N}}$$

where χ^2 = the chi-square value

N = the total of all the values in the contingency table

Step 1. Organize the data into a contingency table as follows. As indicated above, the values in the table represent the number of persons falling into each category.

		Under 3,000	Income Groups (dollars) 3,000— 6,000	6,000— 9,000	Over 9,000
	Under 26	40	30	10	20
Age Group	26 through 60	50	60	20	70
	Over 60	10	20	20	10

Step 2. Add all the numbers in the table to obtain a grand total.

$40 + 30 + \cdots + 20 + 50 + \cdots + 70 + 10 + \cdots + 10 =$ **360**

Step 3. Add the numbers in each row to obtain the row sums.

$40 + 30 + 10 + 20 =$ **100** = sum for Row 1
$50 + 60 + 20 + 70 =$ **200** = sum for Row 2
$10 + 20 + 20 + 10 =$ **60** = sum for Row 3

Step 4. Add the numbers in each column to obtain the column sums.

	Column 1	Column 2	Column 3	Column 4
	40	30	10	20
	50	60	20	70
	+10	+20	+20	+10
Sums:	**100**	**110**	**50**	**100**

Step 5. Make a new table similar to the one in Step 1, but get the new numbers as follows.

Compute the new number for each square in the table by using the row sum from Step 3 and the column sum from Step 4 of the row and column that intersect at the square in which you are interested. For example, to compute the new number that goes into the upper-left square, get the sum for Row 1 from Step 3 and the sum for Column 1 from Step 4; multiply these numbers together, and divide by the result of Step 2.

$$\frac{100 \times 100}{360} = \frac{10,000}{360} = 27.778$$

Continue this operation for each square in the new table.

$$\text{Row 1} \times \text{Column 2:} \ \frac{100 \times 110}{360} = 30.556$$

$$\text{Row 1} \times \text{Column 3:} \ \frac{100 \times 50}{360} = 13.889$$

$$\text{Row 1} \times \text{Column 4:} \ \frac{100 \times 100}{360} = 27.778$$

$$\text{Row 2} \times \text{Column 1:} \ \frac{200 \times 100}{360} = 55.556$$

$$\text{Row 2} \times \text{Column 2:} \ \frac{200 \times 110}{360} = 61.111$$

$$\text{Row 2} \times \text{Column 3:} \ \frac{200 \times 50}{360} = 27.778$$

$$\text{Row 2} \times \text{Column 4:} \ \frac{200 \times 100}{360} = 55.556$$

$$\text{Row 3} \times \text{Column 1:} \ \frac{60 \times 100}{360} = 16.667$$

$$\text{Row 3} \times \text{Column 2:} \ \frac{60 \times 110}{360} = 18.833$$

$$\text{Row 3} \times \text{Column 3:} \frac{60 \times 50}{360} = \mathbf{8.333}$$

$$\text{Row 3} \times \text{Column 4:} \frac{60 \times 100}{360} = \mathbf{16.667}$$

Step 6. Put the numbers from Step 5 into the same tabular form as that of Step 1.

		Income Groups (dollars)			
		Under 3,000	3,000– 6,000	6,000– 9,000	Over 9,000
	Under 26	27.778	30.556	13.889	27.778
Age Group	26 through 60	55.556	61.111	27.778	55.556
	Over 60	16.667	18.333	8.333	16.667

Step 7. The chi-square values can now be computed using the two tables of Step 1 and Step 6. To get these values, take the numbers from the same squares of the two tables, subtract the smaller from the larger, and square that value; then divide by the number that came from the table of Step 6.

For example, the number in the upper-left square of the table of Step 1 is 40; the number from the same square in the table of Step 6 is 27.778. To get the first number for the chi-square, subtract 27.778 from 40, square that difference, and then divide the square by 27.778.

$$\frac{(40 - 27.778)^2}{27.778} = \frac{12.222^2}{27.778} = \frac{149.337}{27.778} = \mathbf{5.378}$$

Continue for each of the other values.
Row 1:

$$\frac{(30.556 - 30)^2}{30.556} = \frac{.556^2}{30.556} = \frac{.309}{30.556} = \mathbf{.01}$$

$$\frac{(13.889 - 10)^2}{13.889} = \frac{3.889^2}{13.889} = \frac{15.124}{13.889} = \mathbf{1.089}$$

$$\frac{(27.778 - 20)^2}{27.778} = \frac{7.778^2}{27.778} = \frac{60.497}{27.778} = \mathbf{2.178}$$

Row 2:

$$\frac{(55.556 - 50)^2}{55.556} = \frac{5.556^2}{55.556} = \frac{30.869}{55.556} = \mathbf{.556}$$

$$\frac{(61.111 - 60)^2}{61.111} = \frac{1.111^2}{61.111} = \frac{1.234}{61.111} = \mathbf{.02}$$

$$\frac{(27.778 - 20)^2}{27.778} = \frac{7.778^2}{27.778} = \frac{60.497}{27.778} = \textbf{2.178}$$

$$\frac{(70 - 55.556)^2}{55.556} = \frac{14.444^2}{55.556} = \frac{208.629}{55.556} = \textbf{3.755}$$

Row 3:

$$\frac{(16.667 - 10)^2}{16.667} = \frac{6.667^2}{16.667} = \frac{44.449}{16.667} = \textbf{2.667}$$

$$\frac{(20 - 18.333)^2}{18.333} = \frac{1.667^2}{18.333} = \frac{2.779}{18.333} = \textbf{.152}$$

$$\frac{(20 - 8.333)^2}{8.333} = \frac{11.667^2}{8.333} = \frac{136.119}{8.333} = \textbf{16.335}$$

$$\frac{(16.667 - 10)^2}{16.667} = \frac{6.667^2}{16.667} = \frac{44.449}{16.667} = \textbf{2.667}$$

Step 8. Chi-square is computed by adding all the values that were computed in Step 7.

$$\chi^2 = 5.378 + .01 + \cdots + 2.667 = \textbf{36.985}$$

Step 9. The degrees of freedom for this chi-square will always be the number of rows minus 1, multiplied by the number of columns minus 1. (In the present example, there are 3 rows and 4 columns.)

$$df = (3 - 1)(4 - 1) = 2 \times 3 = 6$$

A chi-square value larger than 12.6, with 6 degrees of freedom, is significant at the .05 level (see Appendix D).

Step 10. The extent of the relationship between the two variables can be found by first adding the result of Step 8 to the result of Step 2.

$$36.985 + 360 = \textbf{396.985}$$

Step 11. Divide the result of Step 8 by the answer of Step 10.

$$\frac{36.985}{396.985} = \textbf{.093}$$

Then, take the square root of that quotient. This yields the contingency coefficient.

$$C = \sqrt{.093} = \textbf{.305}$$

This contingency coefficient, which is similar in meaning to the correlation coefficient, has already been shown to be significantly different from zero by the test concluded in Step 9. Thus, it is concluded that age and income are significantly related.

SECTION 6.7
Tests for Trends, Runs, and Randomness

Scores often are recorded which represent an event occurring over time. The questions which arise usually relate to the randomness or the extent to which the scores are patterned in some fashion. For example, the temperature recorded at noon daily during the months of February, March, and April would show day-to-day fluctuation, but clearly the trend would be upward as summer approached. Similar nonrandom patterns might be observed in situations as different as the bar-pressing behavior of a rat during successive five-minute periods and the number of letters in the alphabet a child is able to recite on successive days.

The tests described below represent only a small fraction of the tests which can be employed. These were chosen because of their simplicity or their familiarity. The reader who is interested in learning more about the many types and uses of tests of randomness, runs, etc., is referred to the texts listed below.* In addition to those listed below, many statistics books in the areas of economics and finance deal with runs and various forms of trend analyses at great length as a tool required for making financial projections.

Example

Assume that a researcher has recorded the number of times an autistic child speaks a complete sentence during successive half-hour therapy sessions. The number of complete sentences is recorded each day for a twenty-day period. The data are recorded as follows:

Day	1	2	3	4	5	6	7	8	9	10	11	12	13	14	15	16	17	18	19	20
Complete Sentences	6	5	4	5	6	8	9	10	7	6	8	10	11	13	15	12	9	10	13	15

---

*Many textbooks concerned with nonparametric or distribution-free statistics include a section on tests for trends, runs, and randomness. There also are many texts available which deal with analysis of time series. Two texts for consideration by the reader are James V. Bradley, *Distribution-Free Statistical Tests:* Prentice-Hall, 1968, and M. G. Kendall, *Time-Series:* Griffin, London, 1973.

Several questions (hypotheses) will be formulated and answers obtained by use of one or more of the following tests. Each of the computations demonstrated below represents one way of answering a particular question or testing a particular hypothesis.

When dealing with repeated measures of a single event, it is advisable first to graph the data. This is especially true if the number of scores recorded is large.

The graph in Figure 6.1 will prove useful for several of the computations detailed below.

Figure 1

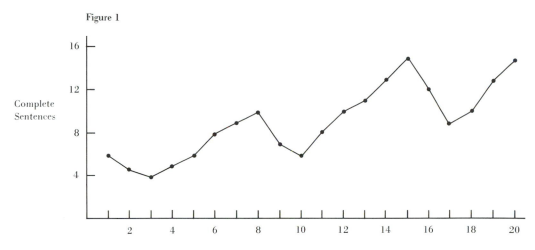

Question 1. If the number of sentences spoken each day shows a consistent tendency to increase or decrease, runs (successive increasing or decreasing scores) would tend to be longer and fewer in number. The specific question to be answered by the following analysis is "Are there fewer runs than would be expected by chance?"

Step 1. The actual number of runs (both up and down) in the data must be counted. The direction of change will be represented by a + when there is an increase from one day to the next and by a − when there is a decrease.

Days	1	2	3	4	5	6	7	8	9	10	11	12	13	14	15	16	17	18	19	20
Complete Sentences	6	5	4	5	6	8	9	10	7	6	8	10	11	13	15	12	9	10	13	15
Change		−	−	+	+	+	+	+	−	−	+	+	+	+	+	−	−	+	+	+

During the twenty-day period used in the example, there were a total of six runs. These occurred between days 1 and 3, 3 and 8, 8 and 10, 10 and 15, 15 and 17, and 17 and 20.

Step 2. Turn to Appendix O, which gives the probability of the total number of runs (both up and down) occurring for up to twenty-five observations. In the present example, six or fewer runs would occur in twenty observations by chance alone less than once in ten-thousand times. Therefore, it is concluded that there are fewer runs than would be expected by chance.

When the number of observations exceeds 25, the following formula approximates a z score and can be referred to the normal tables.

$$z = \frac{r - \left(\dfrac{2n - 1}{3}\right)}{\sqrt{\dfrac{16n - 29}{90}}}$$

where r = the total number of runs

n = the number of observations

Question 2. While the above analysis indicates that the total number of runs is significantly fewer than would be expected by chance, a second question relates to the length of the longest run. Specifically, the question is "Is the length of the longest run greater than would be expected by chance alone?"

Step 1. Using the data as reformulated above, it can be noted that the longest run is five (there actually are two runs of length five). To determine the probability of a run of length five, the following formula is employed:

$$P = \frac{2 + 2(n - S)(S + 1)}{(S + 2)!}$$

where n = the number of observations

s = the length of the longest run

$$= \frac{2 + 2(20 - 5)(5 + 1)}{7 \times 6 \times 5 \times 4 \times 3 \times 2 \times 1}$$

$$= \frac{182}{5,040}$$

$$= .036$$

Thus, the probability of a run of five or more signs of the same type is approximately .036. It is thus concluded that a run of five would not be expected by chance. (If the expected direction of the run, + or −, had been hypothesized prior to the conduct of the experiment, the probability of a positive run of five would have been .036/2.)

Question 3. Further inspection of the data indicates that performance seems to increase more frequently than it decreases. A simple analysis of this observation can be undertaken by use of the test for significance of a proportion as presented in Section 6.1.

Referring again to the reformulated data of Question 1, it is noted that of the 19 possible changes up or down, 13 are positive. In terms of proportion, 13/19 or .684 are positive. If there were no pattern or trend shown in the data, the expected proportion of increases would be .50.

Step 1. Turn to Section 6.1 and begin the analysis with Step 2. In the above example, the observed proportion will be equal to .684, and the expected equal to .50

Results of the calculations for the above example show the z for proportion to be equal to 1.60. Thus, it is concluded that the number of positive changes in relation to the total number of changes is not greater than could be expected by chance.

Question 4. Inspection of the data reveals a tendency for the number of complete sentences to increase as days progress. If this is the case, the scores of days 11 through 20 would be expected to be higher than those of days 1 through 10. Adaptation of the Mann-Whitney U-test described in Section 6.3 permits examination of this hypothesis.

Step 1. Arrange the scores recorded over the twenty-day period into two "groups," one representing the first half of the series, the other representing the second half.

Days	Sentences Spoken	Days	Sentences Spoken
1	6	11	8
2	5	12	10
3	4	13	11
4	5	14	13
5	6	15	15
6	8	16	12
7	9	17	9
8	10	18	10
9	7	19	13
10	6	20	15

Step 2. Turn to Section 6.3 and begin the calculation with Step 1.

Analysis of the above data, using the Mann-Whitney U-test, indicates that the distributions of the scores in the two groups are different beyond the .001 probability level. Thus, it is concluded that significantly more sentences were spoken during the last ten days than during the first ten days.

Question 5. Use of the Mann-Whitney U-test indicates whether the distribution of scores on the last half of the series is different from the first half. Another question might be whether the pattern of increase or decrease is similar during the first and second half of the data series. Use of Spearman's *rho* will permit analysis of this question.

Step 1. Arrange the scores in the first and second half of data sequence in the same fashion as for the Mann-Whitney U-test.

Days	Sentences Spoken	Days	Sentences Spoken
1	6	11	8
2	5	12	10
3	4	13	11
4	5	14	13
5	6	15	15
6	8	16	12
7	9	17	9
8	10	18	10
9	7	19	13
10	6	20	15

Then turn to Section 4.2 and begin the analysis with Step 1. For the data used in this example, the calculated *rho* is equal to $-.05$. Thus, it is concluded that while the number of sentences spoken during the second half of the observation period is larger than the number spoken during the first half (results of the Mann-Whitney U-test), the pattern of increase or decrease during the first and second half is not similar.

The above tests represent a few of the tools available to assist in the decision-making process when data on a single event or subject have been collected. As mentioned initially, the particular test chosen depends in large measure on the hypotheses or questions under study.

APPENDIX A
Normal-Curve Areas

To read this table, first find the z value in which you are interested by looking down the first column for the units and tenths values and across the first row for the hundredths value. Whether the sign of z is positive or negative is not important for reading the table since it merely indicates whether the z lies to the left or right of the mean. After locating the z value of interest, find the number in the body of the table representing the proportional part of the area from the mean ($z = 0.00$) to your particular z value. For example, if the z of interest were -1.24, the proportion would be .3925. To compute the proportional part of the area in one half of the distribution that lies beyond the z value of interest, subtract the number found above from .5000. Thus, the area below a z of -1.24 would be $.5000 - .3925 = .1075$.

Appendix A. Normal Curve

z	.00	.01	.02	.03	.04	.05	.06	.07	.08	.09
0.0	.0000	.0040	.0080	.0120	.0160	.0199	.0239	.0279	.0319	.0359
0.1	.0398	.0438	.0478	.0517	.0557	.0596	.0636	.0675	.0714	.0753
0.2	.0793	.0832	.0871	.0910	.0948	.0987	.1026	.1064	.1103	.1141
0.3	.1179	.1217	.1255	.1293	.1331	.1368	.1406	.1443	.1480	.1517
0.4	.1554	.1591	.1628	.1664	.1700	.1736	.1772	.1808	.1844	.1879
0.5	.1915	.1950	.1985	.2019	.2054	.2088	.2123	.2157	.2190	.2224
0.6	.2257	.2291	.2324	.2357	.2389	.2422	.2454	.2486	.2517	.2549
0.7	.2580	.2611	.2642	.2673	.2704	.2734	.2764	.2794	.2823	.2852
0.8	.2881	.2910	.2939	.2967	.2995	.3023	.3051	.3078	.3106	.3133
0.9	.3159	.3186	.3212	.3238	.3264	.3289	.3315	.3340	.3365	.3389
1.0	.3413	.3438	.3461	.3485	.3508	.3531	.3554	.3577	.3599	.3621
1.1	.3643	.3665	.3686	.3708	.3729	.3749	.3770	.3790	.3810	.3830
1.2	.3849	.3869	.3888	.3907	.3925	.3944	.3962	.3980	.3997	.4015
1.3	.4032	.4049	.4066	.4082	.4099	.4115	.4131	.4147	.4162	.4177
1.4	.4192	.4207	.4222	.4236	.4251	.4265	.4279	.4292	.4306	.4319
1.5	.4332	.4345	.4357	.4370	.4382	.4394	.4406	.4418	.4429	.4441
1.6	.4452	.4463	.4474	.4484	.4495	.4505	.4515	.4525	.4535	.4545
1.7	.4554	.4564	.4573	.4582	.4591	.4599	.4608	.4616	.4625	.4633
1.8	.4641	.4649	.4656	.4664	.4671	.4678	.4686	.4693	.4699	.4706
1.9	.4713	.4719	.4726	.4732	.4738	.4744	.4750	.4756	.4761	.4767
2.0	.4772	.4778	.4783	.4788	.4793	.4798	.4803	.4808	.4812	.4817
2.1	.4821	.4826	.4830	.4834	.4838	.4842	.4846	.4850	.4854	.4857
2.2	.4861	.4864	.4868	.4871	.4875	.4878	.4881	.4884	.4887	.4890
2.3	.4893	.4896	.4898	.4901	.4904	.4906	.4909	.4911	.4913	.4916
2.4	.4918	.4920	.4922	.4925	.4927	.4929	.4931	.4932	.4934	.4936
2.5	.4938	.4940	.4941	.4943	.4945	.4946	.4948	.4949	.4951	.4952
2.6	.4953	.4955	.4956	.4957	.4959	.4960	.4961	.4962	.4963	.4964
2.7	.4965	.4966	.4967	.4968	.4969	.4970	.4971	.4972	.4973	.4974
2.8	.4974	.4975	.4976	.4977	.4977	.4978	.4979	.4979	.4980	.4981
2.9	.4981	.4982	.4982	.4983	.4984	.4984	.4985	.4985	.4986	.4986
3.0	.4987	.4987	.4987	.4988	.4988	.4989	.4989	.4989	.4990	.4990
3.1	.49903									
3.2	.49931									
3.3	.49952									
3.4	.49966									
3.5	.49977									
3.6	.49984									
3.7	.49989									
3.8	.49993									
3.9	.49995									
4.0	.50000									

APPENDIX B
Critical Values of "Student's" t Statistic

To read this table, first find the degrees of freedom (df) in the column at the left. The minimum t values for the several alpha levels of significance are given in the columns to the right of the df column. Note that the alpha levels are given for both directional (one-tailed) and nondirectional (two-tailed) tests of significance. Nearly all tests of significance are, however, nondirectional.

To demonstrate the use of this table, assume that a computed t value of 3.25 was obtained with df equal to 15. Entering the table at the row headed by 15 df, read to the right. Note that the t of 3.25 is larger than the .01 value (2.947) for a two-tailed test but less than the .001 value (4.073). Therefore, it is concluded that this t value is significant beyond the .01 level.

Appendix B. *t* Statistic

df	Alpha level of significance for directional (one-tailed) tests					
	.25	.05	.025	.01	.005	.0005
	Alpha level of significance for nondirectional (two-tailed) tests					
	.50	.10	.05	.02	.01	.001
1	1.000	6.314	12.706	31.821	63.657	636.619
2	.816	2.920	4.303	6.965	9.925	31.598
3	.765	2.353	3.182	4.541	5.841	12.941
4	.741	2.132	2.776	3.747	4.604	8.610
5	.727	2.015	2.571	3.365	4.032	6.859
6	.718	1.943	2.447	3.143	3.707	5.959
7	.711	1.895	2.365	2.998	3.499	5.405
8	.706	1.860	2.306	2.896	3.355	5.041
9	.703	1.833	2.262	2.821	3.250	4.781
10	.700	1.812	2.228	2.764	3.169	4.587
11	.697	1.796	2.201	2.718	3.106	4.437
12	.695	1.782	2.179	2.681	3.055	4.318
13	.694	1.771	2.160	2.650	3.012	4.221
14	.692	1.761	2.145	2.624	2.977	4.140
15	.691	1.753	2.131	2.602	2.947	4.073
16	.690	1.746	2.120	2.583	2.921	4.015
17	.689	1.740	2.110	2.567	2.898	3.965
18	.688	1.734	2.101	2.552	2.878	3.922
19	.688	1.729	2.093	2.539	2.861	3.883
20	.687	1.725	2.086	2.528	2.845	3.850
21	.686	1.721	2.080	2.518	2.831	3.819
22	.686	1.717	2.074	2.508	2.819	3.792
23	.685	1.714	2.069	2.500	2.807	3.767
24	.685	1.711	2.064	2.492	2.797	3.745
25	.684	1.708	2.060	2.485	2.787	3.725
26	.684	1.706	2.056	2.479	2.779	3.707
27	.684	1.703	2.052	2.473	2.771	3.690
28	.683	1.701	2.048	2.467	2.763	3.674
29	.683	1.699	2.045	2.462	2.756	3.659
30	.683	1.697	2.042	2.457	2.750	3.646
40	.681	1.684	2.021	2.423	2.704	3.551
60	.679	1.671	2.000	2.390	2.660	3.460
120	.677	1.658	1.980	2.358	2.617	3.373
∞	.674	1.645	1.960	2.326	2.576	3.291

SOURCE: Appendix B is taken from Table III of Fisher & Yates[1]: *Statistical Tables for Biological, Agricultural and Medical Research* published by Longman Group UK Ltd., London (previously published by Oliver and Boyd Ltd., Edinburgh). Reprinted by permission of the authors and publishers.

APPENDIX C
Critical Values for Sandler's *A* Statistic

To read this table, first find the degrees of freedom (pairs of scores minus one) in the column at the left. The *maximum A* values for the several alpha levels of significance are given in the columns to the right of the *df* column. Note that values for both one-tailed and two-tailed tests are given.

To demonstrate the use of this table, assume that a computed *A* value of .261 was obtained with *df* equal to 18. Entering the table at the row for 18 *df*, read to the right. The computed *A* of .261 is *smaller* than the .05 value (.267) but larger than the .02 value for a two-tailed test. Therefore, it is concluded that the difference is significant beyond the .05 level.

Appendix C. Critical Values of A

For any given value of $n - 1$, the table shows the values of A corresponding to various levels of probability. A is significant at a given level if it is equal to or *less than* the value shown in the table.

	Level of significance for one-tailed test				
	.05	.025	.01	.005	.0005
	Level of significance for two-tailed test				
$n - 1$*	.10	.05	.02	.01	.001
1	0.5125	0.5031	0.50049	0.50012	0.5000012
2	0.412	0.369	0.347	0.340	0.334
3	0.385	0.324	0.286	0.272	0.254
4	0.376	0.304	0.257	0.238	0.211
5	0.372	0.293	0.240	0.218	0.184
6	0.370	0.286	0.230	0.205	0.167
7	0.369	0.281	0.222	0.196	0.155
8	0.368	0.278	0.217	0.190	0.146
9	0.368	0.276	0.213	0.185	0.139
10	0.368	0.274	0.210	0.181	0.134
11	0.368	0.273	0.207	0.178	0.130
12	0.368	0.271	0.205	0.176	0.126
13	0.368	0.270	0.204	0.174	0.124
14	0.368	0.270	0.202	0.172	0.121
15	0.368	0.269	0.201	0.170	0.119
16	0.368	0.268	0.200	0.169	0.117
17	0.368	0.268	0.199	0.168	0.116
18	0.368	0.267	0.198	0.167	0.114
19	0.368	0.267	9.197	0.166	0.113
20	0.368	0.266	0.197	0.165	0.112
21	0.368	0.266	0.196	0.165	0.111
22	0.368	0.266	0.196	0.164	0.110
23	0.368	0.266	0.195	0.163	0.109
24	0.368	0.265	0.195	0.163	0.108
25	0.368	0.265	0.194	0.162	0.108
26	0.368	0.265	0.194	0.162	0.107
27	0.368	0.265	0.193	0.161	0.107
28	0.368	0.265	0.193	0.161	0.106
29	0.368	0.264	0.193	0.161	0.106
30	0.368	0.264	0.193	0.160	0.105
40	0.368	0.263	0.191	0.158	0.102
60	0.369	0.262	0.189	0.155	0.099
120	0.369	0.261	0.187	0.153	0.095
∞	0.370	0.260	0.185	0.151	0.092

*Number of pairs.

SOURCE: Critical values of A by J. Sandler from *British Journal of Psychology*, 1955, **46**, 225–226. Reprinted by permission of Cambridge University Press.

APPENDIX D
Centile Values of the Chi-Square Statistic

To read this table, first find the degrees of freedom (df) in the column at the left. The minimum chi-square values for the several alpha levels are given in the columns to the right of the df column. The subscript values for each chi-square column indicate the probability of a real difference existing. If you wish to find the probability that the obtained difference is due to chance, simply subtract the given subscript value from 1.00, e.g., $1.00 - .95 = .05$.

To demonstrate the use of this table, assume that a computed chi-square of 25.0 was obtained with 15 df. Entering the table at the row headed by 15 df, read to the right. The value of 25.0 is listed in the .95 column. Thus, it is concluded that the probability is .95 that the difference is real, or that the probability is .05 that the difference is due to chance alone.

Appendix D. Chi-Square Statistic

df	$x^2_{.005}$	$x^2_{.01}$	$x^2_{.025}$	$x^2_{.05}$	$x^2_{.10}$	$x^2_{.25}$	$x^2_{.50}$	$x^2_{.75}$	$x^2_{.90}$	$x^2_{.95}$	$x^2_{.975}$	$x^2_{.99}$	$x^2_{.995}$	$x^2_{.999}$
1	—	—	—	—	.02	.10	.45	1.3	2.7	3.8	5.0	6.6	7.9	10.8
2	.01	.02	.05	.10	.21	.58	1.4	2.8	4.6	6.0	7.4	9.2	10.6	13.8
3	.07	.11	.22	.35	.58	1.21	2.4	4.1	6.3	7.8	9.4	11.3	12.8	16.3
4	.21	.30	.48	.71	1.1	1.92	3.4	5.4	7.8	9.5	11.1	13.3	14.9	18.5
5	.41	.55	.83	1.1	1.6	2.7	4.4	6.6	9.2	11.1	12.8	15.1	16.7	20.5
6	.68	.87	1.2	1.6	2.2	3.5	5.4	7.8	10.6	12.6	14.4	16.8	18.5	22.5
7	.99	1.24	1.7	2.2	2.8	4.3	6.4	9.0	12.0	14.1	16.0	18.5	20.3	24.3
8	1.3	1.65	2.2	2.7	3.5	5.1	7.3	10.2	13.4	15.5	17.5	20.1	22.0	26.1
9	1.7	2.09	2.7	3.3	4.2	5.9	8.3	11.4	14.7	16.9	19.0	21.7	23.6	27.9
10	2.2	2.56	3.2	3.9	4.9	6.7	9.3	12.5	16.0	18.3	20.5	23.2	25.2	29.6
11	2.6	3.05	3.8	4.6	5.6	7.6	10.3	13.7	17.3	19.7	21.9	24.7	26.8	31.3
12	3.1	3.57	4.4	5.2	6.3	8.4	11.3	14.8	18.5	21.0	23.3	26.2	28.3	32.9
13	3.6	4.11	5.0	5.9	7.0	9.3	12.3	16.0	19.8	22.4	24.7	27.7	29.8	34.5
14	4.1	4.66	5.6	6.6	7.8	10.2	13.3	17.1	21.1	23.7	26.1	29.1	31.3	36.1
15	4.6	5.23	6.3	7.3	8.5	11.0	14.3	18.2	22.3	25.0	27.5	30.6	32.8	37.7
16	5.1	5.81	6.9	8.0	9.3	11.9	15.3	19.4	23.5	26.3	28.8	32.0	34.3	39.3
17	5.7	6.41	7.6	8.7	10.1	12.8	16.3	20.5	24.8	27.6	30.2	33.4	35.7	40.8
18	6.3	7.01	8.2	9.4	10.9	13.7	17.3	21.6	26.0	28.9	31.5	34.8	37.2	42.3
19	6.8	7.63	8.9	10.1	11.7	14.6	18.3	22.7	27.2	30.1	32.9	36.2	38.6	43.8
20	7.4	8.26	9.6	10.9	12.4	15.5	19.3	23.8	28.4	31.4	34.2	37.6	40.0	45.3
21	8.0	8.9	10.3	11.6	13.2	16.3	20.3	24.9	29.6	32.7	35.5	38.9	41.4	46.8
22	8.6	9.5	11.0	12.3	14.0	17.2	21.3	26.0	30.8	33.9	36.8	40.3	42.8	48.3
23	9.3	10.2	11.7	13.1	14.8	18.1	22.3	27.1	32.0	35.2	38.1	41.6	44.2	49.7
24	9.9	10.9	12.4	13.8	15.7	19.0	23.3	28.2	33.2	36.4	39.4	43.0	45.6	51.2
25	10.5	11.5	13.1	14.6	16.5	19.9	24.3	29.3	34.4	37.7	40.6	44.3	46.9	52.6
26	11.2	12.2	13.8	15.4	17.3	20.8	25.3	30.4	35.6	38.9	41.9	45.6	48.3	54.0
27	11.8	12.9	14.6	16.2	18.1	21.7	26.3	31.5	36.7	40.1	43.2	47.0	49.6	55.5
28	12.5	13.6	15.3	16.9	18.9	22.7	27.3	32.6	37.9	41.3	44.5	48.3	51.0	56.9
29	13.1	14.3	16.0	17.7	19.8	23.6	28.3	33.7	39.1	42.6	45.7	49.6	52.3	58.3
30	13.8	15.0	16.8	18.5	20.6	24.5	29.3	34.8	40.3	43.8	47.0	50.9	53.3	59.7

Note that median values of chi-square are in $x^2_{.50}$ column.

SOURCE: Reproduced from Table 8 of E. S. Pearson and H. O. Hartley, *Biometrika Tables for Statisticians*, Vol. 1, Third Edition, 1966. Used by permission of the Biometrika Trustees.

APPENDIX E
Per Cent Points in the *F* Distribution

To read this table, the degrees of freedom (*df*) for both mean-square values of the *F* ratio must be located in the table. The *df* for the numerator is located in the top row (df_1). The *df* for the denominator of the *F* ratio is located in the column to the far left of the table (df_2). The values in the block where the row and column intersect represent the *F* values at different alpha levels of significance. Note that in this table, the alpha levels are listed as percentages. To convert to probabilities, simply move the decimal point two places to the left. For example, 2.5% is equal to .025.

To demonstrate the use of this table, assume that a computed *F* ratio of 3.54 was obtained with *df* equal to 5 and 15. Entering the table by the column headed by df_1 equal to 5 and by the row headed by df_2 equal to 15, we find that the values range from 1.68 for the 20% level (.20) to 7.57 for the .1% level (.001). The *F* ratio of 3.54 falls just short of the value required for the 2.5% (.025) level. It is concluded, then, that the *F* of 3.54 with *df* equal to 5 and 15 is significant at the 5% (.05) level.

Appendix E. F Distribution

df_2 \ df_1		1	2	3	4	5	6	8	12	24	∞
1	0.1%	405284	500000	540379	562500	576405	585937	598144	610667	623497	636619
	0.5%	16211	20000	21615	22500	23056	23437	23925	24426	24940	25465
	1 %	4052	4999	5403	5625	5764	5859	5981	6106	6234	6366
	2.5%	647.79	799.50	864.16	899.58	921.85	937.11	956.66	976.71	997.25	1018.30
	5 %	161.45	199.50	215.71	224.58	230.16	233.99	238.88	243.91	249.05	254.32
	10 %	39.86	49.50	53.59	55.83	57.24	58.20	59.44	60.70	62.00	63.33
	20 %	9.47	12.00	13.06	13.73	14.01	14.26	14.59	14.90	15.24	15.58
2	0.1	998.5	999.0	999.2	999.2	999.3	999.3	999.4	999.4	999.5	999.5
	0.5	198.50	199.00	199.17	199.25	199.30	199.33	199.37	199.42	199.46	199.51
	1	98.49	99.00	99.17	99.25	99.30	99.33	99.36	99.42	99.46	99.50
	2.5	38.51	39.00	39.17	39.25	39.30	39.33	39.37	39.42	39.46	39.50
	5	18.51	19.00	19.16	19.25	19.30	19.33	19.37	19.41	19.45	19.50
	10	8.53	9.00	9.16	9.24	9.29	9.33	9.37	9.41	9.45	9.49
	20	3.56	4.00	4.16	4.24	4.28	4.32	4.36	4.40	4.44	4.48
3	0.1	167.5	148.5	141.1	137.1	134.6	132.8	130.6	128.3	125.9	123.5
	0.5	55.55	49.80	47.47	46.20	45.39	44.84	44.13	43.39	42.62	41.83
	1	34.12	30.81	29.46	28.71	28.24	27.91	27.49	27.05	26.60	26.12
	2.5	17.44	16.04	15.44	15.10	14.89	14.74	14.54	14.34	14.12	13.90
	5	10.13	9.55	9.28	9.12	9.01	8.94	8.84	8.74	8.64	8.53
	10	5.54	5.46	5.39	5.34	5.31	5.28	5.25	5.22	5.18	5.13
	20	2.68	2.89	2.94	2.96	2.97	2.97	2.98	2.98	2.98	2.98
4	0.1	74.14	61.25	56.18	53.44	51.71	50.53	49.00	47.41	45.77	44.05
	0.5	31.33	26.28	24.26	23.16	22.46	21.98	21.35	20.71	20.03	19.33
	1	21.20	18.00	16.69	15.98	15.52	15.21	14.80	14.37	13.93	13.46
	2.5	12.22	10.65	9.98	9.60	9.36	9.20	8.98	8.75	8.51	8.26
	5	7.71	6.94	6.59	6.39	6.26	6.16	6.04	5.91	5.77	5.63
	10	4.54	4.32	4.19	4.11	4.05	4.01	3.95	3.90	3.83	3.76
	20	2.35	2.47	2.48	2.48	2.48	2.47	2.47	2.46	2.44	2.43
5	0.1	47.04	36.61	33.20	31.09	29.75	28.84	27.64	26.42	25.14	23.78
	0.5	22.79	18.31	16.53	15.56	14.94	14.51	13.96	13.38	12.78	12.14
	1	16.26	13.27	12.06	11.39	10.97	10.67	10.29	9.89	9.47	9.02
	2.5	10.01	8.43	7.76	7.39	7.15	6.98	6.76	6.52	6.28	6.02
	5	6.61	5.79	5.41	5.19	5.05	4.95	4.82	4.68	4.53	4.36
	10	4.06	3.78	3.62	3.52	3.45	3.40	3.34	3.27	3.19	3.10
	20	2.18	2.26	2.25	2.24	2.23	2.22	2.20	2.18	2.16	2.13
6	0.1%	35.51	27.00	23.70	21.90	20.81	20.03	19.03	17.99	16.89	15.75
	0.5%	18.64	14.54	12.92	12.03	11.46	11.07	10.57	10.03	9.47	8.88
	1 %	13.74	10.92	9.78	9.15	8.75	8.47	8.10	7.72	7.31	6.88
	2.5%	8.81	7.26	6.60	6.23	5.99	5.82	5.60	5.37	5.12	4.85
	5 %	5.99	5.14	4.76	4.53	4.39	4.28	4.15	4.00	3.84	3.67
	10 %	3.78	3.46	3.29	3.18	3.11	3.05	2.98	2.90	2.82	2.72
	20 %	2.07	2.13	2.11	2.09	2.08	2.06	2.04	2.02	1.99	1.95

SOURCE: Appendix E is taken from Table V of Fisher & Yates[1]: *Statistical Tables for Biological, Agricultural and Medical Research*, published by Longman Group UK Ltd., London (previously published by Oliver and Boyd Ltd., Edinburgh). Reprinted by permission of the authors and publishers.

Appendix E. *F* Distribution *(cont.)*

df_2 \\ df_1		1	2	3	4	5	6	8	12	24	∞
7	0.1	29.22	21.69	18.77	17.19	16.21	15.52	14.63	13.71	12.73	11.69
	0.5	16.24	12.40	10.88	10.05	9.52	9.16	8.68	8.18	7.65	7.08
	1	12.25	9.55	8.45	7.85	7.46	7.19	6.84	6.47	6.07	5.65
	2.5	8.07	6.54	5.89	5.52	5.29	5.12	4.90	4.67	4.42	4.14
	5	5.59	4.74	4.35	4.12	3.97	3.87	3.73	3.57	3.41	3.23
	10	3.59	3.26	3.07	2.96	2.88	2.83	2.75	2.67	2.58	2.47
	20	2.00	2.04	2.02	1.99	1.97	1.96	1.93	1.91	1.87	1.83
8	0.01	25.42	18.49	15.83	14.39	13.49	12.86	12.04	11.19	10.30	9.34
	0.5	14.69	11.04	9.60	8.81	8.30	7.95	7.50	7.01	6.50	5.95
	1	11.26	8.65	7.59	7.01	6.63	6.37	6.03	5.67	5.28	4.86
	2.5	7.57	6.06	5.42	5.05	4.82	4.65	4.43	4.20	3.95	3.67
	5	5.32	4.46	4.07	3.84	3.69	3.58	3.44	3.28	3.12	2.93
	10	3.46	3.11	2.92	2.81	2.73	2.67	2.59	2.50	2.40	2.29
	20	1.95	1.98	1.95	1.92	1.90	1.88	1.86	1.83	1.79	1.74
9	0.1	22.86	16.39	13.90	12.56	11.71	11.13	10.37	9.57	8.72	7.81
	0.5	13.61	10.11	8.72	7.96	7.47	7.13	6.69	6.23	5.73	5.19
	1	10.56	8.02	6.99	6.42	6.06	5.80	5.47	5.11	4.73	4.31
	2.5	7.21	5.71	5.08	4.72	4.48	4.32	4.10	3.87	3.61	3.33
	5	5.12	4.26	3.86	3.63	3.48	3.37	3.23	3.07	2.90	2.71
	10	3.36	3.01	2.81	2.69	2.61	2.55	2.47	2.38	2.28	2.16
	20	1.91	1.94	1.90	1.87	1.85	1.83	1.80	1.76	1.72	1.67
10	0.1	21.04	14.91	12.55	11.28	10.48	9.92	9.20	8.45	7.64	6.76
	0.5	12.83	9.43	8.08	7.34	6.87	6.54	6.12	5.66	5.17	4.64
	1	10.04	7.56	6.55	5.99	5.64	5.39	5.06	4.71	4.33	3.91
	2.5	6.94	5.46	4.83	4.47	4.24	4.07	3.85	3.62	3.37	3.08
	5	4.96	4.10	3.71	3.48	3.33	3.22	3.07	2.91	2.74	2.54
	10	3.28	2.92	2.73	2.61	2.52	2.46	2.38	2.28	2.18	2.06
	20	1.88	1.90	1.86	1.83	1.80	1.78	1.75	1.72	1.67	1.62
11	0.1	19.69	13.81	11.56	10.35	9.58	9.05	8.35	7.63	6.85	6.00
	0.5	12.23	8.91	7.60	6.88	6.42	6.10	5.68	5.24	4.76	4.23
	1	9.65	7.20	6.22	5.67	5.32	5.07	4.74	4.40	4.02	3.60
	2.5	6.72	5.26	4.63	4.28	4.04	3.88	3.66	3.43	3.17	2.88
	5	4.84	3.98	3.59	3.36	3.20	3.09	2.95	2.79	2.61	2.40
	10	3.23	2.86	2.66	2.54	2.45	2.39	2.30	2.21	2.10	1.97
	20	1.86	1.87	1.83	1.80	1.77	1.75	1.72	1.68	1.63	1.57
12	0.1	18.64	12.97	10.80	9.63	8.89	8.38	7.71	7.00	6.25	5.42
	0.5	11.75	8.51	7.23	6.52	6.07	5.76	5.35	4.91	4.43	3.90
	1	9.33	6.93	5.95	5.41	5.06	4.82	4.50	4.16	3.78	3.36
	2.5	6.55	5.10	4.47	4.12	3.89	3.73	3.51	3.28	3.02	2.72
	5	4.75	3.88	3.49	3.26	3.11	3.00	2.85	2.69	2.50	2.30
	10	3.18	2.81	2.61	2.48	2.39	2.33	2.24	2.15	2.04	1.90
	20	1.84	1.85	1.80	1.77	1.74	1.72	1.69	1.65	1.60	1.54
13	0.1	17.81	12.31	10.21	9.07	8.35	7.86	7.21	6.52	5.78	4.97
	0.5	11.37	8.19	6.93	6.23	5.79	5.48	5.08	4.64	4.17	3.65
	1	9.07	6.70	5.74	5.20	4.86	4.62	4.30	3.96	3.59	3.16
	2.5	6.41	4.97	4.35	4.00	3.77	3.60	3.39	3.15	2.89	2.60
	5	4.67	3.80	3.41	3.18	3.02	2.92	2.77	2.60	2.42	2.21
	10	3.14	2.76	2.56	2.43	2.35	2.28	2.20	2.10	1.98	1.85
	20	1.82	1.83	1.78	1.75	1.72	1.69	1.66	1.62	1.57	1.51

df_1 df_2		1	2	3	4	5	6	8	12	24	∞
14	0.1	17.14	11.78	9.73	8.62	7.92	7.43	6.80	6.13	5.41	4.60
	0.5	11.06	7.92	6.68	6.00	5.56	5.26	4.86	4.43	3.96	3.44
	1	8.86	6.51	5.56	5.03	4.69	4.46	4.14	3.80	3.43	3.00
	2.5	6.30	4.86	4.24	3.89	3.66	3.50	3.29	3.05	2.79	2.49
	5	4.60	3.74	3.34	3.11	2.96	2.85	2.70	2.53	2.35	2.13
	10	3.10	2.73	2.52	2.39	2.31	2.24	2.15	2.05	1.94	1.80
	20	1.81	1.81	1.76	1.73	1.70	1.67	1.64	1.60	1.55	1.48
15	0.1	16.59	11.34	9.34	8.25	7.57	7.09	6.47	5.81	5.10	4.31
	0.5	10.80	7.70	6.48	5.80	5.37	5.07	4.67	4.25	3.79	3.26
	1	8.68	6.36	5.42	4.89	4.56	4.32	4.00	3.67	3.29	2.87
	2.5	6.20	4.77	4.15	3.80	3.58	3.41	3.20	2.96	2.70	2.40
	5	4.54	3.68	3.29	3.06	2.90	2.79	2.64	2.48	2.29	2.07
	10	3.07	2.70	2.49	2.36	2.27	2.21	2.12	2.02	1.90	1.76
	20	1.80	1.79	1.75	1.71	1.68	1.66	1.62	1.58	1.53	1.46
16	0.1%	16.12	10.97	9.00	7.94	7.27	6.81	6.19	5.55	4.85	4.06
	0.5%	10.58	7.51	6.30	5.64	5.21	4.91	4.52	4.10	3.64	3.11
	1 %	8.53	6.23	5.29	4.77	4.44	4.20	3.89	3.55	3.18	2.75
	2.5%	6.12	4.69	4.08	3.73	3.50	3.34	3.12	2.89	2.63	2.32
	5 %	4.49	3.63	3.24	3.01	2.85	2.74	2.59	2.42	2.24	2.01
	10 %	3.05	2.67	2.46	2.33	2.24	2.18	2.09	1.99	1.87	1.72
	20 %	1.79	1.78	1.74	1.70	1.67	1.64	1.61	1.56	1.51	1.43
17	0.1	15.72	10.66	8.73	7.68	7.02	6.56	5.96	5.32	4.63	3.85
	0.5	10.38	7.35	6.16	5.50	5.07	4.78	4.39	3.97	3.51	2.98
	1	8.40	6.11	5.18	4.67	4.34	4.10	3.79	3.45	3.08	2.65
	2.5	6.04	4.62	4.01	3.66	3.44	3.28	3.06	2.82	2.56	2.25
	5	4.45	3.59	3.20	2.96	2.81	2.70	2.55	2.38	2.19	1.96
	10	3.03	2.64	2.44	2.31	2.22	2.15	2.06	1.96	1.84	1.69
	20	1.78	1.77	1.72	1.68	1.65	1.63	1.59	1.55	1.49	1.42
18	0.1	15.38	10.39	8.49	7.46	6.81	6.35	5.76	5.13	4.45	3.67
	0.5	10.22	7.21	6.03	5.37	4.96	4.66	4.28	3.86	3.40	2.87
	1	8.28	6.01	5.09	4.58	4.25	4.01	3.71	3.37	3.00	2.57
	2.5	5.98	4.56	3.95	3.61	3.38	3.22	3.01	2.77	2.50	2.19
	5	4.41	3.55	3.16	2.93	2.77	2.66	2.51	2.34	2.15	1.92
	10	3.01	2.62	2.42	2.29	2.20	2.13	2.04	1.93	1.81	1.66
	20	1.77	1.76	1.71	1.67	1.64	1.62	1.58	1.53	1.48	1.40
19	0.1	15.08	10.16	8.28	7.26	6.61	6.18	5.59	4.97	4.29	3.52
	0.5	10.07	7.09	5.92	5.27	4.85	4.56	4.18	3.76	3.31	2.78
	1	8.18	5.93	5.01	4.50	4.17	3.94	3.63	3.30	2.92	2.49
	2.5	5.92	4.51	3.90	3.56	3.33	3.17	2.96	2.72	2.45	2.13
	5	4.38	3.52	3.13	2.90	2.74	2.63	2.48	2.31	2.11	1.88
	10	2.99	2.61	2.40	2.27	2.18	2.11	2.02	1.91	1.79	1.63
	20	1.76	1.75	1.70	1.66	1.63	1.61	1.57	1.52	1.46	1.39
20	0.1	14.82	9.95	8.10	7.10	6.46	6.02	5.44	4.82	4.15	3.38
	0.5	9.94	6.99	5.82	5.17	4.76	4.47	4.09	3.68	3.22	2.69
	1	8.10	5.85	4.94	4.43	4.10	3.87	3.56	3.23	2.86	2.42
	2.5	5.87	4.46	3.86	3.51	3.29	3.13	2.91	2.68	2.41	2.09
	5	4.35	3.49	3.10	2.87	2.71	2.60	2.45	2.28	2.08	1.84
	10	2.97	2.59	2.38	2.25	2.16	2.09	2.00	1.89	1.77	1.61
	20	1.76	1.75	1.70	1.65	1.62	1.60	1.56	1.51	1.45	1.37

304

Appendix E. *F* Distribution *(cont.)*

df_2	df_1	1	2	3	4	5	6	8	12	24	∞
21	0.1	14.59	9.77	7.94	6.95	6.32	5.88	5.31	4.70	4.03	3.26
	0.5	9.83	6.89	5.73	5.09	4.68	4.39	4.01	3.60	3.15	2.61
	1	8.02	5.78	4.87	4.37	4.04	3.81	3.51	3.17	2.80	2.36
	2.5	5.83	4.42	3.82	3.48	3.25	3.09	2.87	2.64	2.37	2.04
	5	4.32	3.47	3.07	2.84	2.68	2.57	2.42	2.25	2.05	1.81
	10	2.96	2.57	2.36	2.23	2.14	2.08	1.98	1.89	1.75	1.59
	20	1.75	1.74	1.69	1.65	1.61	1.59	1.55	1.50	1.44	1.36
22	0.1	14.38	9.61	7.80	6.81	6.19	5.76	5.19	4.58	3.92	3.15
	0.5	9.73	6.81	5.65	5.02	4.61	4.32	3.94	3.54	3.08	2.55
	1	7.94	5.72	4.82	4.31	3.99	3.76	3.45	3.12	2.75	2.31
	2.5	5.79	4.38	3.78	3.44	3.22	3.05	2.84	2.60	2.33	2.00
	5	4.30	3.44	3.05	2.82	2.66	2.55	2.40	2.23	2.03	1.78
	10	2.95	2.56	2.35	2.22	2.13	2.06	1.97	1.86	1.73	1.57
	20	1.75	1.73	1.68	1.64	1.61	1.58	1.54	1.49	1.43	1.35
23	0.1	14.19	9.47	7.67	6.69	6.08	5.65	5.09	4.48	3.82	3.05
	0.5	9.63	6.73	5.58	4.95	4.54	4.26	3.88	3.47	3.02	2.48
	1	7.88	5.66	4.76	4.26	3.94	3.71	3.41	3.07	2.70	2.26
	2.5	5.75	4.35	3.75	3.41	3.18	3.02	2.81	2.57	2.30	1.97
	5	4.28	3.42	3.03	2.80	2.64	2.53	2.38	2.20	2.00	1.76
	10	2.94	2.55	2.34	2.21	2.11	2.05	1.95	1.84	1.72	1.55
	20	1.74	1.73	1.68	1.63	1.60	1.57	1.53	1.49	1.42	1.34
24	0.1	14.03	9.34	7.55	6.59	5.98	5.55	4.99	4.39	3.74	2.97
	0.5	9.55	6.66	5.52	4.89	4.49	4.20	3.83	3.42	2.97	2.43
	1	7.82	5.61	4.72	4.22	3.90	3.67	3.36	3.03	2.66	2.21
	2.5	5.72	4.32	3.72	3.38	3.15	2.99	2.78	2.54	2.27	1.94
	5	4.26	3.40	3.01	2.78	2.62	2.51	2.36	2.18	1.98	1.73
	10	2.93	2.54	2.33	2.19	2.10	2.04	1.94	1.83	1.70	1.53
	20	1.74	1.72	1.67	1.63	1.59	1.57	1.53	1.48	1.42	1.33
25	0.1	13.88	9.22	7.45	6.49	5.88	5.46	4.91	4.31	3.66	2.89
	0.5	9.48	6.60	5.46	4.84	4.43	4.15	3.78	3.37	2.92	2.38
	1	7.77	5.57	4.68	4.18	3.86	3.63	3.32	2.99	2.62	2.17
	2.5	5.69	4.29	3.69	3.35	3.13	2.97	2.75	2.51	2.24	1.91
	5	4.24	3.38	2.99	2.76	2.60	2.49	2.34	2.16	1.96	1.71
	10	2.92	2.53	2.32	2.18	2.09	2.02	1.93	1.82	1.69	1.52
	20	1.73	1.72	1.66	1.62	1.59	1.56	1.52	1.47	1.41	1.32
26	0.1%	13.74	9.12	7.36	6.41	5.80	5.38	4.83	4.24	3.59	2.82
	0.5%	9.41	6.54	5.41	4.79	4.38	4.10	3.73	3.33	2.87	2.33
	1 %	7.72	5.53	4.64	4.14	3.82	3.59	3.29	2.96	2.58	2.13
	2.5%	5.66	4.27	3.67	3.33	3.10	2.94	2.73	2.49	2.22	1.88
	5 %	4.22	3.37	2.98	2.74	2.59	2.47	2.32	2.15	1.95	1.69
	10 %	2.91	2.52	2.31	2.17	2.08	2.01	1.92	1.81	1.68	1.50
	20 %	1.73	1.71	1.66	1.62	1.58	1.56	1.52	1.47	1.40	1.31
27	0.1	13.61	9.02	7.27	6.33	5.73	5.31	4.76	4.17	3.52	2.75
	0.5	9.34	6.49	5.36	4.74	4.34	4.06	3.69	3.28	2.83	2.29
	1	7.68	5.49	4.60	4.11	3.78	3.56	3.26	2.93	2.55	2.10
	2.5	5.63	4.24	3.65	3.31	3.08	2.92	2.71	2.47	2.19	1.85
	5	4.21	3.35	2.96	2.73	2.57	2.46	2.30	2.13	1.93	1.67
	10	2.90	2.51	2.30	2.17	2.07	2.00	1.91	1.80	1.67	1.49
	20	1.73	1.71	1.66	1.61	1.58	1.55	1.51	1.46	1.40	1.30

df_2	df_1	1	2	3	4	5	6	8	12	24	∞
28	0.1	13.50	8.93	7.19	6.25	5.66	5.24	4.69	4.11	3.46	2.70
	0.5	9.28	6.44	5.32	4.70	4.30	4.02	3.65	3.25	2.79	2.25
	1	7.64	5.45	4.57	4.07	3.75	3.53	3.23	2.90	2.52	2.06
	2.5	5.61	4.22	3.63	3.29	3.06	2.90	2.69	2.45	2.17	1.83
	5	4.20	3.34	2.95	2.71	2.56	2.44	2.29	2.12	1.91	1.65
	10	2.89	2.50	2.29	2.16	2.06	2.00	1.90	1.79	1.66	1.48
	20	1.72	1.71	1.65	1.61	1.57	1.55	1.51	1.46	1.39	1.30
29	0.1	13.39	8.85	7.12	6.19	5.59	5.18	4.64	4.05	3.41	2.64
	0.5	9.23	6.40	5.28	4.66	4.26	3.98	3.61	3.21	2.76	2.21
	1	7.60	5.42	4.54	4.04	3.73	3.50	3.20	2.87	2.49	2.03
	2.5	5.59	4.20	3.61	3.27	3.04	2.88	2.67	2.43	2.15	1.81
	5	4.18	3.33	2.93	2.70	2.54	2.43	2.28	2.10	1.90	1.64
	10	2.89	2.50	2.28	2.15	2.06	1.99	1.89	1.78	1.65	1.47
	20	1.72	1.70	1.65	1.60	1.57	1.54	1.50	1.45	1.39	1.29
30	0.1	13.29	8.77	7.05	6.12	5.53	5.12	4.58	4.00	3.36	2.59
	0.5	9.18	6.35	5.24	4.62	4.23	3.95	3.58	3.18	2.73	2.18
	1	7.56	5.39	4.51	4.02	3.70	3.47	3.17	2.84	2.47	2.01
	2.5	5.57	4.18	3.59	3.25	3.03	2.87	2.65	2.41	2.14	1.79
	5	4.17	3.32	2.92	2.69	2.53	2.42	2.27	2.09	1.89	1.62
	10	2.88	2.49	2.28	2.14	2.05	1.98	1.88	1.77	1.64	1.46
	20	1.72	1.70	1.64	1.60	1.57	1.54	1.50	1.45	1.38	1.28
40	0.1	12.61	8.25	6.60	5.70	5.13	4.73	4.21	3.64	3.01	2.23
	0.5	8.83	6.07	4.98	4.37	3.99	3.71	3.35	2.95	2.50	1.93
	1	7.31	5.18	4.31	3.83	3.51	3.29	2.99	2.66	2.29	1.80
	2.5	5.42	4.05	3.46	3.13	2.90	2.74	2.53	2.29	2.01	1.64
	5	4.08	3.23	2.84	2.61	2.45	2.34	2.18	2.00	1.79	1.51
	10	2.84	2.44	2.23	2.09	2.00	1.93	1.83	1.71	1.57	1.38
	20	1.70	1.68	1.62	1.57	1.54	1.51	1.47	1.41	1.34	1.24
60	0.1	11.97	7.76	6.17	5.31	4.76	4.37	3.87	3.31	2.69	1.90
	0.5	8.49	5.80	4.73	4.14	3.76	3.49	3.13	2.74	2.29	1.69
	1	7.08	4.98	4.13	3.65	3.34	3.12	2.82	2.50	2.12	1.60
	2.5	5.29	3.93	3.34	3.01	2.79	2.63	2.41	2.17	1.88	1.48
	5	4.00	3.15	2.76	2.52	2.37	2.25	2.10	1.92	1.70	1.39
	10	2.79	2.39	2.18	2.04	1.95	1.87	1.77	1.66	1.51	1.29
	20	1.68	1.65	1.59	1.55	1.51	1.48	1.44	1.38	1.31	1.18
120	0.1	11.38	7.31	5.79	4.95	4.42	4.04	3.55	3.02	2.40	1.56
	0.5	8.18	5.54	4.50	3.92	3.55	3.28	2.93	2.54	2.09	1.43
	1	6.85	4.79	3.95	3.48	3.17	2.96	2.66	2.34	1.95	1.38
	2.5	5.15	3.80	3.23	2.89	2.67	2.52	2.30	2.05	1.76	1.31
	5	3.92	3.07	2.68	2.45	2.29	2.17	2.02	1.83	1.61	1.25
	10	2.75	2.35	2.13	1.99	1.90	1.82	1.72	1.60	1.45	1.19
	20	1.66	1.63	1.57	1.52	1.48	1.45	1.41	1.35	1.27	1.12
∞	0.1	10.83	6.91	5.42	4.62	4.10	3.74	3.27	2.74	2.13	1.00
	0.5	7.88	5.30	4.28	3.72	3.35	3.09	2.74	2.36	1.90	1.00
	1	6.64	4.60	3.78	3.32	3.02	2.80	2.51	2.18	1.79	1.00
	2.5	5.02	3.69	3.12	2.79	2.57	2.41	2.19	1.94	1.64	1.00
	5	3.84	2.99	2.60	2.37	2.21	2.09	1.94	1.75	1.52	1.00
	10	2.71	2.30	2.08	1.94	1.85	1.77	1.67	1.55	1.38	1.00
	20	1.64	1.61	1.55	1.50	1.46	1.43	1.38	1.32	1.23	1.00

APPENDIX F

Fisher's z-Transformation Function for Pearson's r Correlation Coefficient: $z = \frac{1}{2}[\log_e (1 + r) - \log_e (1 - r)]$

To read this table, simply find the correlation coefficient value in the r column, and then read the corresponding Z value from the adjacent column. For example, if the r value were .46, the Z would be .497.

Appendix F. Fisher's z Transformation

r	Z	r	Z	r	Z	r	Z	r	Z
.000	.000	.200	.203	.400	.424	.600	.693	.800	1.099
.005	.005	.205	.208	.405	.430	.605	.701	.805	1.113
.010	.010	.210	.213	.410	.436	.610	.709	.810	1.127
.015	.015	.215	.218	.415	.442	.615	.717	.815	1.142
.020	.020	.220	.224	.420	.448	.620	.725	.820	1.157
.025	.025	.225	.229	.425	.454	.625	.733	.825	1.172
.030	.030	.230	.234	.430	.460	.630	.741	.830	1.188
.035	.035	.235	.239	.435	.466	.635	.750	.835	1.204
.040	.040	.240	.245	.440	.472	.640	.758	.840	1.221
.045	.045	.245	.250	.445	.478	.645	.767	.845	1.238
.050	.050	.250	.255	.450	.485	.650	.775	.850	1.256
.055	.055	.255	.261	.455	.491	.655	.784	.855	1.274
.060	.060	.260	.266	.460	.497	.660	.793	.860	1.293
.065	.065	.265	.271	.465	.504	.665	.802	.865	1.313
.070	.070	.270	.277	.470	.510	.670	.811	.870	1.333
.075	.075	.275	.282	.475	.517	.675	.820	.875	1.354
.080	.080	.280	.288	.480	.523	.680	.829	.880	1.376
.085	.085	.285	.293	.485	.530	.685	.838	.885	1.398
.090	.090	.290	.299	.490	.536	.690	.848	.890	1.422
.095	.095	.295	.304	.495	.543	.695	.858	.895	1.447
.100	.100	.300	.310	.500	.549	.700	.867	.900	1.472
.105	.105	.305	.315	.505	.556	.705	.877	.905	1.499
.110	.110	.310	.321	.510	.563	.710	.887	.910	1.528
.115	.116	.315	.326	.515	.570	.715	.897	.915	1.557
.120	.121	.320	.332	.520	.576	.720	.908	.920	1.589
.125	.126	.325	.337	.525	.583	.725	.918	.925	1.623
.130	.131	.330	.343	.530	.590	.730	.929	.930	1.658
.135	.136	.335	.348	.535	.597	.735	.940	.935	1.697
.140	.141	.340	.354	.540	.604	.740	.950	.940	1.738
.145	.146	.345	.360	.545	.611	.745	.962	.945	1.783
.150	.151	.350	.365	.550	.618	.750	.973	.950	1.832
.155	.156	.355	.371	.555	.626	.755	.984	.955	1.886
.160	.161	.360	.377	.560	.633	.760	.996	.960	1.946
.165	.167	.365	.383	.565	.640	.765	1.008	.965	2.014
.170	.172	.370	.388	.570	.648	.770	1.020	.970	2.092
.175	.177	.375	.394	.575	.655	.775	1.033	.975	2.185
.180	.182	.380	.400	.580	.662	.780	1.045	.980	2.298
.185	.187	.385	.406	.585	.670	.785	1.058	.985	2.443
.190	.192	.390	.412	.590	.678	.790	1.071	.990	2.647
.195	.198	.395	.418	.595	.685	.795	1.085	.995	2.994

SOURCE: From *Statistical Methods* by Allen L. Edwards, Copyright 1954, © 1967 by Allen L. Edwards. Reprinted by permission of Holt, Rinehart & Winston, Inc.

APPENDIX G

Critical Values of Pearson's r Correlation Coefficient for Five Alpha Significance Levels

To read this table, find the appropriate degrees of freedom for the correlation coefficient in the column headed by $n' - 2$. When this value has been located, read to the right to find the values that represent the minimum r values for significance at the .10, .05, .02, .01, and .001 alpha levels of significance. For example, if the degrees of freedom were 25, an r of at least .3233 would be needed for significance at the .10 level, .3809 at the .05 level, etc.

Appendix G. Pearson's *r*

n′ − 2	.10	.05	.02	.01	.001
1	.98769	.99692	.999507	.999877	.9999988
2	.90000	.95000	.98000	.990000	.99900
3	.8054	.8783	.93433	.95873	.99116
4	.7293	.8114	.8822	.91720	.97406
5	.6694	.7545	.8329	.8745	.95074
6	.6215	.7067	.7887	.8343	.92493
7	.5822	.6664	.7498	.7977	.8982
8	.5494	.6319	.7155	.7646	.8721
9	.5214	.6021	.6851	.7348	.8471
10	.4973	.5760	.6581	.7079	.8233
11	.4762	.5529	.6339	.6835	.8010
12	.4575	.5324	.6120	.6614	.7800
13	.4409	.5139	.5923	.6411	.7603
14	.4259	.4973	.5742	.6226	.7420
15	.4124	.4821	.5577	.6055	.7246
16	.4000	.4683	.5425	.5897	.7084
17	.3887	.4555	.5285	.5751	.6932
18	.3783	.4438	.5155	.5614	.6787
19	.3687	.4329	.5034	.5487	.6652
20	.3598	.4227	.4921	.5368	.6524
25	.3233	.3809	.4451	.4869	.5974
30	.2960	.3494	.4093	.4487	.5541
35	.2746	.3246	.3810	.4182	.5189
40	.2573	.3044	.3578	.3932	.4896
45	.2428	.2875	.3384	.3721	.4648
50	.2306	.2732	.3218	.3541	.4433
60	.2108	.2500	.2948	.3248	.4078
70	.1954	.2319	.2737	.3017	.3799
80	.1829	.2172	.2565	.2830	.3568
90	.1726	.2050	.2422	.2673	.3375
100	.1638	.1946	.2301	.2540	.3211

SOURCE: Appendix G is taken from Table VI of Fisher & Yates[1]: *Statistical Tables for Biological, Agricultural and Medical Research* published by Longman Group UK Ltd., London (previously published by Oliver and Boyd Ltd., Edinburgh). Reprinted by permission of the authors and publishers.

310

APPENDIX H

Critical Values of the *U* Statistic of the Mann-Whitney Test

To read this table, you must locate the number of cases in both groups in the particular subtable that lists the level of significance desired with either a one-tailed or a two-tailed test. Assume that you wish to find the critical value using a two-tailed test at the .05 level of significance. To do this, the first step is to locate n_1 and n_2 in subtable "c." The point at which the n_1 row and the n_2 column intersect represents the maximum critical value for significance at the .05 level. If, for example, n_1 equals 10, n_2 equals 15, and the computed value equals 23, it is concluded that the difference between the groups is significant—i.e., the tabled critical value for $n_1 = 10$ and $n_2 = 15$ is 39, and 23 is obviously smaller than 39.

Appendix H. *U* Statistic

n_1 \ n_2	(a) Critical Values of *U* for a One-Tailed Test at .001 or for a Two-Tailed Test at .002											
	9	10	11	12	13	14	15	16	17	18	19	20
1												
2												
3									0	0	0	0
4		0	0	0	1	1	1	2	2	3	3	3
5	1	1	2	2	3	3	4	5	5	6	7	7
6	2	3	4	4	5	6	7	8	9	10	11	12
7	3	5	6	7	8	9	10	11	13	14	15	16
8	5	6	8	9	11	12	14	15	17	18	20	21
9	7	8	10	12	14	15	17	19	21	23	25	26
10	8	10	12	14	17	19	21	23	25	27	29	32
11	10	12	15	17	20	22	24	27	29	32	34	37
12	12	14	17	20	23	25	28	31	34	37	40	42
13	14	17	20	23	26	29	32	35	38	42	45	48
14	15	19	22	25	29	32	36	39	43	46	50	54
15	17	21	24	28	32	36	40	43	47	51	55	59
16	19	23	27	31	35	39	43	48	52	56	60	65
17	21	25	29	34	38	43	47	52	57	61	66	70
18	23	27	32	37	42	46	51	56	61	66	71	76
19	25	29	34	40	45	50	55	60	66	71	77	82
20	26	32	37	42	48	54	59	65	70	76	82	88

n_1 \ n_2	(b) Critical Values of *U* for a One-Tailed Test at .01 or for a Two-Tailed Test at .02											
	9	10	11	12	13	14	15	16	17	18	19	20
1												
2					0	0	0	0	0	0	1	1
3	1	1	1	2	2	2	3	3	4	4	4	5
4	3	3	4	5	5	6	7	7	8	9	9	10
5	5	6	7	8	9	10	11	12	13	14	15	16
6	7	8	9	11	12	13	15	16	18	19	20	22
7	9	11	12	14	16	17	19	21	23	24	26	28
8	11	13	15	17	20	22	24	26	28	30	32	34
9	14	16	18	21	23	26	28	31	33	36	38	40
10	16	19	22	24	27	30	33	36	38	41	44	47
11	18	22	25	28	31	34	37	41	44	47	50	53
12	21	24	28	31	35	38	42	46	49	53	56	60
13	23	27	31	35	39	43	47	51	55	59	63	67
14	26	30	34	38	43	47	51	56	60	65	69	73
15	28	33	37	42	47	51	56	61	66	70	75	80
16	31	36	41	46	51	56	61	66	71	76	82	87
17	33	38	44	49	55	60	66	71	77	82	88	93
18	36	41	47	53	59	65	70	76	82	88	94	100
19	38	44	50	56	63	69	75	82	88	94	101	107
20	40	47	53	60	67	73	80	87	93	100	107	114

SOURCE: Adapted and abridged from Tables 1, 3, 5, and 7 of D. Aube, "Extended Tables for the Mann-Whitney Statistic," *Bulletin of the Institute of Educational Research of Indiana University*, 1953, 1, No. 2. Reproduced from S. Siegel, *Nonparametric Statistics for the Behavioral Sciences.* New York: McGraw-Hill Book Company, 1956. Reprinted by permission of the Institute of Educational Research and McGraw-Hill Book Company.

Appendix H. *U* Statistic *(cont.)*

n_1	(c) Critical Values of U for a One-Tailed Test at .025 or for a Two-Tailed Test at .05											
n_1	9	10	11	12	13	14	15	16	17	18	19	20
1												
2	0	0	1	1	1	1	1	1	2	2	2	2
3	2	3	3	4	4	5	5	6	6	7	7	8
4	4	5	6	7	8	9	10	11	11	12	13	13
5	7	8	9	11	12	13	14	15	17	18	19	20
6	10	11	13	14	16	17	19	21	22	24	25	27
7	12	14	16	18	20	22	24	26	28	30	32	34
8	15	17	19	22	24	26	29	31	34	36	38	41
9	17	20	23	26	28	31	34	37	39	42	45	48
10	20	23	26	29	33	36	39	42	45	48	52	55
11	23	26	30	33	37	40	44	47	51	55	58	62
12	26	29	33	37	41	45	49	53	57	61	65	69
13	28	33	37	41	45	50	54	59	63	67	72	76
14	31	36	40	45	50	55	59	64	67	74	78	83
15	34	39	44	49	54	59	64	70	75	80	85	90
16	37	42	47	53	59	64	70	75	81	86	92	98
17	39	45	51	57	63	67	75	81	87	93	99	105
18	42	48	55	61	67	74	80	86	93	99	106	112
19	45	52	58	65	72	78	85	92	99	106	113	119
20	48	55	62	69	76	83	90	98	105	112	119	127

n_1	(d) Critical Values of U for a One-Tailed Test at .05 or for a Two-Tailed Test at .10											
n_1	9	10	11	12	13	14	15	16	17	18	19	20
1											0	0
2	1	1	1	2	2	2	3	3	3	4	4	4
3	3	4	5	5	6	7	7	8	9	9	10	11
4	6	7	8	9	10	11	12	14	15	16	17	18
5	9	11	12	13	15	16	18	19	20	22	23	25
6	12	14	16	17	19	21	23	25	26	28	30	32
7	15	17	19	21	24	26	28	30	33	35	37	39
8	18	20	23	26	28	31	33	36	39	41	44	47
9	21	24	27	30	33	36	39	42	45	48	51	54
10	24	27	31	34	37	41	44	48	51	55	58	62
11	27	31	34	38	42	46	50	54	57	61	65	69
12	30	34	38	42	47	51	55	60	64	68	72	77
13	33	37	42	47	51	56	61	65	70	75	80	84
14	36	41	46	51	56	61	66	71	77	82	87	92
15	39	44	50	55	61	66	72	77	83	88	94	100
16	42	48	54	60	65	71	77	83	89	95	101	107
17	45	51	57	64	70	77	83	89	96	102	109	115
18	48	55	61	68	75	82	88	95	102	109	116	123
19	51	58	65	72	80	87	94	101	109	116	123	130
20	54	62	69	77	84	92	100	107	115	123	130	138

APPENDIX I

Critical Values for Hartley's Maximum F-Ratio Significance
Test for Homogeneity of Variances

Since the sample size in all groups is the same, the df column represents the value for each group. To use this table, simply locate the appropriate df row heading and the column headed by the number of variances that are being compared; then read the value from the point at which the row and column intersect. For example, assume that the df of each sample is 20, that six variances were compared, and that the F_{max} ratio is 4.2. Entering the table at the row headed by 20 df and reading across to the $k = 6$ column, it is noted that .05 value is 3.76 and the .01 value is 4.9. Therefore, the $F_{max} = 4.2$ is considered significant at the .05 level.

Appendix I. F_{max} Statistic

Alpha = .05 and .01 (in italics)

df \ k	2	3	4	5	6	7	8	9	10	11	12
2	39.0	87.5	142.	202.	266.	333.	403.	475.	550.	626.	704.
	199.	*448.*	*729.*	*1036.*	*1362.*	*1705.*	*2063.*	*2432.*	*2813.*	*3204.*	*3605.*
3	15.4	27.8	39.2	50.7	62.0	72.9	83.5	93.9	104.	114.	124.
	47.5	*85.*	*120.*	*151.*	*184.*	*216.**	*249.**	*281.**	*310.**	*337.**	*361.**
4	9.60	15.5	20.6	25.2	29.5	33.6	37.5	41.1	44.6	48.0	51.4
	23.2	*37.*	*49.*	*59.*	*69.*	*79.*	*89.*	*97.*	*106.*	*113.*	*120.*
5	7.15	10.8	13.7	16.3	18.7	20.8	22.9	24.7	26.5	28.2	29.9
	14.9	*22.*	*28.*	*33.*	*38.*	*42.*	*46.*	*50.*	*54.*	*57.*	*60.*
6	5.82	8.38	10.4	12.1	13.7	15.0	16.3	17.5	18.6	19.7	20.7
	11.1	*15.5*	*19.1*	*22.*	*25.*	*27.*	*30.*	*32.*	*34.*	*36.*	*37.*
7	4.99	6.94	8.44	9.70	10.8	11.8	12.7	13.5	14.3	15.1	15.8
	8.89	*12.1*	*14.5*	*16.5*	*18.4*	*20.*	*22.*	*23.*	*24.*	*26.*	*27.*
8	4.43	6.00	7.18	8.12	9.03	9.78	10.5	11.1	11.7	12.2	12.7
	7.50	*9.9*	*11.7*	*13.2*	*14.5*	*15.8*	*16.9*	*17.9*	*18.9*	*19.8*	*21.*
9	4.03	5.34	6.31	7.11	7.80	8.41	8.95	9.45	9.91	10.3	10.7
	6.54	*8.5*	*9.9*	*11.1*	*12.1*	*13.1*	*13.9*	*14.7*	*15.3*	*16.0*	*16.6*
10	3.72	4.85	5.67	6.34	6.92	7.42	7.87	8.28	8.66	9.01	9.34
	5.85	*7.4*	*8.6*	*9.6*	*10.4*	*11.1*	*11.8*	*12.4*	*12.9*	*13.4*	*13.9*
12	3.28	4.16	4.79	5.30	5.72	6.09	6.42	6.72	7.00	7.25	7.48
	4.91	*6.1*	*6.9*	*7.6*	*8.2*	*8.7*	*9.1*	*9.5*	*9.9*	*10.2*	*10.6*
15	2.86	3.54	4.01	4.37	4.68	4.95	5.19	5.40	5.59	5.77	5.93
	4.07	*4.9*	*5.5*	*6.0*	*6.4*	*6.7*	*7.1*	*7.3*	*7.5*	*7.8*	*8.0*
20	2.46	2.95	3.29	3.54	3.76	3.94	4.10	4.24	4.37	4.49	4.59
	3.32	*3.8*	*4.3*	*4.6*	*4.9*	*5.1*	*5.3*	*5.5*	*5.6*	*5.8*	*5.9*
30	2.07	2.40	2.61	2.78	2.91	3.02	3.12	3.21	3.29	3.36	3.39
	2.63	*3.0*	*3.3*	*3.4*	*3.6*	*3.7*	*3.8*	*3.9*	*4.0*	*41*	*4.2*
60	1.67	1.85	1.96	2.04	2.11	2.17	2.22	2.26	2.30	2.33	2.36
	1.96	*2.2*	*2.3*	*2.4*	*2.4*	*2.5*	*2.5*	*2.6*	*2.6*	*2.7*	*2.7*
∞	1.00	1.00	1.00	1.00	1.00	1.00	1.00	1.00	1.00	1.00	1.00
	1.00	*1.00*	*1.00*	*1.00*	*1.00*	*1.00*	*1.00*	*1.00*	*1.00*	*1.00*	*1.00*

*Values in the column $k = 2$ and in the rows $df = 2$ and ∞ are exact. Elsewhere the third digit may be in error by a few units for $F_{.95}$ and several units for $F_{.99}$. The third digit figures of values marked by an asterisk are the most uncertain.

SOURCE: Reproduced from Table 31 of E. S. Pearson and H. O. Hartley, *Biometrika Tables for Statisticians*, Vol. 1, Third Edition, 1966. Used by permission of the Biometrika Trustees.

APPENDIX J
Significant Studentized Ranges for Duncan's New Multiple-Range Test

To use these tables, the *df* and the range of the pairs of groups being compared must be located in the particular subtable which lists the level of significance desired. Assume that you wish to find the critical value using a two-tailed test at the .05 level of significance. To do this, first locate the *df* and *r* values in the subtable for alpha equal to .05. The point at which the *df* row and the *r* column intersect represents the minimum significant value (*k*) at the .05 level. This is then multiplied by the standard error of the means to obtain the minimum critical differences for each of the several ranges. If, for example, the *df* are equal to 15, the range of the pair being compared is equal to 6, and the standard error of the means is equal to 1.50, then the critical difference would be equal to 3.36 × 1.50, or 5.04. For the next pair, with a range of 5, the critical difference would be 3.31 × 1.50, or 4.96. The same procedure would be followed for the remainder of the range comparisons.

316

Appendix J. Duncan's New Multiple-Range Test

Error df	Protection Level	r = number of means for range being tested					
		2	3	4	5	6	7
1	.05	18.0	18.0	18.0	18.0	18.0	18.0
	.01	90.0	90.0	90.0	90.0	90.0	90.0
2	.05	6.09	6.09	6.09	6.09	6.09	6.09
	.01	14.0	14.0	14.0	14.0	14.0	14.0
3	.05	4.50	4.50	4.50	4.50	4.50	4.50
	.01	8.26	8.5	8.6	8.7	8.8	8.9
4	.05	3.93	4.01	4.02	4.02	4.02	4.02
	.01	6.51	6.8	6.9	7.0	7.1	7.1
5	.05	3.64	3.74	3.79	3.83	3.83	3.83
	.01	5.70	5.96	6.11	6.18	6.26	6.33
6	.05	3.46	3.58	3.64	3.68	3.68	3.68
	.01	5.24	5.51	5.65	5.73	5.81	5.88
7	.05	3.35	3.47	3.54	3.58	3.60	3.61
	.01	4.95	5.22	5.37	5.45	5.53	5.61
8	.05	3.26	3.39	3.47	3.52	3.55	3.56
	.01	4.74	5.00	5.14	5.23	5.32	5.40
9	.05	3.20	3.34	3.41	3.47	3.50	3.52
	.01	4.60	4.86	4.99	5.08	5.17	5.25
10	.05	3.15	3.30	3.37	3.43	3.46	3.47
	.01	4.48	4.73	4.88	4.96	5.06	5.13
11	.05	3.11	3.27	3.35	3.39	3.43	3.44
	.01	4.39	4.63	4.77	4.86	4.94	5.01
12	.05	3.08	3.23	3.33	3.36	3.40	3.42
	.01	4.32	4.55	4.68	4.76	4.84	4.92
13	.05	3.06	3.21	3.30	3.35	3.38	3.41
	.01	4.26	4.48	4.62	4.69	4.74	4.84
14	.05	3.03	3.18	3.27	3.33	3.37	3.39
	.01	4.21	4.42	4.55	4.63	4.70	4.78
15	.05	3.01	3.16	3.25	3.31	3.36	3.38
	.01	4.17	4.37	4.50	4.58	4.64	4.72

SOURCE: Abridged from D. B. Duncan, Multiple range and multiple F tests, *Biometrics*, 1955, **11**, 1–41. Reprinted by permission of the Biometric Society.

8	9	10	12	14	16	18	20
18.0	18.0	18.0	18.0	18.0	18.0	18.0	18.0
90.0	90.0	90.0	90.0	90.0	90.0	90.0	90.0
6.09	6.09	6.09	6.09	6:09	6.09	6.09	6.09
14.0	14.0	14.0	14.0	14.0	14.0	14.0	14.0
4.50	4.50	4.50	4.50	4.50	4.50	4.50	4.50
8.9	9.0	9.0	9.0	9.1	9.2	9.3	9.3
4.02	4.02	4.02	4.02	4.02	4.02	4.02	4.02
7.2	7.2	7.3	7.3	7.4	7.4	7.5	7.5
3.83	3.83	3.83	3.83	3.83	3.83	3.83	3.83
6.40	6.44	6.5	6.6	6.6	6.7	6.7	6.8
3.68	3.68	3.68	3.68	3.68	3.68	3.68	3.68
5.95	6.00	6.0	6.1	6.2	6.2	6.3	6.3
3.61	3.61	3.61	3.61	3.61	3.61	3.61	3.61
5.69	5.73	5.8	5.8	5.9	5.9	6.0	6.0
3.56	3.56	3.56	3.56	3.56	3.56	3.56	3.56
5.47	5.51	5.5	5.6	5.7	5.7	5.8	5.8
3.52	3.52	3.52	3.52	3.52	3.52	3.52	3.52
5.32	5.36	5.4	5.5	5.5	5.6	5.7	5.7
3.47	3.47	3.47	3.47	3.47	3.47	3.47	3.48
5.20	5.24	5.28	5.36	5.42	5.48	5.54	5.55
3.45	3.46	3.46	3.46	3.46	3.46	3.47	3.48
5.06	5.12	5.15	5.24	5.28	5.34	5.38	5.39
3.44	3.44	3.46	3.46	3.46	3.46	3.47	3.48
4.96	5.02	5.07	5.13	5.17	5.22	5.24	5.26
3.42	3.44	3.45	3.45	3.46	3.46	3.47	3.47
4.88	4.94	4.98	5.04	5.08	5.13	5.14	5.15
3.41	3.42	3.44	3.45	3.46	3.46	3.47	3.47
4.83	4.87	4.91	4.96	5.00	5.04	5.06	5.07
3.40	3.42	3.43	3.44	3.45	3.46	3.47	3.47
4.77	4.81	4.84	4.90	4.94	4.97	4.99	5.00

Appendix J. Duncan's New Multiple-Range Test *(cont.)*

Error df	Protection Level	r = number of means for range being tested						
		2	3	4	5	6	7	8
16	.05	3.00	3.15	3.23	3.30	3.34	3.37	3.39
	.01	4.13	4.34	4.45	4.54	4.60	4.67	4.72
17	.05	2.98	3.13	3.22	3.28	3.33	3.36	3.38
	.01	4.10	4.30	4.41	4.50	4.56	4.63	4.68
18	.05	2.97	3.12	3.21	3.27	3.32	3.35	3.37
	.01	4.07	4.27	4.38	4.46	4.53	4.59	4.64
19	.05	2.96	3.11	3.19	3.26	3.31	3.35	3.37
	.01	4.05	4.24	4.35	4.43	4.50	4.56	4.61
20	.05	2.95	3.10	3.18	3.25	3.30	3.34	3.36
	.01	4.02	4.22	4.33	4.40	4.47	4.53	4.58
22	.05	2.93	3.08	3.17	3.24	3.29	3.32	3.35
	.01	3.99	4.17	4.28	4.36	4.42	4.48	4.53
24	.05	2.92	3.07	3.15	3.22	3.28	3.31	3.34
	.01	3.96	4.14	4.24	4.33	4.39	4.44	4.49
26	.05	2.91	3.06	3.14	3.21	3.27	3.30	3.34
	.01	3.93	4.11	4.21	4.30	4.36	4.41	4.46
28	.05	2.90	3.04	3.13	3.20	3.26	3.30	3.33
	.01	3.91	4.08	4.18	4.28	4.34	4.39	4.43
30	.05	2.89	3.04	3.12	3.20	3.25	3.29	3.32
	.01	3.89	4.06	4.16	4.22	4.32	4.36	4.41
40	.05	2.86	3.01	3.10	3.17	3.22	3.27	3.30
	.01	3.82	3.99	4.10	4.17	4.24	4.30	4.34
60	.05	2.83	2.98	3.08	3.14	3.20	3.24	3.28
	.01	3.76	3.92	4.03	4.12	4.17	4.23	4.27
100	.05	2.80	2.95	3.05	3.12	3.18	3.22	3.26
	.01	3.71	3.86	3.93	4.06	4.11	4.17	4.21
∞	.05	2.77	2.92	3.02	3.09	3.15	3.19	3.23
	.01	3.64	3.80	3.80	3.98	4.04	4.09	4.14

9	10	12	14	16	18	20
3.41	3.43	3.44	3.45	3.46	3.47	3.47
4.76	4.79	4.84	4.88	4.91	4.93	4.94
3.40	3.42	3.44	3.45	3.46	3.47	3.47
4.72	4.75	4.80	4.83	4.86	4.88	4.89
3.39	3.41	3.43	3.45	3.46	3.47	3.47
4.68	4.71	4.76	4.79	4.82	4.84	4.85
3.39	3.41	3.43	3.44	3.46	3.47	3.47
4.64	4.67	4.72	4.76	4.79	4.81	4.82
3.38	3.40	3.43	3.44	3.46	3.46	3.47
4.61	4.65	4.69	4.73	4.76	4.78	4.79
3.37	3.39	3.42	3.44	3.45	3.46	3.47
4.57	4.60	4.65	4.68	4.71	4.74	4.75
3.37	3.38	3.41	3.44	3.45	3.46	3.47
4.53	4.57	4.62	4.64	4.67	4.70	4.72
3.36	3.38	3.41	3.43	3.45	3.46	3.47
4.50	4.53	4.58	4.62	4.65	4.67	4.69
3.35	3.37	3.40	3.43	3.45	3.46	3.47
4.47	4.51	4.56	4.60	4.62	4.65	4.67
3.35	3.37	3.40	3.43	3.44	3.46	3.47
4.45	4.48	4.54	4.58	4.61	4.63	4.65
3.33	3.35	3.39	3.42	3.44	3.46	3.47
4.37	4.41	4.46	4.51	4.54	4.57	4.59
3.31	3.33	3.37	3.40	3.43	3.45	3.47
4.31	4.34	4.39	4.44	4.47	4.50	4.58
3.29	3.32	3.36	3.40	3.42	3.45	3.47
4.25	4.29	4.35	4.38	4.42	4.45	4.48
3.26	3.29	3.34	3.38	3.41	3.44	3.47
4.17	4.20	4.26	4.31	4.34	4.38	4.44

APPENDIX K

Significant Studentized Ranges for the Newman-Keuls' and Tukey Multiple-Comparison Tests

To use these tables, the df and the range (r) of the pairs of groups being compared must be located in the table where the desired level of significance is listed. Assume that you wish to find a significant value at the .05 level. To do this, first locate the df and the r values in the table for alpha equal to .05. The point at which the df row and the r column intersect (q) represents the minimum significant value at the .05 level. This value is then multiplied by the standard error of the means to obtain the minimum critical value.

Appendix K. Significant Studentized Ranges (Two-tailed)

Error df	α	r = number of means or number of steps between ordered means									
		2	3	4	5	6	7	8	9	10	11
5	.05	3.64	4.60	5.22	5.67	6.03	6.33	6.58	6.80	6.99	7.17
	.01	5.70	6.98	7.80	8.42	8.91	9.32	9.67	9.97	10.24	10.48
6	.05	3.46	4.34	4.90	5.30	5.63	5.90	6.12	6.32	6.49	6.65
	.01	5.24	6.33	7.03	7.56	7.97	8.32	8.61	8.87	9.10	9.30
7	.05	3.34	4.16	4.68	5.06	5.36	5.61	5.82	6.00	6.16	6.30
	.01	4.95	5.92	6.54	7.01	7.37	7.68	7.94	8.17	8.37	8.55
8	.05	3.26	4.04	4.53	4.89	5.17	5.40	5.60	5.77	5.92	6.05
	.01	4.75	5.64	6.20	6.62	6.96	7.24	7.47	7.68	7.86	8.03
9	.05	3.20	3.95	4.41	4.76	5.02	5.24	5.43	5.59	5.74	5.87
	.01	4.60	5.43	5.96	6.35	6.66	6.91	7.13	7.33	7.49	7.65
10	.05	3.15	3.88	4.33	4.65	4.91	5.12	5.30	5.46	5.60	5.72
	.01	4.48	5.27	5.77	6.14	6.43	6.67	6.87	7.05	7.21	7.36
11	.05	3.11	3.82	4.26	4.57	4.82	5.03	5.20	5.35	5.49	5.61
	.01	4.39	5.15	5.62	5.97	6.25	6.48	6.67	6.84	6.99	7.13
12	.05	3.08	3.77	4.20	4.51	4.75	4.95	5.12	5.27	5.39	5.51
	.01	4.32	5.05	5.50	5.84	6.10	6.32	6.51	6.67	6.81	6.94
13	.05	3.06	3.73	4.15	4.45	4.69	4.88	5.05	5.19	5.32	5.43
	.01	4.26	4.96	5.40	5.73	5.98	6.19	6.37	6.53	6.67	6.79
14	.05	3.03	3.70	4.11	4.41	4.64	4.83	4.99	5.13	5.25	5.36
	.01	4.21	4.89	5.32	5.63	5.88	6.08	6.26	6.41	6.54	6.66
15	.05	3.01	3.67	4.08	4.37	4.59	4.78	4.94	5.08	5.20	5.31
	.01	4.17	4.84	5.25	5.56	5.80	5.99	6.16	6.31	6.44	6.55

SOURCE: Reproduced from Table 29 of E. S. Pearson and H. O. Hartley, *Biometrika Tables for Statisticans*, Vol. 1, Third Edition, 1966. Used by permission of the Biometrika Trustees.

If, for example, the tables are being used for the Newman-Keuls' test, where the *df* are equal to 15, alpha = .05, the range of the first pair to be compared is 6 and the standard error of the means is equal to 1.50, then a minimum critical value of 6.88 would be needed for significance (4.59 × 1.50). For the next comparison, with a range of 5, the critical value would have to be equal to or larger than 6.56 (4.37 × 1.50). For a range of 4, the critical value would have to be at least 6.12, etc.

Using the tables for the Tukey test, only the value for the largest range is needed. Thus, where the *df* are equal to 15, alpha = .05, the largest range is equal to 6 and the standard error of the means is equal to 1.50, a critical value larger than 6.88 would be needed for all comparisons.

Error df	α	\multicolumn{10}{c}{r = number of means or number of steps between ordered means}
		2	3	4	5	6	7	8	9	10	11
16	.05	3.00	3.65	4.05	4.33	4.56	4.74	4.90	5.03	5.15	5.26
	.01	4.13	4.79	5.19	5.49	5.72	5.92	6.08	6.22	6.35	6.46
17	.05	2.98	3.63	4.02	4.30	4.52	4.70	4.86	4.99	5.11	5.21
	.01	4.10	4.74	5.14	5.43	5.66	5.85	6.01	6.15	6.27	6.38
18	.05	2.97	3.61	4.00	4.28	4.49	4.67	4.82	4.96	5.07	5.17
	.01	4.07	4.70	5.09	5.38	5.60	5.79	5.94	6.08	6.20	6.31
19	.05	2.96	3.59	3.98	4.25	4.47	4.65	4.70	4.92	5.04	5.14
	.01	4.05	4.67	5.05	5.33	5.55	5.73	6.89	6.02	6.14	6.25
20	.05	2.95	3.58	3.96	4.23	4.45	4.62	4.77	4.90	5.01	5.11
	.01	4.02	4.64	5.02	5.29	5.51	5.69	5.84	5.97	6.09	6.19
24	.05	2.92	3.53	3.90	4.17	4.37	4.54	4.68	4.81	4.92	5.01
	.01	3.96	4.55	4.91	5.17	5.37	5.54	5.69	5.81	5.92	6.02
30	.05	2.89	3.49	3.85	4.10	4.30	4.46	4.60	4.72	4.82	4.92
	.01	3.89	4.45	4.80	5.05	5.24	5.40	5.54	5.65	5.76	5.85
40	.05	2.86	3.44	3.79	4.04	4.23	4.39	4.52	4.63	4.73	4.82
	.01	3.82	4.37	4.70	4.93	5.11	5.26	5.39	5.50	5.60	5.69
60	.05	2.83	3.40	3.74	3.98	4.16	4.31	4.44	4.55	4.65	4.73
	.01	3.76	4.28	4.59	4.82	4.99	5.13	5.25	5.36	5.45	5.53
120	.05	2.80	3.36	3.68	3.92	4.10	4.24	4.36	4.47	4.56	4.64
	.01	3.70	4.20	4.50	4.71	4.87	5.01	5.12	5.21	5.30	5.37
∞	.05	2.77	3.31	3.63	3.86	4.03	4.17	4.29	4.39	4.47	4.55
	.01	3.64	4.12	4.40	4.60	4.76	4.88	4.99	5.08	5.16	5.23

Appendix K. Significant Studentized Ranges *(cont.)*

Error df	α	\multicolumn{9}{c}{r = number of means or number of steps between ordered means}
		12	13	14	15	16	17	18	19	20
5	.05	7.32	7.47	7.60	7.72	7.83	7.93	8.03	8.12	8.21
	.01	10.70	10.89	11.08	11.24	11.40	11.55	11.68	11.81	11.93
6	.05	6.79	6.92	7.03	7.14	7.24	7.34	7.43	7.51	7.59
	.01	9.48	9.65	9.81	9.95	10.08	10.21	10.32	10.43	10.54
7	.05	6.43	6.55	6.66	6.76	6.85	6.94	7.02	7.10	7.17
	.01	8.71	8.86	9.00	9.12	9.24	9.35	9.46	9.55	9.65
8	.05	6.18	6.29	6.39	6.48	6.57	6.65	6.73	6.80	6.87
	.01	8.18	8.31	8.44	8.55	8.66	8.76	8.85	8.94	9.03
9	.05	5.98	6.09	6.19	6.28	6.36	6.44	6.51	6.58	6.64
	.01	7.78	7.91	8.03	8.13	8.23	8.33	8.41	8.49	8.57
10	.05	5.83	5.93	6.03	6.11	6.19	6.27	6.34	6.40	6.47
	.01	7.49	7.60	7.71	7.81	7.91	7.99	8.08	8.15	8.23
11	.05	5.71	5.81	5.90	5.98	6.06	6.13	6.20	6.27	6.33
	.01	7.25	7.36	7.46	7.56	7.65	7.73	7.81	7.88	7.95
12	.05	5.61	5.71	5.80	5.88	5.95	6.02	6.09	6.15	6.21
	.01	7.06	7.17	7.26	7.36	7.44	7.52	7.59	7.66	7.73
13	.05	5.53	5.63	5.71	5.79	5.86	5.93	5.99	6.05	6.11
	.01	6.90	7.01	7.10	7.19	7.27	7.35	7.42	7.48	7.55
14	.05	5.46	5.55	5.64	5.71	5.79	5.85	5.91	5.97	6.03
	.01	6.77	6.87	6.96	7.05	7.13	7.20	7.27	7.33	7.39
15	.05	5.40	5.49	5.57	5.65	5.72	5.78	5.85	5.90	5.96
	.01	6.66	6.76	6.84	6.93	7.00	7.07	7.14	7.20	7.26

Error df	α	r = number of means or number of steps between ordered means								
		12	13	14	15	16	17	18	19	20
16	.05	5.35	5.44	5.52	5.59	5.66	5.73	5.79	5.84	5.90
	.01	6.56	6.66	6.74	6.82	6.90	6.97	7.03	7.09	7.15
17	.05	5.31	5.39	5.47	5.54	5.61	5.67	5.73	5.79	5.84
	.01	6.48	6.57	6.66	6.73	6.81	6.87	6.94	7.00	7.05
18	.05	5.27	5.35	5.43	5.50	5.57	5.63	5.69	5.74	5.79
	.01	6.41	6.50	6.58	6.65	6.73	6.79	6.85	6.91	6.97
19	.05	5.23	5.31	5.39	5.46	5.53	5.59	5.65	5.70	5.75
	.01	6.34	6.43	6.51	6.58	6.65	6.72	6.78	6.84	6.89
20	.05	5.20	5.28	5.36	5.43	5.49	5.55	5.61	5.66	5.71
	.01	6.28	6.32	6.45	6.52	6.59	6.65	6.71	6.77	6.82
24	.05	5.10	5.18	5.25	5.32	5.38	5.44	5.49	5.55	5.59
	.01	6.11	6.19	6.26	6.33	6.39	6.45	6.51	6.56	6.61
30	.05	5.00	5.08	5.15	5.21	5.27	5.33	5.38	5.43	5.47
	.01	5.93	6.01	6.08	6.14	6.20	6.26	6.31	6.36	6.41
40	.05	4.90	4.98	5.04	5.11	5.16	5.22	5.27	5.31	5.36
	.01	5.76	5.83	5.90	5.96	6.02	6.07	6.12	6.16	6.21
60	.05	4.81	4.88	4.94	5.00	5.06	5.11	5.15	5.20	5.24
	.01	5.60	5.67	5.73	5.78	5.84	5.89	5.93	5.97	6.01
120	.05	4.71	4.78	4.84	4.90	4.95	5.00	5.04	5.09	5.13
	.01	5.44	5.50	5.56	5.61	5.66	5.71	5.75	5.79	5.83
∞	.05	4.62	4.68	4.74	4.80	4.85	4.89	4.93	4.97	5.01
	.01	5.29	5.35	5.40	5.45	5.49	5.54	5.57	5.61	5.65

APPENDIX L

Dunnett's Test: Comparison of Treatment Means with a Control

To use this table, the df and the range (r) of the pairs of groups being compared must be located in the table where the desired level of significance is listed. Assume that you wish to compare four means with

Appendix L. Comparison of Treatment Means with a Control (Two-tailed)

Error df	α	r = number of treatment means, including control								
		2	3	4	5	6	7	8	9	10
5	.05	2.57	3.03	3.29	3.48	3.62	3.73	3.82	3.90	3.97
	.01	4.03	4.63	4.98	5.22	5.41	5.56	5.69	5.80	5.89
6	.05	2.45	2.86	3.10	3.26	3.39	3.49	3.57	3.64	3.71
	.01	3.71	4.21	4.51	5.71	4.87	5.00	5.10	5.20	5.28
7	.05	2.36	2.75	2.97	3.12	3.24	3.33	3.41	3.47	3.53
	.01	3.50	3.95	4.21	4.39	4.53	4.64	4.74	4.82	4.89
8	.05	2.31	2.67	2.88	3.02	3.13	3.22	3.29	3.35	3.41
	.01	3.36	3.77	4.00	4.17	4.29	4.40	4.48	4.56	4.62
9	.05	2.26	2.61	2.81	2.95	3.05	3.14	3.20	3.26	3.32
	.01	3.25	3.63	3.85	4.01	4.12	4.22	4.30	4.37	4.43
10	.05	2.23	2.57	2.76	2.89	2.99	3.07	3.14	3.19	3.24
	.01	3.17	3.53	3.74	3.88	3.99	4.08	4.16	4.22	4.28
11	.05	2.20	2.53	2.72	2.84	2.94	3.02	3.08	3.14	3.19
	.01	3.11	3.45	3.65	3.79	3.89	3.98	4.05	4.11	4.16
12	.05	2.18	2.50	2.68	2.81	2.90	2.98	3.04	3.09	3.14
	.01	3.05	3.39	3.58	3.71	3.81	3.89	3.96	4.02	4.07
13	.05	2.16	2.48	2.65	2.78	2.87	2.94	3.00	3.06	3.10
	.01	3.01	3.33	3.52	3.65	3.74	3.82	3.89	3.94	3.99
14	.05	2.14	2.46	2.63	2.75	2.84	2.91	2.97	3.02	3.07
	.01	2.98	3.29	3.47	3.59	3.69	3.76	3.83	3.88	3.93
15	.05	2.13	2.44	2.61	2.73	2.82	2.89	2.95	3.00	3.04
	.01	2.95	3.25	3.43	3.55	3.64	3.71	3.78	3.83	3.88

SOURCE: Table reproduced from C. W. Dunnett, New tables for multiple comparisons with a control, *Biometrics*, 1964, **20**, 482–491. Reprinted by permission of the Biometric Society.

the control group. To do this, first locate the *df* and the *r* in the table for alpha equal to .05. The point at which the *df* and *r* intersect (*d*) represents the minimum significant value. This value is then multiplied by the standard error of the means to obtain the minimum critical value. If, for example, the *df* are equal to 15, the maximum range of the means being compared is equal to 4 and the standard error of the means is equal to 1.70, then the critical difference for all comparisons would be equal to 2.61 × 1.70, or 4.44.

Error df	α	\multicolumn{9}{c}{r = number of treatment means, including control}
		2	3	4	5	6	7	8	9	10
16	.05	2.12	2.42	2.59	2.71	2.80	2.87	2.92	2.97	3.02
	.01	2.92	3.22	3.39	3.51	3.60	3.67	3.73	3.78	3.83
17	.05	2.11	2.41	2.58	2.69	2.78	2.85	2.90	2.95	3.00
	.01	2.90	3.19	3.36	3.47	3.56	3.63	3.69	3.74	3.79
18	.05	2.10	2.40	2.56	2.68	2.76	2.83	2.89	2.94	2.98
	.01	2.88	3.17	3.33	3.44	3.53	3.60	3.66	3.71	3.75
19	.05	2.09	2.39	2.55	2.66	2.75	2.81	2.87	2.92	2.96
	.01	2.86	3.15	3.31	3.42	3.50	3.57	3.63	3.68	3.72
20	.05	2.09	2.38	2.54	2.65	2.73	2.80	2.86	2.90	2.95
	.01	2.85	3.13	3.29	3.40	3.48	3.55	3.60	3.65	3.69
24	.05	2.06	2.35	2.51	2.61	2.70	2.76	2.81	2.86	2.90
	.01	2.80	3.07	3.22	3.32	3.40	3.47	3.52	3.57	3.61
30	.05	2.04	2.32	2.47	2.58	2.66	2.72	2.77	2.82	2.86
	.01	2.75	3.01	3.15	3.25	3.33	3.39	3.44	3.49	3.52
40	.05	2.02	2.29	2.44	2.54	2.62	2.68	2.73	2.77	2.81
	.01	2.70	2.95	3.09	3.19	3.26	3.32	3.37	3.41	3.44
60	.05	2.00	2.27	2.41	2.51	2.58	2.64	2.69	2.73	2.77
	.01	2.66	2.90	3.03	3.12	3.19	3.25	3.29	3.33	3.37
120	.05	1.98	2.24	2.38	2.47	2.55	2.60	2.65	2.69	2.73
	.01	2.62	2.85	2.97	3.06	3.12	3.18	3.22	3.26	3.29
∞	.05	1.96	2.21	2.35	2.44	2.51	2.57	2.61	2.65	2.69
	.01	2.58	2.79	2.92	3.00	3.06	3.11	3.15	3.19	3.22

APPENDIX M

Critical Values of Wilcoxon's *T* Statistic for the Matched-Pairs Signed-Ranks Test

To use this table, first locate the number of *pairs* of scores in the n' column. The critical values for the several levels of significance are listed in the columns to the right. For example, if n' were 20 and the computed value 25, it would be concluded that since 25 is less than 38, this value is significant beyond the .01 level of significance for a two-tailed test.

	Level of significance for one-tailed test		
	.025	.01	.005
	Level of significance for two-tailed test		
n'	.05	.02	.01
6	1	—	—
7	2	0	—
8	4	2	0
9	6	3	2
10	8	5	3
11	11	7	5
12	14	10	7
13	17	13	10
14	21	16	13
15	25	20	16
16	30	24	19
17	35	28	23
18	40	33	28
19	46	38	32
20	52	43	37
21	59	49	43
22	66	56	49
23	73	62	55
24	81	69	61
25	90	77	68

Note that n' is the number of matched pairs.

SOURCE: Adapted from Table 2 of F. Wilcoxon and R. A. Wilcox, *Some Rapid Approximate Statistical Procedures*, rev. ed. Pearl River, New York: American Cyanamid Company, 1964. Reproduced from S. Siegel, *Nonparametric Statistics for the Behavioral Sciences*, New York: McGraw-Hill Book Company, 1956. Reprinted by permission of the American Cyanamid Company and McGraw-Hill Book Company.

APPENDIX N
Coefficients for Orthogonal Polynomials

In this table, the column headed by k indicates the number of components being compared. The columns headed by the ten X values indicate the actual coefficients to be used. Thus, if $k = 3$ groups were being analyzed, the coefficients for analysis of the linear component would be -1 for group 1, 0 for Group 2, and 1 for Group 3. The quadratic coefficients would be 1, -2, and 1 for the three respective groups.

k	Polynomial	$X = 1$	2	3	4	5	6	7	8	9	10
3	Linear	-1	0	1							
	Quadratic	1	-2	1							
4	Linear	-3	-1	1	3						
	Quadratic	1	-1	-1	1						
	Cubic	-1	3	-3	1						
5	Linear	-2	-1	0	1	2					
	Quadratic	2	-1	-2	-1	2					
	Cubic	-1	2	0	-2	1					
	Quartic	1	-4	6	-4	1					
6	Linear	-5	-3	-1	1	3	5				
	Quadratic	5	-1	-4	-4	-1	5				
	Cubic	-5	7	4	-4	-7	5				
	Quartic	1	-3	2	2	-3	1				
7	Linear	-3	-2	-1	0	1	2	3			
	Quadratic	5	0	-3	-4	-3	0	5			
	Cubic	-1	1	1	0	-1	-1	1			
	Quartic	3	-7	1	6	1	-7	3			
8	Linear	-7	-5	-3	-1	1	3	5	7		
	Quadratic	7	1	-3	-5	-5	-3	1	7		
	Cubic	-7	5	7	3	-3	-7	-5	7		
	Quartic	7	-13	-3	9	9	-3	-13	7		
	Quintic	-7	23	-17	-15	15	17	-23	7		
9	Linear	-4	-3	-2	-1	0	1	2	3	4	
	Quadratic	28	7	-8	-17	-20	-17	-8	7	28	
	Cubic	-14	7	13	9	0	-9	-13	-7	14	
	Quartic	14	-21	-11	9	18	9	-11	-21	14	
	Quintic	-4	11	-4	-9	0	9	4	-11	4	
10	Linear	-9	-7	-5	-3	-1	1	3	5	7	9
	Quadratic	6	2	-1	-3	-4	-4	-3	-1	2	6
	Cubic	-42	14	35	31	12	-12	-31	-35	-14	42
	Quartic	18	-22	-17	3	18	18	3	-17	-22	18
	Quintic	-6	14	-1	-11	-6	6	11	1	-14	6

SOURCE: Abridged from Table B.10 of *Statistical Principles in Experimental Design* by B. J. Winer, Copyright 1962 by McGraw-Hill Book Company. Used by permission of McGraw-Hill Book Company.

APPENDIX O
Cumulative Probability Distribution for r, the Total Number of Runs Up or Down

To use this table, the number of observations (n) and the number of runs (r') must be known. Enter the table according to the number of observations (n) and runs (r'). For example, if there are twenty observations and six runs, the probability of six or fewer runs occurring by chance alone is less than one in ten thousand.

Appendix O. Cumulative Probability Distribution for r, the Total Number of Runs Up and Down

$P(r \leq r')$
Number of observations, n

Number of runs r'	1	2	3	4	5	6	7	8	9	10	11	12
1		1.0000	.3333	.0833	.0167	.0028	.0004	.0000	.0000	.0000	.0000	.0000
2			1.0000	.5833	.2500	.0861	.0250	.0063	.0014	.0003	.0001	.0000
3				1.0000	.7333	.4139	.1909	.0749	.0257	.0079	.0022	.0005
4					1.0000	.8306	.5583	.3124	.1500	.0633	.0239	.0082
5						1.0000	.8921	.6750	.4347	.2427	.1196	.0529
6							1.0000	.9313	.7653	.5476	.3438	.1918
7								1.0000	.9563	.8329	.6460	.4453
8									1.0000	.9722	.8823	.7280
9										1.0000	.9823	.9179
10											1.0000	.9887
11												1.0000
12												
13												
14												
15												
16												
17												
18												
19												
20												
21												
22												
23												
24												

SOURCE: From "Probability Table for Number of Runs of Signs of First Differences in Ordered Series" by Eugene S. Edgington, *Journal of the American Statistical Association*, **56**(1961), 156–159. Copyright © 1961 by the American Statistical Association. Reprinted by permission.

Appendix O. Cumulative Probability Distribution for r (cont.)

$P(r \leq r')$
Number of runs r'
Number of observations, n

13	14	15	16	17	18	19	20	21	22	23	24	25
.0000	.0000	.0000	.0000	.0000	.0000	.0000	.0000	.0000	.0000	.0000	.0000	.0000
.0000	.0000	.0000	.0000	.0000	.0000	.0000	.0000	.0000	.0000	.0000	.0000	.0000
.0001	.0000	.0000	.0000	.0000	.0000	.0000	.0000	.0000	.0000	.0000	.0000	.0000
.0026	.0007	.0027	.0001	.0003	.0001	.0000	.0000	.0000	.0000	.0000	.0000	.0000
.0213	.0079	.0186	.0009	.0026	.0009	.0003	.0001	.0000	.0000	.0000	.0000	.0000
.0964	.0441	.0782	.0072	.0160	.0065	.0025	.0009	.0003	.0001	.0000	.0000	.0000
.2749	.1534	.2216	.0367	.0638	.0306	.0137	.0058	.0023	.0009	.0003	.0001	.0000
.5413	.3633	.4520	.1238	.1799	.1006	.0523	.0255	.0117	.0050	.0021	.0008	.0003
.7942	.6278	.7030	.2975	.3770	.2443	.1467	.0821	.0431	.0213	.0099	.0044	.0018
.9432	.8464	.8866	.5369	.6150	.4568	.3144	.2012	.1202	.0674	.0356	.0177	.0084
.9928	.9609	.9733	.7665	.8188	.6848	.5337	.3873	.2622	.1661	.0988	.0554	.0294
1.0000	.9954	.9971	.9172	.9400	.8611	.7454	.6055	.4603	.3276	.2188	.1374	.0815
····	1.0000	1.0000	.9818	.9877	.9569	.8945	.7969	.6707	.5312	.3953	.2768	.1827
····	····	····	.9981	.9988	.9917	.9692	.9207	.8398	.7286	.5980	.4631	.3384
····	····	····	1.0000	1.0000	.9992	.9944	.9782	.9409	.8749	.7789	.6595	.5292
····	····	····	····	····	1.0000	.9995	.9962	.9846	.9563	.9032	.8217	.7148
····	····	····	····	····	····	1.0000	.9997	.9975	.9892	.9679	.9258	.8577
····	····	····	····	····	····	····	1.0000	.9998	.9983	.9924	.9765	.9436
····	····	····	····	····	····	····	····	1.0000	.9999	.9989	.9947	.9830
····	····	····	····	····	····	····	····	····	1.0000	.9999	.9993	.9963
····	····	····	····	····	····	····	····	····	····	1.0000	1.0000	.9995
····	····	····	····	····	····	····	····	····	····	····	1.0000	1.0000
····	····	····	····	····	····	····	····	····	····	····	····	1.0000

APPENDIX P
Sample Computer Programs for Various Analyses

The following computer programs demonstrate how the step-by-step approach of this book can be applied to the use of BASIC for computer programming. These programs were written from the computational examples presented in this book and were verified on several different kinds of computers. They should work, with hardly any modification, on all computers that process BASIC. Only the most fundamental features of the BASIC language were used so that the programs would be as widely applicable as possible.

The programs should prove especially useful to the beginning programmer and also to the experienced programmer who is not familiar with statistical procedures. The programs demonstrate the basic computational steps that are part of all statistical analyses. While more efficient and elaborate programs could be written, the programs presented here demonstrate solutions to most types of problems that might be encountered in statistical programming.

P–1   Sums and sums of squares—the most common statistical computations.

```
10 PRINT "SUMS AND SUMS OF SQUARES"
20 PRINT :PRINT "DEMO OF THE FOR LOOP"
60 PRINT :PRINT "HOW MANY NUMBERS DO YOU HAVE?":INPUT N
75 DIM X(N),L$(10)
80 REM BEFORE STARTING TO ACCUMULATE THE NUMBERS, IT IS
90 REM NECCESSARY TO SET THE ACCUMULATERS TO ZERO.
100 SUM=0
110 SQ=0
120 PRINT "ENTER THE NUMBERS"
130 FOR I=1 TO N
142 INPUT X
150 SUM=SUM+X
160 SQ=SQ+X^2
170 NEXT I
175 PRINT :PRINT
180 PRINT "THE SUM =  ";SUM
185 PRINT
190 PRINT "THE SUM OF SQUARES =   ";SQ
195 PRINT
230 END
```

P-2   Chi-squared for goodness of fit.

```
10 PRINT "CHI-SQUARED TEST FOR GOODNESS OF FIT"
20 PRINT :PRINT "HOW MANY CATEGORIES?"
30 INPUT C
50 DIM O(C),E(C),A$(10)
60 PRINT :D=0
80 PRINT "ENTER THE OBSERVED VALUES"
82 PRINT
84 FOR I=1 TO C
98 INPUT O:O(I)=O:NEXT I
103 PRINT :PRINT "ENTER THE EXPECTED VALUES"
110 FOR I=1 TO C
122 INPUT E:E(I)=E:NEXT I
130 FOR I=1 TO C
140 D=D+(O(I)-E(I))^2/E(I)
150 NEXT I
160 DF=C-1
170 PRINT "CHI SQUARED = ";D
175 PRINT :PRINT "DF = ";DF
200 END
```

P-3   Mean and standard deviation   Section 1.2.

```
10 PRINT :PRINT "MEAN AND SD"
60 PRINT :PRINT "HOW MANY NUMBERS DO YOU HAVE?"
70 INPUT N
75 DIM X(N),L$(10)
80 REM BEFORE STARTING TO ACCUMULATE THE NUMBERS, IT IS
90 REM NECCESSARY TO SET THE ACCUMULATORS TO ZERO.
100 SUM=0
110 SQ=0
120 PRINT "ENTER THE NUMBERS"
121 PRINT
130 FOR I=1 TO N
144 INPUT X
145 X(I)=X
150 SUM=SUM+X(I)
160 SQ=SQ+X^2
170 NEXT I
180 PRINT "THE SUM = ";SUM
190 PRINT :PRINT "THE SUM OF SQUARES = ";SQ
192 MEAN=SUM/N
194 SD=SQR((SQ-SUM^2/N)/N)
196 PRINT :PRINT "THE MEAN = ";MEAN
198 PRINT :PRINT "THE STANDARD DEVIATION = ";SD
230 END
```

P-4   Chi-squared for independence.   Section 5.6.

```
10 PRINT "CHI SQUARED TEST FOR INDEPENDENCE"
20 PRINT :PRINT "ENTER THE NUMBER OF ROWS"
30 INPUT N1
40 PRINT :PRINT "ENTER THE NUMBER OF COLUMNS"
50 INPUT N2
60 DIM L$(10),O(N1,N2),R(N1),E(N1,N2),C(N2)
70 FOR I=1 TO N1:R(I)=0:FOR J=1 TO N2:O(I,J)=0:E(I,J)=0:NEXT J:NEXT I
75 FOR J=1 TO N2:C(J)=0:NEXT J
```

```
80 PRINT :DF=(N1-1)*(N2-1)
85 PRINT "ENTER THE VALUE FOR"
90 FOR I=1 TO N1:FOR J=1 TO N2
100 PRINT "ROW ";I;" COL ";J
110 INPUT O:O(I,J)=O:NEXT J:NEXT I
120 FOR I=1 TO N1:FOR J=1 TO N2
130 R(I)=R(I)+O(I,J):C(J)=C(J)+O(I,J):T1=T1+O(I,J)
140 NEXT J:NEXT I
150 FOR I=1 TO N1:FOR J=1 TO N2:E(I,J)=R(I)*C(J)/T1:NEXT J:NEXT I
160 FOR I=1 TO N1:FOR J=1 TO N2:CH=CH+(O(I,J)-E(I,J))^2/E(I,J)
170 NEXT J:NEXT I
180 PRINT :PRINT "CHI SQUARED = ";CH;"  DF = ";DF
190 END
```

P-5 Grouped frequency distribution, used with large distributions.

```
10 PRINT :PRINT "FREQUENCY DISTRIBUTION"
40 PRINT :PRINT "HOW MANY SCORES ARE THERE?"
50 INPUT N
70 DIM X(N+1),M(100),F(100),D(100),E(100),L$(10)
80 PRINT :PRINT "ENTER THE NUMBERS"
82 PRINT
84 FOR I=1 TO N
102 INPUT X
105 X(I)=X
110 NEXT I
115 S=0:REM RANK ORDER THE NUMBERS
120 FOR I=1 TO N-1
130 IF X(I)<=X(I+1) THEN 180
140 S=S+1
150 Z=X(I)
160 X(I)=X(I+1)
170 X(I+1)=Z
180 NEXT I
190 IF S>0 THEN 115
200 SUM=0
210 RAN=X(N)-X(1)+2
220 SQ=0
230 RA=INT(SQR(RAN))
240 IF RA/2<>INT(RA/2) THEN 270
250 G=RA-1
260 GOTO 280
270 G=RA:REM SELECT INTERVAL WIDTH
280 C=INT(RAN/G):REM AND NUMBER OF INTERVALS
290 PRINT
300 IF C*G<RAN-1 THEN C=C+1
320 FOR I=1 TO N
330 SUM=SUM+X(I)
340 SQ=SQ+X(I)^2
350 NEXT I
360 MEAN=SUM/N
363 PRINT "INTERVAL WIDTH = ";G
365 PRINT :PRINT "NUMBER OF INTERVALS = ";C
370 SD=SQR(ABS(SQ-SUM^2/N)/N)
375 PRINT
380 PRINT "THE MEAN = ";MEAN
390 PRINT :PRINT "THE SD = ";SD
400 T=X(1)
410 REM PREPARE INTERVAL END-POINTS
420 FOR I=0 TO C-1
430 D(I+1)=T+G*I
440 E(I+1)=T+G*(I+1)-1
```

```
450 F(I)=0
460 NEXT I
470 L=1
480 F(C)=0
490 FOR J=1 TO C
500 FOR I=L TO N
510 IF X(I)<=E(J) THEN F(J)=F(J)+1
520 NEXT I
530 L=L+F(J)
540 NEXT J
550 PRINT
560 PRINT "THE GROUPED FREQUENCY TABLE."
570 PRINT
580 PRINT "   INTERVALS   MIDPOINTS   FREQUENCIES"
600 FOR I=1 TO C
610 M(I)=D(I)+INT(G/2):REM COMPUTE MIDPOINTS
615 NEXT I
617 FOR I=1 TO C
618 IF E(I)>=10 AND M(I)<10 THEN 637
620 IF D(I)<10 AND E(I)<10 THEN 635
625 IF D(I)<10 THEN 633
630 PRINT "    ";D(I);" - ";E(I);"         ";M(I);"           ";F(I)
632 GOTO 640
633 PRINT "     ";D(I);" - ";E(I);"         ";M(I);"          ";F(I)
634 GOTO 640
635 PRINT "     ";D(I);" - ";E(I);"          ";M(I);"           ";F(I)
636 GOTO 640
637 PRINT "     ";D(I);" - ";E(I);"          ";M(I);"           ";F(I)
640 NEXT I
680 END
```

P-6 Rank ordered numbers, from small to large.

```
5 PRINT "RANK ORDERED DISTRIBUTION"
7 PRINT
10 PRINT "HOW MANY NUMBERS ARE THERE?"
20 INPUT N
30 DIM X(N+1),R(N+1),L$(10)
40 PRINT "ENTER THE SCORES"
42 PRINT
50 FOR I=1 TO N
66 INPUT X:X(I)=X
70 NEXT I
80 S=0
90 FOR I=1 TO N-1
100 IF X(I)<=X(I+1) THEN 130
110 S=S+1:Z=X(I)
120 X(I)=X(I+1):X(I+1)=Z
130 NEXT I
140 IF S>0 THEN 80
150 D=0
160 M=0
170 FOR I=1 TO N
180 IF X(1)<X(2) THEN R(1)=1
190 IF X(I)<>X(I+1) AND I>1 AND X(I)>X(I-1) THEN R(I)=I:GOTO 250
200 D=D+I:M=M+1
210 IF X(I)=X(I+1) THEN 250
220 R(I)=D/M:FOR J=I-(M-1) TO I:R(J)=R(I):NEXT J
230 IF X(I)=X(I+1) THEN 250
240 D=0:M=0
250 NEXT I
```

```
255 IF X(N)>X(N-1) THEN R(N)=N
270 PRINT "SCORE    RANK"
280 PRINT
290 FOR I=1 TO N
300 PRINT X(I),R(I)
310 NEXT I
350 END
```

P-7   Mann-Whitney U test.   Section 5.3.

```
10 PRINT "MANN-WHITNEY U TEST"
20 PRINT :PRINT "ENTER N FOR THE SMALLER GROUP"
40 INPUT N1
50 PRINT :PRINT "ENTER N FOR THE LARGER GROUP"
60 INPUT N2
70 DIM X(N1+1),Y(N1+N2+1),R(N1+N2+1),R1(N2+1),R2(N2+1),L(N1+N2+1),L$(10)
80 PRINT :PRINT "ENTER THE SCORES FOR GROUP 1"
82 PRINT
84 FOR I=1 TO N1
98 INPUT X:X(I)=X
102 NEXT I
104 PRINT :PRINT "ENTER THE SCORES FOR GROUP 2"
105 PRINT
107 FOR I=N1+1 TO N1+N2
124 INPUT Y:Y(I)=Y
129 NEXT I
131 FOR I=1 TO N1:L(I)=X(I):NEXT I
135 FOR I=N1+1 TO N1+N2:L(I)=Y(I):NEXT I
140 S=0
150 FOR I=1 TO N1+N2-1
160 IF L(I)<=L(I+1) THEN 190
170 S=S+1:Z=L(I)
180 L(I)=L(I+1):L(I+1)=Z
190 NEXT I
195 IF S>0 THEN 140
200 D=0:M=0
230 FOR I=1 TO N1+N2
240 IF L(1)<L(2) THEN R(1)=1
250 IF L(I)<>L(I+1) AND I>1 AND L(I)>L(I-1) THEN R(I)=I:GOTO 310
260 D=D+I:M=M+1
270 IF L(I)=L(I+1) THEN 310
280 R(I)=D/M:FOR J=I-(M-1) TO I:R(J)=R(I):NEXT J
290 IF L(I)=L(I+1) THEN 310
300 D=0:M=0
310 NEXT I
320 FOR I=1 TO N1:FOR J=1 TO N1+N2
330 IF X(I)=L(J) THEN R1(I)=R(J)
340 NEXT J
350 NEXT I
360 PRINT :PRINT
370 P1=N1*N2
380 P2=((N1+1)*N1)/2
390 FOR I=1 TO N1:P3=P3+R1(I):NEXT I
400 U=P1+P2-P3
410 U1=P1-U
415 IF N1>20 OR N2>20 THEN 460
420 PRINT " U=";U;" U1=";U1
430 PRINT :PRINT "TAKE THE SMALLER OF U AND U1 AND LOOK IN APPENDIX H TO EVALUAT
E"
440 GOTO 580
460 Q1=P1/2
```

```
470 Q2=N1+N2+1
480 Q3=P1*Q2
490 Q4=SQR(Q3/12)
500 Q5=Q1/Q4
510 PRINT
520 PRINT "SINCE ONE N IS LARGER THAN 20 USE Z TO EVALUATE"
525 PRINT :PRINT "Z = ";Q5
527 PRINT
530 PRINT "LOOK IN APPENDIX A TO EVALUATE"
580 END
```

P–8   T-test for independent groups.   Section 1.6.

```
10 PRINT "THE INDEPENDENT T-TEST"
30 PRINT
50 PRINT "WHAT IS THE N FOR GROUP 1?"
60 INPUT N1
70 PRINT
80 PRINT "WHAT IS THE N FOR GROUP 2?"
90 INPUT N2
95 DIM X1(N1),X2(N2),P$(10)
100 S1=0:S2=0:S3=0:S4=0:PRINT
150 PRINT "ENTER DATA FOR GROUP 1"
160 PRINT :FOR I=1 TO N1:INPUT X1
170 S1=S1+X1:S2=S2+X1^2:NEXT I
230 PRINT :PRINT "ENTER THE DATA FOR GROUP 2"
240 PRINT :FOR I=1 TO N2:INPUT X2
260 S3=S3+X2:S4=S4+X2^2:NEXT I
290 Q3=S2-(S1^2/N1)
300 G3=S4-(S3^2/N2)
310 T1=Q3+G3
320 D5=N1+N2-2
330 T2=T1/D5
340 D=T2*((1/N1)+(1/N2))
350 D=SQR(D)
360 M1=S1/N1:M2=S3/N2
380 D1=M1-M2:D2=D1/D
410 PRINT :PRINT "MEAN FOR GROUP 1 = ";M1
415 PRINT :PRINT "MEAN FOR GROUP 2 = ";M2
420 PRINT
430 PRINT "THE T VALUE = ";D2
440 PRINT :PRINT "THE DF = ";D5
450 END
```

P–9   T-test for matched groups.   Section 1.7.

```
10 PRINT "T-RATIO FOR RELATED DATA"
15 DIM S$(10)
30 PRINT
40 PRINT "HOW MANY PAIRS OF MEASURES?"
50 INPUT K
60 PRINT
70 PRINT "ENTER THE PAIRS OF NUMBERS, BOTH ON THE SAME LINE"
80 PRINT :PRINT "PUT A COMMA BETWEEN THE NUMBERS"
90 REM SET THE ACCUMULATORS TO ZERO.
100 D=0:DS=0:DSQ=0:X1S=0:X2S=0
128 FOR I=1 TO K
153 INPUT X1,X2
170 D=X1-X2:DS=DS+D:DSQ=DSQ+D^2
```

```
200 X1S=X1S+X1
210 X2S=X2S+X2
220 NEXT I
230 M1=X1S/K:M2=X2S/K:NSD=K*DSQ
260 SQD=DS^2:DIF=NSD-SQD:DF=K-1
290 RAT=DIF/DF
300 RA=SQR(RAT)
310 T=DS/RA
320 PRINT :PRINT "THE T-VALUE = ";T,"DF = ";DF
330 PRINT :PRINT "THE GROUP 1 MEAN = ";M1
335 PRINT :PRINT "THE GROUP 2 MEAN = ";M2
340 END
```

P-10   The correlation coefficient.   Section 4.1.

```
10 PRINT "CORRELATION PROGRAM."
20 PRINT
30 PRINT "HOW MANY PAIRS OF NUMBERS DO YOU HAVE?"
40 INPUT N
50 REM INITIALIZE THE SUMMATION REGISTERS.
60 A=0:B=0:A1=0:B1=0:AB=0
110 PRINT "ENTER THE PAIRS OF NUMBERS; ON THE SAME LINE; WITH A COMMA BETWEEN TH
EM"
120 PRINT :FOR I=1 TO N:INPUT X,Y
150 A=A+X:B=B+Y:A1=A1+X^2
180 B1=B1+Y^2:AB=AB+X*Y:NEXT I
210 AB=N*AB:A3=AB-A*B:B3=A1*N-A^2
250 A4=B1*N-B^2:B4=B3*A4:A5=SQR(B4)
270 R=A3/A5
280 PRINT "THE CORRELATION = ";R
290 PRINT
300 T=R*SQR((N-2)/(1-R^2))
310 DF=N-2
320 PRINT "T = ";T;"  DF = ";DF
330 END
```

P-11   Intercorrelations—many variables.

```
10 PRINT "INTERCORRELATIONS - MANY VARIABLES"
20 PRINT :PRINT "INPUT N - THE NUMBER OF PEOPLE"
30 INPUT N
40 PRINT :PRINT "INPUT K - THE NUMBER OF VARIABLES"
50 INPUT K
80 DIM A(K),S(K),R(K,K)
90 DIM X(K),L$(10)
100 FOR I=1 TO K
110 A(I)=0:S(I)=0
130 FOR J=1 TO K
140 R(I,J)=0:NEXT J:NEXT I
170 PRINT :PRINT "ENTER THE SCORES"
180 FOR I=1 TO N
190 FOR J=1 TO K
210 PRINT :PRINT "DATA FOR S ";I
215 FOR J=1 TO K
220 INPUT X:X(J)=X:A(J)=A(J)+X
250 S(J)=S(J)+X^2:NEXT J
270 FOR J=1 TO K:FOR M=J TO K
290 R(J,M)=R(J,M)+X(J)*X(M)
300 NEXT M:NEXT J:NEXT I
```

```
330 FOR I=1 TO K
340 A(I)=A(I)/N
350 S(I)=SQR(R(I,I)/N-(A(I)^2))
360 NEXT I:PRINT
380 FOR I=1 TO K:FOR J=I TO K
400 IF S(I)*S(J)=0 THEN 420
410 R(J,I)=((R(I,J)/N)-A(I)*A(J))/(S(I)*S(J))
420 R(I,J)=R(J,I):NEXT J
440 R(I,I)=1:NEXT I
460 FOR I=1 TO K:FOR J=I TO K
480 R(I,J)=INT(R(I,J)*100)
490 R(I,J)=R(I,J)/100
500 NEXT J:NEXT I
520 PRINT "THE CORRELATIONS"
530 PRINT
540 FOR I=1 TO K
550 FOR J=I TO K
555 IF I=J THEN 570
560 PRINT "R ";I;" ";J;" = ";R(I,J):?
570 NEXT J:NEXT I
580 END
```

P–12   Percentile ranks and z-scores.   Common descriptive statistics.

```
10 PRINT "PERCENTILE RANKS AND Z SCORES"
50 PRINT
60 PRINT "ENTER N, THE NUMBER OF NUMBERS."
70 INPUT N
80 DIM X(N+1),P(N),Z(N),L$(10)
90 PRINT :PRINT "ENTER THE NUMBERS"
100 FOR I=1 TO N:INPUT X:X(I)=X
122 NEXT I
124 S=0
129 FOR I=1 TO N-1
130 IF X(I)<=X(I+1) THEN 180
140 S=S+1:Z=X(I):X(I)=X(I+1)
170 X(I+1)=Z
180 NEXT I
190 IF S>0 THEN 124
200 SUM=0:SQ=0:FOR I=1 TO N:SUM=SUM+X(I)
240 SQ=SQ+X(I)^2:NEXT I
260 FOR I=1 TO N-1:A=1:B=I-1
290 IF X(I)=X(I+1) THEN 320
300 P(I)=INT((B/N+0.5*(A/N))*100)
310 GO TO 430
320 L=I:Q=I
340 A=A+1:L=L+1
360 IF L=N THEN 390
370 Q=Q+1
380 IF X(L)=X(L+1) THEN 340
390 FOR L=I TO Q
400 P(L)=INT((B/N+0.5*(A/N))*100)
410 NEXT L:I=Q
430 NEXT I
440 IF X(N-1)<>X(N) THEN A=1
450 IF X(N-1)<>X(N) THEN B=N-1
460 P(N)=INT((B/N+0.5*(A/N))*100)
470 MEAN=SUM/N
480 SD=SQR((SQ-SUM*2/N)/N)
490 PRINT
500 PRINT "THE MEAN = ";MEAN
```

```
502 PRINT :PRINT "THE STANDARD DEVIATION = ";SD
510 FOR I=1 TO N
520 Z(I)=(X(I)-MEAN)/SD
530 NEXT I:PRINT :PRINT
540 PRINT
560 PRINT
570 PRINT "  X     PERCENTILE     Z-SCORE"
600 PRINT :FOR I=1 TO N
605 IF X(I)=X(I-1) THEN 620
610 PRINT "  ";X(I),P(I),Z(I)
620 NEXT I:END
```

P-13   Completely randomized design.   Section 2.1.

```
10 PRINT "SIMPLE RANDOM ANOVA"
20 PRINT
40 PRINT "HOW MANY GROUPS (TREATMENTS)?"
50 INPUT NG
55 DIM N(NG),T(NG),AM(NG),L$(10)
60 FOR I=1 TO NG:T(I)=0:NEXT I
90 SB=0:TOT=0:SQ=0:SS=0:PRINT
140 PRINT "ENTER  THE GROUP N VALUES"
150 FOR I=1 TO NG:INPUT N:N(I)=N:NEXT I:PRINT
200 PRINT "ENTER THE DATA"
210 FOR I=1 TO NG
220 PRINT :PRINT "DATA FOR GROUP ";I
240 L=N(I):FOR J=1 TO L
250 SS=SS+1:INPUT X:T(I)=T(I)+X
270 SQ=SQ+X^2:NEXT J:NEXT I
300 FOR I=1 TO NG
310 TOT=TOT+T(I):NT=NT+N(I):NEXT I
340 C=TOT^2/NT:SST=SQ-C:FOR I=1 TO NG
370 SB=SB+T(I)^2/N(I):AM(I)=T(I)/N(I):NEXT I
400 SSB=SB-C:SSW=SST-SSB:DFT=NT-1:DFB=NG-1:DFW=DFT-DFB:MSB=SSB/DFB
460 MSW=SSW/DFW:F=MSB/MSW:Y=DFB:Y1=DFW:Y2=F:GOSUB 690:PRINT
470 F=MSB/MSW
480 Y=DFB
490 Y1=DFW
530 PRINT "THE SUMMARY TABLE"
540 PRINT
550 PRINT "TOTAL SS = ";SST;"  DF = ";DFT
560 PRINT
570 PRINT "TMTS  SS = ";SSB;"  DF = ";DFB
575 PRINT :PRINT " MS = ";MSB;"  F = ";F
577 PRINT :PRINT " PROB = ";P
580 PRINT
590 PRINT "ERROR SS = ";SSW;"  DF = ";DFW
595 PRINT :PRINT " MS = ";MSW
600 PRINT :PRINT
610 PRINT "THE GROUP MEANS"
620 PRINT
630 FOR I=1 TO NG
640 PRINT "  ";AM(I);
650 NEXT I
655 PRINT
660 GOTO 880
670 PRINT
680 REM PROBABILITY COMPUTATION
690 P=1
700 IF Y*Y1*Y2=0 THEN 860
710 IF Y2<1 THEN 760
720 Q=Y:U=Y1:F1=Y2
```

```
750 GOTO 790
760 Q=Y1:U=Y:F1=1/Y2
790 U1=2/(9*Q)
800 B9=2/(9*U)
810 Z=ABS(((((1-B9)*F1^0.333333)-1+U1)/SQR((B9*F1^0.666667)+U1))
820 IF U<4 THEN Z=Z*(1+(0.08*Z^4/U^3))
830 P=0.5/(1+Z*(0.196854+Z*(0.115194+Z*(3.44E-04+Z*0.019527)))))^4
840 IF F1<1 THEN P=1-P
850 GOTO 870
860 PRINT "THERE IS A PROBLEM WITH DEGREES OF FREEDOM OR F VALUES."
870 RETURN
880 END
```

P-14   Factorial design—two factors.   Section 2.2.

```
10 PRINT "A X B ANOVA"
50 PRINT
60 PRINT "HOW MANY ROW TREATMENTS?"
70 INPUT A
80 PRINT "HOW MANY COLUMN TREATMENTS?"
90 INPUT B
100 DIM Q(A,B),R1(B),C1(A)
110 DIM N1(A,B),N2(A),N3(B),P(B),P1(A),P2(A,B),L$(10)
130 FOR I=1 TO A:C1(I)=0:N2(I)=0:FOR J=1 TO B:Q(I,J)=0:R1(J)=0:N3(J)=0:NEXT J:NE
XT I
240 C=0:T=0:M=0:C5=0:H1=0:H2=0
284 PRINT :PRINT "ENTER THE CELL N VALUES"
290 PRINT :FOR I=1 TO A:FOR J=1 TO B
297 PRINT "FOR CELL ";I;" ";J
320 INPUT N1:N1(I,J)=N1:NEXT J:NEXT I
325 PRINT :NN=0
330 FOR I=1 TO A:FOR J=1 TO B:NN=NN+N1(I,J):NEXT J:NEXT I
340 PRINT "ENTER THE DATA"
350 DIM X(NN):DD=0
360 FOR I=1 TO A:FOR J=1 TO B
370 PRINT "FOR CELL ";I;" ";J
380 K=N1(I,J):FOR L3=1 TO K:DD=DD+1:INPUT X:M=M+1
440 Q(I,J)=Q(I,J)+X:C=C+X:T=T+X^2:NEXT L3
490 C5=C5+Q(I,J)^2/N1(I,J)
500 N2(I)=N2(I)+N1(I,J)
510 N3(J)=N3(J)+N1(I,J)
512 NEXT J:NEXT I
548 FOR I=1 TO B
550 H2=H2+R1(I)^2/N3(I)
552 NEXT I
556 FOR I=1 TO A:FOR J=1 TO B
560 C1(I)=C1(I)+Q(I,J)
570 R1(J)=R1(J)+Q(I,J):NEXT J
590 H1=H1+C1(I)^2/N2(I):NEXT I
610 FOR J=1 TO B
620 H2=H2+R1(J)^2/N3(J):NEXT J
640 C=C^2/M:T=T-C:A1=H2-C
670 IF B=1 THEN GOTO 690
680 B1=H1-C
690 C5=C5-C
700 IF B=1 THEN GOTO 730
710 L=C5-A1-B1
720 GOTO 740
730 L=C5-A1
740 E=T-C5:D1=M-1:D2=A-1:D3=B-1:D4=D2*D3:D5=D1-D2-D3-D4:M2=A1/D2
810 IF B=1 THEN GOTO 840
820 M3=B1/D3:M4=L/D4
```

```
840 M5=E/D5
850 F8=M2/M5:Y=D2:Y1=D5:Y2=F8:GOSUB 1470:PP1=G
910 IF B=1 THEN GOTO 1050
920 F2=M3/M5:Y=D3:Y1=D5:Y2=F2:GOSUB 1470:PP2=G
980 F3=M4/M5:Y=D4:Y1=D5:Y2=F3:GOSUB 1470:PP3=G
1050 PRINT
1060 PRINT :PRINT "THE SUMMARY TABLE OF THE RESULTS"
1100 PRINT :PRINT :PRINT "TOTAL SS = ";T;"  DF = ";D1
1110 PRINT
1120 PRINT "ROWS  SS = ";B1;"  DF = ";D3
1127 PRINT " MS = ";M3;"  F = ";F2
1128 PRINT " P = ";PP2
1130 PRINT
1160 PRINT "COLS  SS = ";A1;"  DF = ";D2
1165 PRINT " MS = ";M2;"  F = ";F8
1167 PRINT " P = ";PP1
1180 PRINT :PRINT "R C  SS = ";L;"   DF = ";D4
1185 PRINT " MS = ";M4;"  F = ";F3
1187 PRINT " P = ";PP3
1200 PRINT :PRINT "ERROR SS = ";E;"   DF = ";D5
1205 PRINT " MS = ";M5
1210 PRINT
1212 PRINT "PRESS RETURN FOR THE MEANS."
1214 INPUT L$
1230 PRINT :PRINT "THE COL MEANS":PRINT
1260 FOR I=1 TO B:P(I)=R1(I)/N3(I):NEXT I
1280 FOR I=1 TO B:PRINT P(I);" ";
1285 NEXT I
1287 PRINT :PRINT
1290 IF B=1 THEN 1450
1300 PRINT "THE ROW MEANS"
1320 PRINT :FOR J=1 TO A
1330 P1(J)=C1(J)/N2(J)
1340 PRINT P1(J);"  ";
1360 NEXT J:PRINT :PRINT
1365 IF B=1 THEN 1450
1370 PRINT "THE CELL MEANS"
1380 PRINT
1400 FOR I=1 TO A:FOR J=1 TO B
1410 P2(I,J)=Q(I,J)/N1(I,J)
1420 PRINT P2(I,J);"  ";
1430 NEXT J:PRINT :NEXT I:GOTO 1660
1460 REM PROBABILITY FUNCTION.
1470 G=1:IF Y2=0 THEN 1650:IF Y2<1 THEN 1540:Q1=Y
1510 Q1=Y:U=Y1:F1=Y2
1530 GOTO 1570
1540 Q1=Y1:U=Y:F1=1/Y2
1570 U1=2/(9*Q1)
1580 B9=1/(9*U)
1590 Z=ABS((((1-B9)*F1^0.333333)-1+U1)/SQR((B9*F1^0.666667)+U1))
1600 IF U<4 THEN Z=Z*(1+(0.08*Z^4/U^3))
1610 G=0.5/(1+Z*(0.196854+Z*(0.115194+Z*(3.44E-04+Z*0.019527))))^4
1620 IF F1<1 THEN G=1-G
1630 GOTO 1650
1650 RETURN
1660 END
```

P–15   Treatments-by-treatments-by-subjects design.   Section 2.6.

```
10 PRINT "T BY T BY S ANOVA"
30 PRINT
40 PRINT "HOW MANY TREATMENTS ON THE FIRST VARIABLE?"
50 INPUT N1
60 PRINT "HOW MANY TREATMENTS ON THE SECOND VARIABLE?"
70 INPUT N2
80 PRINT "HOW MANY SUBJECTS?"
90 INPUT N3
100 DIM X(N1,N2),S(N3),MA(N1),B(N2),C(N1,N2),A(N1)
110 DIM A1(N1,N3),B1(N2,N3),M2(N2),M1(N1),M3(N3)
115 DIM L$(10),E(N1*N2)
120 DIM M4(N1,N2),M5(N2,N3),M6(N1,N3)
130 FOR I=1 TO N1:FOR J=1 TO N2:FOR K=1 TO N3
160 A(I)=0:B(J)=0:S(K)=0:C(I,J)=0:A1(I,K)=0:B1(J,K)=0:NEXT K:NEXT J:NEXT I
250 S10=0:S9=0:S12=0:S13=0:S14=0:S15=0:S16=0:S17=0
340 PRINT :PRINT "ENTER THE DATA"
360 G=N1*N2
370 DIM R(G)
380 FOR K=1 TO N3
385 PRINT :PRINT "SUBJECT ";K
390 FOR E=1 TO G:INPUT R:R(E)=R:NEXT E
430 L=0
440 FOR I=1 TO N1:FOR J=1 TO N2:L=L+1
470 X(I,J)=R(L):NEXT J:NEXT I
510 FOR I=1 TO N1:FOR J=1 TO N2
530 S10=S10+X(I,J)
540 S9=S9+X(I,J)^2
550 A(I)=A(I)+X(I,J)
560 B(J)=B(J)+X(I,J)
570 C(I,J)=C(I,J)+X(I,J)
580 S(K)=S(K)+X(I,J)
590 A1(I,K)=A1(I,K)+X(I,J)
600 B1(J,K)=B1(J,K)+X(I,J)
610 NEXT J:NEXT I:NEXT K
640 REM COMPUTE THE MEANS.
650 FOR I=1 TO N1
660 M1(I)=A(I)/(N2*N3)
670 FOR J=1 TO N2
680 M4(I,J)=C(I,J)/N3:NEXT J:NEXT I
710 FOR J=1 TO N2
720 M2(J)=B(J)/(N1*N3):NEXT J
740 REM COMPUTE THE SUMS OF SQUARES.
750 FOR K=1 TO N3
760 S12=S12+S(K)^2
770 FOR I=1 TO N1
780 S16=S16+A1(I,K)^2:NEXT I:NEXT K
810 FOR J=1 TO N2
820 S13=S13+B(J)^2
830 FOR I=1 TO N1
840 S15=S15+C(I,J)^2:NEXT I:NEXT J
880 FOR I=1 TO N1:S14=S14+A(I)^2:NEXT I
900 FOR K=1 TO N3:FOR J=1 TO N2
920 S17=S17+B1(J,K)^2:NEXT J:NEXT K
950 C1=S10^2/(N1*N2*N3):T1=S9-C1
970 Q1=(S12/(N2*N1))-C1
980 Q2=(S13/(N3*N1))-C1
990 Q3=(S14/(N3*N2))-C1
1000 Q4=(S15/N3)-C1-Q2-Q3
1010 Q5=(S16/N2)-C1-Q3-Q1
1020 Q6=(S17/N1)-C1-Q2-Q1
1030 Q7=T1-Q1-Q2-Q3-Q4-Q5-Q6
1040 REM COMPUTE THE DF AND MEAN SQUARES.
```

```
1050 D1=N1-1:D2=N2-1:D3=N3-1:D4=D1*D2:D5=D1*D3:D6=D2*D3
1110 D7=D6*D1:Y1=Q3/D1:Y2=Q2/D2:Y3=Q1/D3:Y4=Q4/D4:Y5=Q5/D5:Y6=Q6/D6:Y7=Q7/D7
1190 REM COMPUTE THE F RATIONS
1200 F1=Y1/Y5:F2=Y2/Y6:F4=Y4/Y7
1230 REM COMPUTE PROBABILITIES.
1240 E=D1:E1=D5:E2=F1:GOSUB 1770:P1=G7
1290 E=D2:E1=D6:E2=F2:GOSUB 1770:P2=G7
1340 E=D4:E1=D7:E2=F4:GOSUB 1770:P3=G7:PRINT :PRINT
1410 PRINT "THE SUMMARY TABLE OF THE RESULTS"
1430 PRINT
1435 D11=N1*N2*N3-1
1440 PRINT "TOTAL  SS= ";T1;" DF= ";D11
1450 PRINT :PRINT "  S BET  SS= ";Q1;" DF= ";D3
1470 PRINT :PRINT "VAR 1  SS= ";Q3;" DF= ";D1;" MS= ";Y1;" F= ";F1;" P= ";P1
1472 PRINT
1475 PRINT "  VAR 2  SS= ";Q2;" DF= ";D2;" MS= ";Y2;" F= ";F2;" P= ";P2
1477 PRINT
1480 PRINT "V1xV2  SS= ";Q4;" DF= ";D4;" MS= ";Y4;" F= ";F4;" P= ";P3
1482 PRINT
1485 PRINT "E1 SS= ";Q5;" DF= ";D5;" MS= ";Y5
1487 PRINT
1490 PRINT "E2 SS= ";Q6;" DF= ";D6;" MS= ";Y6
1492 PRINT
1495 PRINT "E12 SS= ";Q7;" DF= ";D7;" MS= ";Y7
1496 PRINT
1500 PRINT "PRESS RETURN FOR THE MEANS"
1505 INPUT L$
1560 PRINT "THE MEANS FOR THE FIRST VARIABLE"
1580 PRINT :FOR I=1 TO N1
1590 PRINT M1(I),
1600 NEXT I
1620 PRINT :PRINT :PRINT "THE MEANS FOR THE SECOND VARIABLE":PRINT
1640 FOR I=1 TO N2
1650 PRINT M2(I),
1660 NEXT I
1680 PRINT :PRINT :PRINT "THE CELL MEANS":PRINT
1700 FOR I=1 TO N1:FOR J=1 TO N2
1720 PRINT M4(I,J),
1730 NEXT J:PRINT :NEXT I
1750 GOTO 1960
1760 REM PROBABILITY FUNCTION.
1770 G7=1
1780 IF E*E1*E2=0 THEN 1940
1790 IF E2<1 THEN 1840
1800 H1=E:H=E1:H2=E2
1830 GOTO 1870
1840 H1=E1:H=E:H2=1/E2
1870 H3=2/(9*H1)
1880 B9=2/(9*H)
1890 Z=ABS((((1-B9)*H2^0.333333)-1+H3)/SQR((B9*H2^0.666667)+H3))
1900 IF H<4 THEN Z=Z*(1+(0.08*Z^4/H^3))
1910 G7=0.5/(1+Z*(0.196854+Z*(0.115194+Z*(3.44E-04+Z*0.019527))))^4
1920 IF H2<1 THEN G7=1-G7
1930 GOTO 1950
1940 PRINT "THERE IS A PROBLEM WITH DF OR F VALUES."
1950 RETURN
1960 END
```

P–16   Factorial design: Three factors.   Section 2.3.

```
10 PRINT "ABC ANOVA"
90 PRINT
100 PRINT "HOW MANY TREATMENTS ON THE OUTSIDE VAR?"
110 INPUT A1
120 PRINT "HOW MANY TREATMENTS ON THE MIDDLE VAR?"
130 INPUT A2
140 PRINT "HOW MANY TREATMENTS ON THE INSIDE VAR?"
150 INPUT A3
155 L=A1*A2*A3
160 DIM S2(A1),S3(A2),S4(A3),S5(A2,A3),S6(A1,A3),S7(A1,A2)
170 DIM S8(L),N(L),N1(A1),N2(A2),N3(A3),N4(A2,A3),L$(10)
180 DIM N5(A1,A3),N6(A1,A2),G1(L)
190 REM INITIALIZE THE SUMMATION REGISTERS.
200 FOR K=1 TO A1:S2(K)=0:N1(K)=0:FOR J=1 TO A2:N2(J)=0:S3(J)=0:FOR I=1 TO A3:N3
(I)=0:S4(I)=0:N4(J,I)=0
300 S5(J,I)=0:N5(K,I)=0:S6(K,I)=0:N6(K,J)=0:S7(K,J)=0:NEXT I:NEXT J:NEXT K
320 N8=A1*A2*A3
380 FOR L=1 TO N8:G1(L)=0:NEXT L
410 T=0:B1=0:B2=0:B3=0:B4=0:B5=0:Q4=0:B6=0:B7=0
510 PRINT "ENTER THE CELL N VALUES"
512 PRINT
530 FOR I=1 TO N8
534 PRINT "CELL ";I
540 INPUT N:N(I)=N:NEXT I
570 FOR L=1 TO N8:Q4=Q4+N(L):NEXT L
610 L=0:FOR K=1 TO A1:FOR J=1 TO A2:FOR I=1 TO A3
650 L=L+1:N1(K)=N1(K)+N(L):N2(J)=N2(J)+N(L):N3(I)=N3(I)+N(L)
680 N4(J,I)=N4(J,I)+N(L)
690 N5(K,I)=N5(K,I)+N(L)
700 N6(K,J)=N6(K,J)+N(L):NEXT I:NEXT J:NEXT K
760 L=0:LLL=0:DIM X(Q4)
770 FOR K=1 TO A1:FOR J=1 TO A2:FOR I=1 TO A3
810 L=L+1:IF L>1 THEN 840
820 PRINT :PRINT "ENTER THE DATA"
840 A4=N(L)
860 PRINT :PRINT "CELL ";L
880 FOR H=1 TO A4
881 LLL=LLL+1
900 INPUT X:X(H)=X
910 S1=S1+X(H)^2
920 S2(K)=S2(K)+X(H)
930 S3(J)=S3(J)+X(H)
940 S4(I)=S4(I)+X(H)
950 S5(J,I)=S5(J,I)+X(H)
960 S6(K,I)=S6(K,I)+X(H)
970 S7(K,J)=S7(K,J)+X(H)
980 G1(L)=G1(L)+X(H)
990 NEXT H:NEXT I:NEXT J:NEXT K
1030 REM COMPUTE THE SUMS OF SQUARES.
1040 FOR I=1 TO A1:T=T+S2(I):NEXT I
1070 C=T^2/Q4:T=S1-C
1100 FOR I=1 TO A1:B1=B1+S2(I)^2/N1(I):NEXT I:B1=B1-C
1140 FOR I=1 TO A2:B2=B2+S3(I)^2/N2(I):NEXT I:B2=B2-C
1180 FOR I=1 TO A3:B3=B3+S4(I)^2/N3(I):NEXT I:B3=B3-C
1210 FOR I=1 TO A2:FOR J=1 TO A3
1230 B4=B4+S5(I,J)^2/N4(I,J)
1240 NEXT J:NEXT I:B4=B4-C-B2-B3
1270 FOR I=1 TO A1:FOR J=1 TO A3
1290 B5=B5+S6(I,J)^2/N5(I,J)
1300 NEXT J:NEXT I:B5=B5-C-B1-B3
1330 FOR I=1 TO A1:FOR J=1 TO A2
1340 FOR J=1 TO A2
```

```
1350 B6=B6+S7(I,J)^2/N6(I,J)
1360 NEXT J:NEXT I:B6=B6-C-B1-B2
1390 FOR L=1 TO N8
1400 B7=B7+G1(L)^2/N(L):NEXT L
1420 B7=B7-C-B1-B2-B3-B4-B5-B6
1430 B8=T-B1-B2-B3-B4-B5-B6-B7
1440 REM COMPUTE THE DEGREES OF FREEDOM.
1450 D1=Q4-1:D2=A1-1:D3=A2-1:D4=A3-1:D5=D3*D4:D6=D2*D4
1510 D7=D2*D3:D8=D2*D3*D4
1530 D9=D1-(D2+D3+D4+D5+D6+D7+D8)
1540 REM COMPUTE THE MEAN SQUARES AND F VALUES
1550 M1=B1/D2:M9=B8/D9:M2=B2/D3:M3=B3/D4:M4=B4/D5:M5=B5/D6
1610 M6=B6/D7:M7=B7/D8:F8=M1/M9
1640 Y=D2:Y1=D9:Y2=F8:GOSUB 2830:P1=P
1690 F2=M2/M9:Y=D3:Y1=D9:Y2=F2:GOSUB 2830:P2=P
1750 F3=M3/M9:Y=D4:Y1=D9:Y2=F3:GOSUB 2830:P3=P
1810 F4=M4/M9:Y=D5:Y1=D9:Y2=F4:GOSUB 2830:P4=P
1870 F5=M5/M9:Y=D6:Y1=D9:Y2=F5:GOSUB 2830:P5=P
1930 F6=M6/M9:Y=D7:Y1=D9:Y2=F6:GOSUB 2830:P6=P
1990 F7=M7/M9:Y=D8:Y1=D9:Y2=F7:GOSUB 2830:P7=P
2050 REM COMPUTE ROW, COLUMN,SLICE AND CELL MEANS.
2060 FOR J=1 TO A1
2070 S2(J)=S2(J)/N1(J):NEXT J
2090 FOR I=1 TO A2
2100 S3(I)=S3(I)/N2(I):NEXT I
2120 FOR I=1 TO A3
2130 S4(I)=S4(I)/N3(I):NEXT I
2150 FOR L=1 TO N8
2160 G1(L)=G1(L)/N(L):NEXT L
2210 PRINT "THE A MEANS"
2220 PRINT
2230 FOR I=1 TO A1
2240 PRINT S2(I),
2250 NEXT I:PRINT :PRINT
2280 PRINT "THE B MEANS"
2300 PRINT :FOR I=1 TO A2
2310 PRINT S3(I),
2320 NEXT I:PRINT :PRINT
2350 PRINT "THE C MEANS"
2370 PRINT :FOR I=1 TO A3
2380 PRINT S4(I),
2390 NEXT I:PRINT
2410 PRINT
2420 DIM G2(A1,A2),VV$(20)
2430 PRINT "PRESS RETURN WHEN YOU WANT MORE"
2440 L=0
2450 INPUT VV$
2460 FOR I=1 TO A3
2480 PRINT "THE BC CELL MEANS FOR A",I
2490 FOR K=1 TO A1:FOR J=1 TO A2
2520 L=L+1:G2(K,J)=G1(L):NEXT J
2530 NEXT K:PRINT
2560 FOR R=1 TO A1:FOR N=1 TO A2
2580 PRINT G2(R,N),
2590 NEXT N:PRINT
2610 PRINT :NEXT R:NEXT I
2650 PRINT :PRINT "PRESS RETURN WHEN YOU ARE READY FOR THE SUMMARY TABLE."
2660 INPUT VV$
2680 PRINT "SUMMARY TABLE FOR THE ANOVA - ABC"
2700 PRINT :PRINT "TOTAL SS= ";T;" DF= ";D1
2710 PRINT "A     SS= ";B1;" DF= ";D2;" MS= ";M1;" F= ";F8;" PROB= ";P1
2720 PRINT "B     SS= ";B2;" DF= ";D3;" MS= ";M2;" F= ";F2;" PROB= ";P2
2730 PRINT "C     SS= ";B3;" DF= ";D4;" MS= ";M3;" F= ";F3;" PROB= ";P3
2735 PRINT
2740 PRINT "A x B SS= ";B6;" DF= ";D7;" MS= ";M6;" F= ";F6;" PROB= ";P6
```

```
2750 PRINT "A x C SS= ";B5;" DF= ";D6;" MS= ";M5;" F= ";F5;" PROB= ";P5
2760 PRINT "B x C SS= ";B4;" DF= ";D5;" MS= ";M4;" F= ";F4;" PROB= ";P4
2770 PRINT "ABC    SS= ";B7;" DF= ";D8;" MS= ";M7;" F= ";F7;" PROB= ";P7
2775 PRINT
2780 PRINT "ERROR SS= ";B8;" DF= ";D9;" MS= ";M9
2810 GOTO 3020
2820 REM PROBABILITY FUNCTION.
2830 P=1
2840 IF Y*Y1*Y2=0 THEN 3000
2850 IF Y2<1 THEN 2900
2860 A=Y:B=Y1:F1=Y2
2890 GOTO 2930
2900 A=Y1:B=Y:F1=1/Y2
2930 AA=2/(9*A)
2940 B9=2/(9*B)
2950 Z=ABS(((1-B9)*F1^0.333333)-1+AA)/SQR((B9*F1^0.666667)+AA))
2960 IF B<4 THEN Z=Z*(1+(0.08*Z^4/B^3))
2970 P=0.5/(1+Z*(0.196854+Z*(0.115194+Z*(3.44E-04+Z*0.019527))))^4
2980 IF F1<1 THEN P=1-P
2990 GOTO 3010
3000 PRINT "THERE IS A PROBLEM WITH DEGREES OF FREEDOM OR F VALUES."
3010 RETURN
3020 END
```

P-17   Type 1: Two factor mixed design.   Section 2.7.

```
10 PRINT "TYPE 1 ANOVA; OR T BY S"
80 PRINT
90 PRINT "HOW MANY GROUPS?  ENTER 1 IF THIS IS A T x S DESIGN."
100 PRINT
130 PRINT
140 INPUT N1
150 PRINT "HOW MANY SUBJECTS IN THE TOTAL EXPERIMENT?"
160 INPUT N2
170 PRINT "HOW MANY TREATMENTS FOR EACH SUBJECT?  (OR ITEMS FOR ALPHA)"
180 INPUT N3
190 DIM X(N3),S1(N1,N3),S2(N2),S3(N3),N(N1),S4(N1),B(N3)
200 PRINT "ENTER THE GROUP N VALUES"
220 DIM A1(N1,N3),A(N1),L$(10)
230 FOR J=1 TO N1
234 PRINT "GROUP      ";J
250 INPUT N:N(J)=N:NEXT J
270 REM -  INITIALIZE THE SUMMATION REGISTERS.
280 FOR J=1 TO N1:S4(J)=0
300 FOR I=1 TO N3:S3(I)=0:S1(J,I)=0:NEXT I:NEXT J
350 FOR I=1 TO N2:S2(I)=0:NEXT I:T=0
390 A2=0:B2=0:S=0:A6=0:S5=0:L=1:L1=N(1)
480 FOR J=1 TO N1
485 PRINT "ENTER DATA FOR GROUP ";J
490 FOR K=L TO L1
495 PRINT :PRINT "SUBJECT ";K
500 PRINT :FOR I=1 TO N3:INPUT X
530 S4(J)=S4(J)+X
540 S5=S5+X^2
550 S2(K)=S2(K)+X
560 S3(I)=S3(I)+X
570 S1(J,I)=S1(J,I)+X
580 NEXT I:NEXT K:L=L1+1
610 IF J=N1 THEN 660
630 M=J+1:L1=L1+N(M):NEXT J
660 D1=N3*N2-1:D2=N2-1:D3=D1-D2:D4=N1-1:D5=N3-1
710 D6=D4*D5:D7=D2-D4:D8=D3-D5-D6:D9=D2*D5
```

```
750 REM COMPUTE SUMS OF SQUARES.
770 FOR I=1 TO N3:T=T+S3(I)
780 B2=B2+S3(I)^2/N2:NEXT I
800 C=T^2/(N2*N3):T=S5-C:B1=B2-C
830 IF N1=1 THEN 940
840 FOR J=1 TO N1
850 A2=A2+S4(J)^2/(N3*N(J)):NEXT J
870 A2=A2-C
880 FOR J=1 TO N1
890 FOR I=1 TO N3
900 A6=A6+S1(J,I)^2/N(J)
910 NEXT I:NEXT J
930 A6=A6-A2-B1-C
940 FOR K=1 TO N2
950 S=S+S2(K)^2/N3:NEXT K
970 S=S-C
980 REM -   COMPUTE MEAN SQUARES AND F VALUES.
990 IF N1>1 THEN 1040
1000 D8=D9
1010 E2=T-S-B1
1020 C1=E2/D8
1030 GOTO 1290
1040 S9=T-S
1050 E1=S-A2
1060 E2=T-S-B1-A6
1070 C1=E2/D8
1080 C2=A2/D4
1090 C3=A6/D6
1100 C4=E1/D7
1110 F2=C2/C4:Y=D4:Y1=D7:Y2=F2:GOSUB 1860:P1=P
1170 F3=C3/C1:Y=D6:Y1=D8:Y2=F3:GOSUB 1860:P2=P
1230 FOR J=1 TO N1
1240 FOR I=1 TO N3
1250 A1(J,I)=S1(J,I)/N(J)
1260 NEXT I:NEXT J
1280 REM -   COMPUTE ROW, COL, AND CELL MEANS.
1290 B5=B1/D5:F4=B5/C1:Y=B5:Y1=D8:Y2=F4:GOSUB 1860:P3=P
1360 FOR J=1 TO N1
1370 A(J)=S4(J)/(N3*N(J)):NEXT J
1390 FOR I=1 TO N3
1400 B(I)=S3(I)/N2:NEXT I
1420 PRINT "THE MEANS OF INTEREST ARE"
1440 PRINT :IF N1=1 THEN 1500
1450 PRINT "FOR GROUPS"
1460 PRINT :FOR I=1 TO N1
1470 PRINT A(I),
1480 NEXT I
1500 PRINT :PRINT :PRINT "FOR TREATMENTS"
1510 PRINT :FOR I=1 TO N3
1520 PRINT B(I),
1530 NEXT I:PRINT
1540 IF N1=1 THEN 1625
1560 PRINT :PRINT "FOR THE CELLS"
1565 PRINT
1570 FOR I=1 TO N1:FOR J=1 TO N3
1590 PRINT A1(I,J),
1600 NEXT J:PRINT :PRINT :NEXT I:PRINT
1625 PRINT "PRESS RETURN FOR THE SUMMARY TABLE"
1627 INPUT L$
1630 PRINT "SUMMARY TABLE FOR THE ANOVA."
1640 PRINT
1650 PRINT "TOTAL SS= ";T;" DF= ";D1
1655 PRINT
1660 PRINT "SUB B SS= ";S;" DF= ";D2
1665 PRINT
```

```
1700 IF N1=1 THEN 1750
1710 PRINT "B VAR SS= ";A2;" DF= ";D4;" MS= ";C2;" F= ";F2;" P= ";P1
1715 PRINT
1720 PRINT "ERR B SS= ";E1;" DF= ";D7;" MS= ";C4
1730 PRINT "_____"
1740 PRINT "SUB W SS= ";S9;" DF= ";D3
1745 PRINT
1750 PRINT "W VAR SS= ";B1;" DF= ";D5;" MS= ";B5;" F= ";F4;" P= ";P3
1760 IF N1=1 THEN 1775
1765 PRINT
1770 PRINT "B x W SS= ";A6;" DF= ";D6;" MS= ";C3;" F= ";F3;" P= ";P2
1775 PRINT
1780 PRINT "ERR W SS= ";E2;" DF= ";D8;" MS= ";C1
1800 KR=(S/D2-C1)/(S/D2)
1820 GOTO 2030
1840 REM FUNCTION PRBF
1850 PRINT
1860 LET P=1
1870 IF Y*Y1*Y2=0 THEN 2020
1880 IF Y2<1 THEN 1930
1890 Q=Y:U=Y1:F1=Y2
1920 GOTO 1960
1930 Q=Y1:U=Y:F1=1/Y2
1960 U1=2/(9*Q)
1970 B9=1/(9*U)
1980 Z=ABS(((((1-B9)*F1^0.333333)-1+U1)/SQR((B9*F1^0.666667)+U1))
1990 IF U<4 THEN Z=Z*(1+(0.08*Z^4/U^3))
2000 P=0.5/(1+Z*(0.196854+Z*(0.115194+Z*(3.44E-04+Z*0.019527))))^4
2010 IF F1<1 THEN P=1-P
2020 RETURN
2030 END
```

P-18 Type 3: Three factor mixed design—two betweens. Section 2.8.

```
10 PRINT "TYPE 3 ANOVA"
30 PRINT
40 PRINT "HOW MANY TREATMENTS ON THE OUTSIDE BETWEEN VARIABLE (A)?"
50 INPUT N1
60 PRINT "HOW MANY TREATMENTS ON THE INSIDE BETWEEN-VARIABLE (B)?"
70 INPUT N2
80 PRINT "HOW MANY TREATMENTS ON THE WITHIN-VARIABLE?"
90 INPUT N3
95 DIM L$(10)
100 PRINT "HOW MANY SUBJECTS IN THE TOTAL EXPERIMENT?"
110 INPUT N4
130 PRINT :N7=N1*N2*N3
140 DIM N5(N1,N2)
150 PRINT "ENTER THE GROUP N VALUES"
170 FOR I=1 TO N1:FOR J=1 TO N2
174 PRINT "GROUP ";I;" ";J
210 INPUT N:N5(I,J)=N:NEXT J:NEXT I
220 DIM S9(N4),B1(N1),B2(N1),B3(N2),B4(N2),B5(N1,N2)
230 DIM B6(N3),B7(N1,N3),B8(N2,N3),E1(N7),Q4(N1,N2)
240 DIM Q5(N1),Q6(N2),A1(N1),A2(N1,N2),C1(N2)
250 DIM C2(N2,N3),C3(N3),C4(N1,N3),CC5(N7),CC6(N2,N3)
270 X1=0:S=0:S1=0:S2=0:S3=0:S4=0:S5=0:S6=0:S7=0:S8=0:T=0
360 FOR K=1 TO N4
370 S9(K)=0:NEXT K
390 FOR I=1 TO N1
400 B1(I)=0:B2(I)=0:FOR J=1 TO N2:B3(J)=0:B4(J)=0:B5(I,J)=0:FOR M=1 TO N3
```

```
470 T=T+1:B6(M)=0:B7(I,M)=0:B8(J,M)=0:E1(T)=0:NEXT M:NEXT J:NEXT I
570 L=1:E5=0:L1=N5(1,1)
590 PRINT "ENTER THE DATA"
610 PRINT :GG=0
630 FOR I=1 TO N1:FOR J=1 TO N2:FOR K=L TO L1:GG=GG+1
670 PRINT "GROUP ";I;" ";J;" SUBJECT ";GG
677 FOR M=1 TO N3:FOR P=1 TO N3
690 INPUT X:X1=X1+X^2
710 S=S+X:B1(I)=B1(I)+X:B3(J)=B3(J)+X
740 S9(K)=S9(K)+X:B7(I,M)=B7(I,M)+X
760 B5(I,J)=B5(I,J)+X:B8(J,M)=B8(J,M)+X
780 B6(M)=B6(M)+X:T=M+E5:E1(T)=E1(T)+X
810 NEXT M:NEXT K:L=L1+1:E5=E5+N3
850 IF J=N2 THEN 880
860 M=J+1:L1=L1+N5(I,M)
880 NEXT J
890 IF I=N1 THEN 930
900 J=1:N=I+1:L1=L1+N5(I,J)
930 NEXT I
940 D1=N1-1:D2=N2-1:D3=N3-1:D4=N4-1:D5=D1*D2:D6=D1*D3
990 D7=D2*D3:D8=D6*D2:D9=N3*N4-1:Q1=D9-D4:Q2=D4-D1-D2-D5
1000 Q3=Q1-D3-D6-D7-D8
1010 REM COMPUTE THE VARIOUS N VALUES.
1030 FOR I=1 TO N1:FOR J=1 TO N2
1040 B2(I)=B2(I)+N5(I,J)
1050 Q4(I,J)=N5(I,J)*N3
1060 B4(J)=B4(J)+N5(I,J)
1080 NEXT J:NEXT I
1090 FOR I=1 TO N1
1100 Q5(I)=B2(I)*N3:NEXT I
1120 FOR J=1 TO N2
1130 Q6(J)=B4(J)*N3:NEXT J
1150 REM COMPUTE THE MEANS.
1160 FOR I=1 TO N1
1170 A1(I)=B1(I)/Q5(I)
1180 FOR J=1 TO N2
1190 A2(I,J)=B5(I,J)/Q4(I,J):NEXT J
1210 NEXT I
1220 FOR J=1 TO N2
1230 C1(J)=B3(J)/Q6(J)
1240 FOR K=1 TO N3
1250 C2(J,K)=B8(J,K)/B4(J):NEXT K:NEXT J
1280 FOR K=1 TO N3
1290 C3(K)=B6(K)/N4
1300 FOR I=1 TO N1
1310 C4(I,K)=B7(I,K)/B2(I):NEXT I:NEXT K
1340 C=S^2/(N3*N4):G=X1-C
1360 FOR I=1 TO N1
1370 S1=S1+B1(I)^2/Q5(I)
1380 FOR J=1 TO N2
1390 S3=S3+B5(I,J)^2/(N5(I,J)*N3)
1410 NEXT J:NEXT I
1420 FOR M=1 TO N3
1430 S4=S4+B6(M)^2/N4:NEXT M
1460 FOR I=1 TO N1:FOR M=1 TO N3
1470 S5=S5+B7(I,M)^2/B2(I)
1480 NEXT M:NEXT I:T=0
1510 FOR J=1 TO N2
1520 S2=S2+B3(J)^2/Q6(J)
1530 FOR M=1 TO N3
1540 S6=S6+B8(J,M)^2/B4(J)
1550 FOR I=1 TO N1
1570 T=T+1:S7=S7+E1(T)^2/N5(I,J)
1580 CC5(T)=E1(T)/N5(I,J)
1600 NEXT I:NEXT M:NEXT J
```

```
1620 FOR K=1 TO N4
1630 S8=S8+S9(K)^2/N3:NEXT K
1710 S1=S1-C:S2=S2-C:S3=S3-C-S1-S2:S4=S4-C:S5=S5-C-S1-S4:S6=S6-C-S2-S4:S7=S7-C-S
1-S2-S3-S4-S5-S6
1750 S8=S8-C:C5=G-S8:C6=S8-S1-S2-S3:C7=C5-S4-S5-S6-S7
1760 REM COMPUTE MS, F, PROBS.
1860 SS1=S1/D1:SS2=S2/D2:SS3=S3/D5:SS4=C6/Q2:SS5=S4/D3:SS6=S5/D6:SS7=S6/D7:SS8=S
7/D8:SS9=C7/Q3:S20=SS1/SS4
1910 Y=D1:Y1=Q2:Y2=S20:GOSUB 2970:S21=P
1920 S22=SS2/SS4:Y=D2:Y1=Q2:Y2=S22:GOSUB 2970:S23=P
1980 S24=SS3/SS4:Y=D5:Y1=Q2:Y2=S24:GOSUB 2970:S25=P
2040 S26=SS5/SS9:Y=D3:Y1=Q3:Y2=S26:GOSUB 2970:S27=P
2100 S28=SS6/SS9:Y=D4:Y1=Q3:Y2=S28:GOSUB 2970:S29=P
2160 S30=SS7/SS9:Y=D7:Y1=Q3:Y2=S30:GOSUB 2970:S31=P
2220 S32=SS8/SS9:Y=D8:Y1=Q3:Y2=S32:GOSUB 2970:S33=P
2290 PRINT "SUMMARY TABLE FOR THE TYPE 3 ANOVA"
2330 PRINT :PRINT :PRINT "TOTAL    SS=";G;"           DF=";D9
2340 PRINT :PRINT "SUB B     SS=";S8;"           DF=";D4
2350 PRINT :PRINT "A         SS=";S1;"  MS=";SS1;"  DF=";D1;"  F=";S20;"  P=";S21
2360 PRINT :PRINT "B         SS=";S2;"  MS=";SS2;"  DF=";D2;"  F=";S22;"  P=";S23
2370 PRINT :PRINT "A x B     SS=";S3;"  MS=";SS3;"  DF=";D5;"  F=";S24;"  P=";S25
2380 PRINT :PRINT "ERR B     SS=";C6;"  MS=";SS4;"  DF=";Q2
2390 PRINT "- - - - - - - - - - -"
2395 PRINT "PRESS RETURN FOR MORE"
2396 INPUT L$
2400 PRINT "SUB W     SS=";C5;"           DF=";Q1
2410 PRINT :PRINT "W VAR     SS=";S4;"  MS=";SS5;"  DF=";D3;"  F=";S26;"  P=";S27
2420 PRINT :PRINT "A x W     SS=";S5;"  MS=";SS6;"  DF=";D6;"  F=";S28;"  P=";S29
2430 PRINT :PRINT "B x W     SS=";S6;"  MS=";SS7;"  DF=";D7;"  F=";S30;"  P=";S31
2440 PRINT :PRINT "ABW       SS=";S7;"  MS=";SS8;"  DF=";D8;"  F=";S32;"  P=";S33
2450 PRINT :PRINT "ERR W     SS=";C7;"  MS=";SS9;"  DF=";Q3
2460 PRINT
2470 PRINT "PRESS RETURN FOR THE MEANS"
2480 INPUT L$
2500 PRINT "THE MEANS OF INTEREST"
2520 PRINT :PRINT "THE A MEANS"
2540 PRINT :FOR I=1 TO N1
2550 PRINT A1(I),
2590 NEXT I:PRINT :PRINT :PRINT "THE B MEANS"
2600 PRINT
2610 PRINT :FOR I=1 TO N2
2620 PRINT C1(I),
2630 NEXT I:PRINT :PRINT
2660 PRINT "THE C MEANS"
2680 PRINT :FOR I=1 TO N3
2690 PRINT C3(I),
2700 NEXT I
2710 PRINT
2720 PRINT
2730 PRINT "PRESS RETURN FOR THE CELL MEANS"
2740 INPUT L$:T=0
2770 FOR L=1 TO N1:FOR J=1 TO N2:FOR K=1 TO N3
2810 T=T+1:CC6(J,K)=CC5(T)
2840 NEXT K:NEXT J:PRINT
2850 PRINT "THE BC MEANS FOR A",L
2860 PRINT
2880 FOR I=1 TO N2:FOR J=1 TO N3
2890 PRINT CC6(I,J),
2900 NEXT J:PRINT :NEXT I:NEXT L
2950 GOTO 3180
2960 REM PROBABILITIES
2970 P=1
2990 IF Y*Y1*Y2=0 THEN 3160
3000 IF Y2<1 THEN 3060
3010 PRINT :Q=Y:U=Y1:F1=Y2
```

```
3050 GOTO 3090
3060 Q=Y1:U=Y:F1=1/Y2
3090 U1=2/(9*Q)
3100 B9=2/(9*U)
3110 Z=ABS((((1-B9)*F1^0.333333)-1+U1)/SQR((B9*F1^0.666667)+U1))
3120 IF U<4 THEN Z=Z*(1+(0.08*Z^4/U^3))
3130 P=0.5/(1+Z*(0.196854+Z*(0.115194+Z*(3.44E-04+Z*0.019527))))^4
3140 IF F1<1 THEN P=1-P
3150 GOTO 3170
3160 PRINT "THERE IS A PROBLEM WITH DEGREES OF FREEDOM OR F VALUES."
3170 RETURN
3180 END
```

P-19   Type 6: Three factor mixed design—two withins.   Section 2.9.

```
10 PRINT "TYPE 6 ANOVA"
40 PRINT
50 PRINT "HOW MANY GROUPS"
60 INPUT G
70 PRINT "HOW MANY TMTS ON THE OUTSIDE WITHIN-VARIABLE?"
80 INPUT A1
90 PRINT "HOW MANY TMTS ON THE INSIDE WITHIN-VARIABLE?"
100 INPUT B1
110 PRINT "HOW MANY SUBJECTS IN THE ENTIRE EXPERIMENT?"
120 INPUT T
130 C11=A1*B1*G
140 DIM N(G),CE(C11),S(T),SW1(A1,T),SW2(B1,T),L$(10)
150 DIM GP(G),W1(A1),W2(B1),W1G(G,A1),W2G(G,B1),TT(A1,B1),W1W2(A1,B1),S1(T)
160 DIM GM(G),G1M(G,A1),W1M(A1),G2M(G,B1),W2M(B1),W12M(A1,B1),WM(A1,B1)
163 DIM X(A1,B1)
170 PRINT "ENTER THE GROUP N VALUES"
184 FOR I=1 TO G:PRINT "GROUP      ";I:INPUT N:N(I)=N:NEXT I
220 FOR I=1 TO G:GP(I)=0:FOR W=1 TO B1:W2G(I,W)=0:NEXT W:NEXT I
230 FOR J=1 TO A1:W1(J)=0:FOR W=1 TO B1:W1W2(J,W)=0:TT(J,W)=0
240 NEXT W:NEXT J
250 FOR W=1 TO B1:W2(W)=0:FOR I=1 TO T:SW2(W,I)=0:NEXT I:NEXT W
260 FOR I=1 TO G:FOR J=1 TO A1:W1G(I,J)=0:NEXT J:NEXT I
270 FOR I=1 TO C11:CE(I)=0:NEXT I
280 FOR J=1 TO T:S1(J)=0:FOR I=1 TO A1:SW1(I,J)=0:NEXT I:NEXT J:S=0
300 V=0:GGG=0:S3=0:SSG=0:S1W=0:S2W=0
310 GW1=0:GW2=0:SSW1=0:SSW2=0:SUB=0:CELL=0:W12=0
320 PRINT "ENTER THE DATA"
340 FOR I=1 TO G:GG=N(I)
370 FOR L=1 TO GG:GGG=GGG+1:PRINT
380 PRINT "GROUP ";I;" SUBJECT ";GGG
394 FOR J=1 TO A1:FOR K=1 TO B1
396 DDD=DDD+1
415 INPUT X:X(J,K)=X
420 S=S+X(J,K):S3=S3+X(J,K)^2
430 GP(I)=GP(I)+X(J,K):W1(J)=W1(J)+X(J,K):W2(K)=W2(K)+X(J,K)
440 W1G(I,J)=W1G(I,J)+X(J,K):W2G(I,K)=W2G(I,K)+X(J,K)
450 TT(J,K)=TT(J,K)+X(J,K):W1W2(J,K)=W1W2(J,K)+X(J,K):S1(GGG)=S1(GGG)+X(J,K)
460 SW1(J,GGG)=SW1(J,GGG)+X(J,K):SW2(K,GGG)=SW2(K,GGG)+X(J,K)
470 NEXT K:NEXT J:NEXT L
510 FOR D=1 TO A1:FOR E=1 TO B1
520 V=V+1:CE(V)=TT(D,E)
530 NEXT E:NEXT D
560 FOR D=1 TO A1:FOR E=1 TO B1
570 TT(D,E)=0:NEXT E:NEXT D:NEXT I
610 REM COMPUTE SUMS OF SQUARES
620 C1=S^2/(A1*B1*T):T1=S3-C1
640 FOR I=1 TO G
```

```
650 SSG=SSG+GP(I)~2/(N(I)*A1*B1)
670 NEXT I:SSG=SSG-C1
680 FOR I=1 TO A1
690 SSW1=SSW1+W1(I)~2/(B1*T)
710 NEXT I:SSW1=SSW1-C1
720 FOR I=1 TO B1
730 SSW2=SSW2+W2(I)~2/(A1*T)
740 FOR J=1 TO T:S2W=S2W+SW2(I,J)~2/A1:NEXT J:NEXT I
760 SSW2=SSW2-C1
780 FOR I=1 TO G:FOR J=1 TO A1
790 GW1=GW1+W1G(I,J)~2/(N(I)*B1)
820 NEXT J:NEXT I:GW1=GW1-C1-SSG-SSW1
840 FOR I=1 TO G:FOR J=1 TO B1
850 GW2=GW2+W2G(I,J)~2/(N(I)*A1)
870 NEXT J:NEXT I
880 GW2=GW2-C1-SSG-SSW2
900 FOR I=1 TO A1:FOR J=1 TO B1
910 W12=W12+W1W2(I,J)~2/T
930 NEXT J:NEXT I
940 W12=W12-C1-SSW1-SSW2
950 FOR I=1 TO T
960 SUB=SUB+S1(I)~2/(A1*B1)
970 FOR J=1 TO A1:S1W=S1W+SW1(J,I)~2/B1:NEXT J:NEXT I:SUB=SUB-C1
1000 S1W=S1W-C1-SUB-SSW1-GW1
1010 S2W=S2W-C1-SUB-SSW2-GW2:D=0
1040 FOR I=1 TO G:FOR J=1 TO A1*B1
1060 D=D+1:CELL=CELL+CE(D)~2/N(I)
1080 NEXT J:NEXT I
1090 CELL=CELL-C1-SSG-SSW1-SSW2-GW1-GW2-W12
1100 SSWS=T1-SUB:ERB=SUB-SSG
1110 REM
1120 REM COMPUTE DF AND MEAN SQUARES.
1130 REM
1140 DFG=G-1:DFW1=A1-1:DFW2=B1-1:DFGW1=DFG*DFW1:DFGW2=DFG*DFW2
1150 DFW12=DFW1*DFW2:DFCELL=DFW12*DFG:DFTOT=T*A1*B1-1:DFSUB=T-1
1160 DFWS=DFTOT-DFSUB:DFERB=DFSUB-DFG
1170 DFWIN=(T-G)*(A1*B1-1):DFE1=(A1-1)*(T-G):DFE2=(B1-1)*(T-G)
1180 DFE12=DFWIN-DFE1-DFE2
1190 WIN=SSWS-CELL-SSW1-SSW2-GW1-GW2-W12
1200 WE1=S1W:WE2=S2W:WE12=WIN-WE1-WE2
1210 MSG=SSG/DFG:MSEB=ERB/DFERB
1220 MS1=SSW1/DFW1:MS2=SSW2/DFW2:MSG1=GW1/DFGW1:MSG2=GW2/DFGW2
1230 MS12=W12/DFW12:MSCELL=CELL/DFCELL:MSE1=WE1/DFE1
1240 MSE2=WE2/DFE2:MSE12=WE12/DFE12
1250 FG=MSG/MSEB:F8=MS1/MSE1:F2=MS2/MSE2:F12=MS12/MSE12
1260 FG1=MSG1/MSE1:FG2=MSG2/MSE2:FG12=MSCELL/MSE12
1270 Y=DFG:Y11=DFERB:Y2=FG:GOSUB 1880:P1=P
1280 Y=DFW1:Y11=DFE1:Y2=F8:GOSUB 1880:P2=P
1290 Y=DFW2:Y11=DFE2:Y2=F2:GOSUB 1880:P3=P
1300 Y=DFW12:Y11=DFE12:Y2=F12:GOSUB 1880:P6=P
1310 Y=DFGW1:Y11=DFE1:Y2=FG1:GOSUB 1880:P4=P
1320 Y=DFGW2:Y11=DFE2:Y2=FG2:GOSUB 1880:P5=P
1330 Y=DFCELL:Y11=DFE12:Y2=FG12:GOSUB 1880:P7=P
1340 FOR I=1 TO G:GM(I)=GP(I)/(N(I)*A1*B1):FOR J=1 TO A1
1350 G1M(I,J)=W1G(I,J)/(N(I)*B1):NEXT J:NEXT I
1360 FOR J=1 TO A1:W1M(J)=W1(J)/(T*B1):FOR K=1 TO B1
1370 W12M(J,K)=W1W2(J,K)/T:NEXT K:NEXT J
1380 FOR K=1 TO B1:W2M(K)=W2(K)/(T*A1):FOR I=1 TO G
1390 G2M(I,K)=W2G(I,K)/(N(I)*A1):NEXT I:NEXT K
1400 PRINT "THE GROUP MEANS"
1408 PRINT :FOR I=1 TO G
1410 PRINT GM(I),
1420 NEXT I:PRINT
1430 PRINT :PRINT "THE MEANS FOR THE FIRST WITHIN VARIABLE"
1440 PRINT :FOR I=1 TO A1:PRINT W1M(I),:NEXT I
```

```
1460 PRINT :PRINT :PRINT "THE MEANS FOR THE SECOND WITHIN  VARIABLE"
1470 PRINT :PRINT :FOR I=1 TO B1:PRINT W2M(I),:NEXT I
1500 PRINT :PRINT :PRINT "PRESS RETURN FOR MORE MEANS"
1510 INPUT L$:PRINT
1520 PRINT "THE GROUPS x FIRST WITHIN VAR MEANS"
1530 PRINT :FOR I=1 TO G:FOR J=1 TO A1:PRINT G1M(I,J),
1540 NEXT J:PRINT :NEXT I:PRINT :PRINT :PRINT "THE GROUPS x SECOND WITHIN VAR ME
ANS"
1550 PRINT :FOR I=1 TO G:FOR J=1 TO B1:PRINT G2M(I,J),
1560 NEXT J:PRINT :NEXT I:PRINT :PRINT :PRINT "THE W1 x W2 MEANS":PRINT
1570 FOR I=1 TO A1:FOR J=1 TO B1:PRINT W12M(I,J),
1580 NEXT J:PRINT :NEXT I:PRINT :PRINT
1590 PRINT "PRESS RETURN FOR MORE MEANS"
1600 INPUT L$:PRINT
1610 D=0:FOR I=1 TO G:PRINT "THE W1 x W2 MEANS FOR GROUP";I
1620 FOR J=1 TO A1:FOR K=1 TO B1:D=D+1:WM(J,K)=CE(D)/N(I)
1630 NEXT K:NEXT J
1640 PRINT :FOR J=1 TO A1:FOR K=1 TO B1
1650 PRINT WM(J,K),:NEXT K:PRINT :NEXT J:PRINT
1660 NEXT I:PRINT
1670 PRINT :PRINT "PRESS RETURN FOR THE SUMMARY TABLE OF THE ANOVA"
1680 INPUT L$:PRINT
1700 PRINT "TOTAL SS = ";T1;" DF = ";DFTOT
1710 PRINT :PRINT
1730 PRINT "SUB B SS = ";SUB;" DF = ";DFSUB
1740 PRINT :PRINT "GRPS  SS = ";SSG;" DF = ";DFG;" MS = ";MSG;" F = ";FG;" PROB
= ";P1
1750 PRINT :PRINT "ERR B SS = ";ERB;" DF = ";DFERB;" MS = ";MSEB
1760 PRINT :PRINT "- - - - - - - - "
1770 PRINT "SUB W SS = ";SSWS;" DF = ";DFWS
1780 PRINT :PRINT "W1     SS = ";SSW1;" DF = ";DFW1;" MS = ";MS1;" F = ";F8;" PRO
B = ";P2
1790 PRINT :PRINT "W2     SS = ";SSW2;" DF = ";DFW2;" MS = ";MS2;" F = ";F2;" PRO
B = ";P3
1792 PRINT :PRINT "PRESS RETURN FOR MORE"
1794 INPUT L$
1800 PRINT :PRINT "G W1  SS = ";GW1;" DF = ";DFGW1;" MS = ";MSG1;" F = ";FG1;" P
ROB = ";P4
1810 PRINT :PRINT "G W2  SS = ";GW2;" DF = ";DFGW2;" MS = ";MSG2;" F = ";FG2;" P
ROB = ";P5
1820 PRINT :PRINT "W1 W2 SS = ";W12;" DF = ";DFW12;" MS = ";MS12;" F = ";F12;" P
ROB = ";P6
1830 PRINT :PRINT "TRIP  SS = ";CELL;" DF = ";DFCELL;" MS = ";MSCELL;" F = ";FG1
2;" PROB = ";P7
1840 PRINT :PRINT "E1 G1 SS = ";WE1;" DF = ";DFE1;" MS = ";MSE1
1850 PRINT :PRINT "E2 G2 SS = ";WE2;" DF = ";DFE2;" MS = ";MSE2
1860 PRINT :PRINT "E12 G SS = ";WE12;" DF = ";DFE12;" MS = ";MSE12
1870 GOTO 2070
1880 P=1
1890 IF Y*Y11*Y2<=0 THEN 2070
1900 IF Y2<1 THEN 1950
1930 QQ=Y:U=Y11:F1=Y2
1940 GOTO 1980
1950 QQ=Y11:U=Y:F1=1/Y2
1980 U1=2/(9*QQ)
1990 B9=2/(9*U)
2000 Z=ABS(((((1-B9)*F1^0.333333)-1+U1)/SQR((B9*F1^0.666667)+U1))
2010 IF U<4 THEN Z=Z*(1+(0.08*Z^4/U^3))
2020 P=0.5/(1+Z*(0.196854+Z*(0.115194+Z*(3.44E-04+Z*0.019527))))^4
2030 IF F1<1 THEN P=1-P
2040 GOTO 2060
2050 PRINT "THERE IS A PROBLEM WITH DEGREES OF FREEDOM OR F VALUES."
2060 RETURN
2070 END
```

P-20   Covariance analyses.   Sections 4.10 and 4.11.

```
10 PRINT "COVARIANCE ANALYSIS"
50 PRINT :PRINT "HOW MANY TREATMENTS ON THE FIRST VARIABLE?"
60 INPUT NA
80 PRINT "HOW MANY TREATMENTS ON THE OTHER VARIABLE?  ENTER 1 IF THIS IS A ONE-V
ARIABLE DESIGN"
100 INPUT NB
110 PRINT
120 DIM N(NB,NA),SCX(NB,NA),SCY(NB,NA),N1(NB),N2(NA),SBX(NB),SBY(NB),SAY(NA)
130 DIM SAX(NA),XRM(NB),XCLM(NB,NA),XCM(NA),YRM(NB),YCLM(NB,NA),YCM(NA)
140 DIM YRM1(NB),YCLM1(NB,NA),YCM1(NA)
145 DIM L$(10)
150 PRINT "ENTER THE CELL N VALUES"
160 FOR I=1 TO NB:FOR J=1 TO NA:PRINT "CELL ";I;"  ";J
164 INPUT N
170 N(I,J)=N:NEXT J:NEXT I
190 AX=0:AY=0:CLX=0:CLY=0:BY=0:BX=0:NT=0:TX=0:TY=0:TXS=0:TYS=0
200 TXY=0:BXY=0:CLXY=0:AXY=0
210 FOR I=1 TO NB:SBX(I)=0:SBY(I)=0:N1(I)=0
220 FOR J=1 TO NA:SCX(I,J)=0:SCY(I,J)=0:NEXT J:NEXT I
230 FOR J=1 TO NA:N2(J)=0:SAX(J)=0:SAY(J)=0:NEXT J
240 REM
250 REM COMPUTE ROW, COL, AND TOTAL N
260 REM
270 FOR I=1 TO NB:FOR J=1 TO NA:NT=NT+N(I,J):N1(I)=N1(I)+N(I,J)
280 N2(J)=N2(J)+N(I,J):NEXT J:NEXT I
282 GG=0
290 PRINT "ENTER THE DATA; BOTH ON A LINE; SEPARATE THE NUMBERS WITH A COMMA"
300 FOR I=1 TO NB:FOR J=1 TO NA:NL=N(I,J)
310 FOR K=1 TO NL
312 GG=GG+1
330 PRINT "S ";GG;" CELL ";I;"  ";J
334 INPUT X,Y
340 TX=TX+X:TY=TY+Y
350 TXS=TXS+X^2:TYS=TYS+Y^2:TXY=TXY+X*Y:SBX(I)=SBX(I)+X
360 SBY(I)=SBY(I)+Y:SAX(J)=SAX(J)+X:SAY(J)=SAY(J)+Y:SCX(I,J)=SCX(I,J)+X
370 SCY(I,J)=SCY(I,J)+Y:NEXT K:NEXT J:NEXT I
380 FOR I=1 TO NB:BX=BX+SBX(I)^2/N1(I):BY=BY+SBY(I)^2/N1(I)
390 BXY=BXY+SBX(I)*SBY(I)/N1(I):FOR J=1 TO NA:CLX=CLX+SCX(I,J)^2/N(I,J)
400 CLY=CLY+SCY(I,J)^2/N(I,J):CLXY=CLXY+SCX(I,J)*SCY(I,J)/N(I,J)
410 NEXT J:NEXT I
420 FOR J=1 TO NA:AX=AX+SAX(J)^2/N2(J):AY=AY+SAY(J)^2/N2(J)
430 AXY=AXY+SAX(J)*SAY(J)/N2(J):NEXT J
440 CX=TX^2/NT:CY=TY^2/NT:STX=TXS-CX:STY=TYS-CY:CXY=(TY*TX)/NT
450 TTXY=TXY-CXY:SAX1=AX-CX:SAY1=AY-CY:SBX1=BX-CX
460 SBY1=BY-CY:SABX=CLX-CX-SAX1-SBX1:SABY=CLY-CY-SAY1-SBY1
470 SXE=TXS-CLX:SYE=TYS-CLY:SAXY=AXY-CXY:SBXY=BXY-CXY
480 SABXY=CLXY-CXY-SAXY-SBXY:SXYE=TXY-CLXY
490 REM
500 REM COMPUTE CORRELATIONS  - RW AND RT.
510 REM
520 RT=TTXY/SQR(STX*STY):RW=SXYE/SQR(SXE*SYE)
530 REM
540 REM COMPUTE THE ADJUSTED Y SUMS OF SQUARES.
550 REM
560 EYA=SYE-(SXYE^2/SXE):YA=(SAY1+SYE)-((SAXY+SXYE)^2/(SAX1+SXE))-EYA
570 YB=(SBY1+SYE)-((SBXY+SXYE)^2/(SBX1+SXE))-EYA
580 YAB=(SABY+SYE)-((SABXY+SXYE)^2/(SABX+SXE))-EYA
590 REM
600 REM COMPUTE ALL DF AND ADJUSTED DF ERROR
610 REM
620 NDT=NT-1:NDA=NA-1:NDE=NT-NA*NB:NDYE=NDE-1:NDYT=NDT-1
630 SMA=YA/NDA:SMYE=EYA/NDYE
640 IF NB=1 THEN GOTO 660
```

```
650 NDB=NB-1:NDAB=NDA*NDB:SMB=YB/NDB:SMAB=YAB/NDAB
660 FA=SMA/SMYE:FB=SMB/SMYE:FAB=SMAB/SMYE
670 Y=NDA:Y1=NDYE:Y2=FA:GOSUB 1280:P1=G
680 IF NB=1 THEN GOTO 740
690 Y=NDB:Y1=NDYE:Y2=FB:GOSUB 1280:P2=G
700 Y=NDAB:Y1=NDYE:Y2=FAB:GOSUB 1280:P3=G
710 REM
720 REM COMPUTE BW, SM, X MEANS, Y MEANS, AND ADJUSTED Y MEANS.
730 REM
740 BW=SXYE/SXE:XM=TX/NT
750 FOR I=1 TO NB:XRM(I)=SBX(I)/N1(I):FOR J=1 TO NA
760 XCLM(I,J)=SCX(I,J)/N(I,J):NEXT J:NEXT I
770 FOR J=1 TO NA:XCM(J)=SAX(J)/N2(J):NEXT J
780 FOR I=1 TO NB:YRM(I)=SBY(I)/N1(I):FOR J=1 TO NA
790 YCLM(I,J)=SCY(I,J)/N(I,J):NEXT J:NEXT I
800 FOR J=1 TO NA:YCM(J)=SAY(J)/N2(J):NEXT J
810 FOR I=1 TO NB:YRM1(I)=YRM(I)-BW*(XRM(I)-XM):FOR J=1 TO NA
820 YCLM1(I,J)=YCLM(I,J)-BW*(XCLM(I,J)-XM):NEXT J:NEXT I
830 FOR J=1 TO NA:YCM1(J)=YCM(J)-BW*(XCM(J)-XM):NEXT J
850 PRINT "THE CORRELATIONS ARE:"
853 PRINT :PRINT "WITHIN = ";RW
855 PRINT :PRINT "TOTAL  = ";RT
860 PRINT
870 PRINT "PRESS RETURN FOR THE ANCOVA SUMMARY TABLE"
880 INPUT L$
900 PRINT "THE ANCOVA FOR THE CORRECTED Y"
910 PRINT
920 IF NB=1 THEN GOTO 1010
930 PRINT
940 PRINT "VAR 1   SS = ";YA
945 PRINT " DF = ";NDA;"    MS = ";SMA
950 PRINT " F = ";FA;"    PROB = ";P1
955 PRINT :PRINT "VAR 2   SS = ";YB
960 PRINT " DF = ";NDB;"    MS = ";SMB
965 PRINT " F = ";FB;"    PROB = ";P2
970 PRINT :PRINT "INTER   SS = ";YAB
975 PRINT " DF = ";NDAB;"    MS = ";SMAB
980 PRINT " F = ";FAB;"    PROB = ";P3
985 PRINT
990 PRINT "ERROR   SS = ";EYA
995 PRINT " DF = ";NDYE;"    MS = ";SMYE
1000 IF NB>1 THEN GOTO 1070
1010 PRINT
1020 PRINT "TREATMENT  SS = ";YA
1022 PRINT :PRINT " DF = ";NDA;"    MS = ";SMA
1024 PRINT :PRINT " F = ";FA;"    PROB = ";P1
1025 PRINT :PRINT
1030 PRINT "ERROR SS = ";EYA
1040 PRINT :PRINT " DF = ";NDYE;"    MS = ";SMYE
1070 PRINT :PRINT "PRESS RETURN FOR THE CORRECTED MEANS"
1080 INPUT L$
1100 IF NB=1 THEN 1180
1110 PRINT "FOR THE FIRST VARIABLE"
1120 PRINT
1130 FOR I=1 TO NB
1140 PRINT YRM1(I);"  ";
1150 NEXT I
1160 PRINT
1180 PRINT :PRINT "FOR THE SECOND VARIABLE":PRINT
1185 PRINT :FOR J=1 TO NA
1190 PRINT YCM1(J);"  ";:NEXT J
1200 PRINT :PRINT
1210 IF NB=1 THEN GOTO 1380
1220 PRINT "FOR THE CELLS":PRINT
1230 FOR I=1 TO NB:FOR J=1 TO NA
1240 PRINT "  ";YCLM1(I,J);:NEXT J
```

```
1245 PRINT :PRINT :NEXT I
1250 PRINT
1260 GOTO 1380
1270 REM PROBABILITY FUNCTION
1280 G=1:IF Y2<=0 THEN 1370
1290 IF Y2<1 THEN 1310
1300 Q1=Y:U=Y1:F1=Y2:GOTO 1320
1310 Q1=Y1:U=Y:F1=1/Y2
1320 U1=2/(9*Q1):B9=2/(9*U)
1330 Z=ABS((((1-B9)*F1^0.333333)-1+U1)/SQR((B9*F1^0.666667)+U1))
1340 IF U<4 THEN Z=Z*(1+(0.08*Z^4/U^3))
1350 G=0.5/(1+Z*(0.196854+Z*(0.115194+Z*(3.44E-04+Z*0.019527))))^4
1360 IF F1<1 THEN G=1-G
1370 RETURN
1380 END
```

P-21 Multiple regression—two predictors.   Section 4.9.

```
10 PRINT "MULTIPLE REGRESSION - TWO PREDICTORS"
20 GOSUB 1020
30 B1=(S(3)*(R(1,3)-R(2,3)*R(1,2)))/(S(1)*(1-R(1,2)^2))
40 B2=(S(3)*(R(2,3)-R(1,3)*R(1,2)))/(S(2)*(1-R(1,2)^2))
50 A=A(3)-B1*A(1)-B2*A(2)
56 A=INT(A*1000):A=A/1000
80 PRINT :PRINT "THE REGRESSION EQUATION"
90 PRINT :PRINT "Y =(X1)(";B1;")+(X2)(";B2;")+(";A;")"
100 R=SQR((R(1,3)^2+R(2,3)^2-2*R(1,3)*R(2,3)*R(1,2))/(1-R(1,2)^2))
102 R=INT(R*1000):R=R/1000
120 F=(R^2/(K-1))/((1-R^2)/(N-K))
122 F=INT(F*1000):F=F/1000
125 DFN=K-1:DFD=N-K
130 PRINT :PRINT "MULT R = ";R;"  F = ";F
140 PRINT :PRINT "DF = ";DFN;" AND ";DFD
150 END
1020 PRINT :PRINT "INPUT N - THE NUMBER OF PEOPLE"
1030 INPUT N
1040 K=3
1055 PRINT
1060 IF N<=(K+1) THEN PRINT "IMPOSSIBLE! - THERE MUST BE AT LEAST    5 PEOPLE.":
? :GOTO 150
1080 DIM A(K),S(K),R(K,K)
1090 DIM X(K)
1100 FOR I=1 TO K
1110 A(I)=0
1120 S(I)=0
1130 FOR J=1 TO K
1140 R(I,J)=0
1150 NEXT J
1160 NEXT I
1170 PRINT "ENTER THE SCORES FOR EACH SUBJECT ON    ONE LINE; SEPARATED BY COMMA
S"
1175 PRINT
1180 FOR I=1 TO N
1190 PRINT "SUBJECT ";I
1200 INPUT X,Y,Z
1225 X(1)=X:X(2)=Y:X(3)=Z
1227 FOR J=1 TO K
1230 A(J)=A(J)+X(J)
1250 S(J)=S(J)+X(J)^2
1280 FOR M=J TO K
1290 R(J,M)=R(J,M)+X(J)*X(M)
1300 NEXT M
1310 NEXT J
1320 NEXT I
1330 FOR I=1 TO K
1340 A(I)=A(I)/N
1350 S(I)=SQR(R(I,I)/N-(A(I)^2))
```

```
1360 NEXT I
1370 PRINT
1380 FOR I=1 TO K
1390 FOR J=I TO K
1400 IF S(I)*S(J)=0 THEN 1420
1410 R(J,I)=((R(I,J)/N)-A(I)*A(J))/(S(I)*S(J))
1420 R(I,J)=R(J,I)
1430 NEXT J
1440 R(I,I)=1
1450 NEXT I
1460 FOR I=1 TO K
1470 FOR J=I TO K
1480 R(I,J)=INT(R(I,J)*100)
1490 R(I,J)=R(I,J)/100
1500 NEXT J
1510 NEXT I
1520 PRINT "THE CORRELATIONS"
1530 PRINT
1540 FOR I=1 TO K
1550 FOR J=I TO K
1555 IF I=J THEN 1570
1557 IF J=3 THEN 1565
1560 PRINT "R X";I;"X";J;" = ";R(I,J):PRINT
1562 GOTO 1570
1565 PRINT "R X";I;"Y";" = ";R(I,J):PRINT
1570 NEXT J:NEXT I
1580 RETURN
1590 END
```

P-22   MANOVA for sample and population.   Section 5.1.

```
20 PRINT "MANOVA FOR A SAMPLE AND POPULATION"
30 PRINT :PRINT "ENTER THE N"
40 INPUT N
42 S1=0:S2=0:SS1=0:SS2=0
44 CP=0
50 DIM L$(10),D(4),CD(4),CDI(4)
60 PRINT "ENTER DATA -- BOTH NUMBERS ON ONE LINE; SEPARATED BY A COMMA"
65 PRINT
70 FOR I=1 TO N
80 PRINT "SUBJECT ";I
130 INPUT T1,T2
160 S1=S1+T1:S2=S2+T2:SS1=SS1+T1^2:SS2=SS2+T2^2:CP=CP+T1*T2
170 NEXT I
180 M1=S1/N:M2=S2/N
190 PRINT "ENTER THE POPULATION MEANS"
203 PRINT "MEAN 1"
210 INPUT PM1
215 PRINT :PRINT "MEAN 2"
220 INPUT PM2
230 D1=M1-PM1:D2=M2-PM2
240 SS1=SS1-(S1)^2/N
250 SS2=SS2-(S2)^2/N
260 CP=CP-S1*S2/N
270 REM COMPUTATION OF THE DISPERSION MATRIX
272 D(1)=SS1:D(2)=CP:D(3)=CP:D(4)=SS2
274 FOR I=1 TO 4:CD(I)=1/(N-1)*D(I):NEXT I
280 CC=1/((CD(1)*CD(4))-(CD(2)*CD(3)))
285 REM COMPUTATION OF THE INVERSE MATRIX
290 CDI(1)=CC*CD(4):CDI(2)=CC*(-CD(2)):CDI(3)=CC*(-CD(3)):CDI(4)=CC*CD(1)
292 TSQ=N*((CDI(1)*D1^2)+(2*CDI(2)*D1*D2)+(CDI(4)*D2^2))
310 F=((N-2)/(2*(N-1)))*TSQ
320 DFN=2:DFD=N-2
340 PRINT "T SQUARED = ";TSQ
350 PRINT :PRINT "F = ";F
360 PRINT :PRINT "DFN = ";DFN;" DFD = ";DFD
370 END
```

P-23 MANOVA for one group—matched pairs.   Section 5.2.

```
20 PRINT "MANOVA FOR ONE SAMPLE - DIFF. SCORES"
30 PRINT :PRINT "ENTER THE N"
40 INPUT N:PRINT
45 DIM P1(N),P2(N),PO1(N),PO2(N)
50 PRINT "ENTER THE DATA; PUT ALL FOUR NUMBERS ON ONE LINE; SEPARATE THEM WITH C
OMMAS":PRINT
60 FOR I=1 TO N
70 PRINT "SUBJECT ";I
130 INPUT P1,PO1,P2,PO2
140 P1(I)=P1:PO1(I)=PO1
150 P2(I)=P2
160 PO2(I)=PO2
170 NEXT I
180 DIM D(N),D1(N),CD(4),CDI(4)
190 FOR I=1 TO N:D(I)=P1(I)-PO1(I):D1(I)=P2(I)-PO2(I):NEXT I
195 T1S=0:T2S=0:CP=0
200 FOR I=1 TO N:T1=T1+D(I):T2=T2+D1(I):T1S=T1S+D(I)~2:T2S=T2S+D1(I)~2:CP=CP+D(I
)*D1(I):NEXT I
210 M1=T1/N:M2=T2/N
220 SSX1=T1S-(T1~2/N):SSX2=T2S-(T2~2/N):SSCP=CP-(T1*T2/N)
230 CD(1)=1/(N-1)*SSX1:CD(2)=1/(N-1)*SSCP:CD(3)=1/(N-1)*SSCP:CD(4)=1/(N-1)*SSX2
240 CC=1/(CD(1)*CD(4)-CD(2)*CD(3))
250 CDI(1)=CC*CD(4):CDI(2)=CC*(-CD(2)):CDI(3)=CC*(-CD(3)):CDI(4)=CC*CD(1)
260 TSQ=N*(CDI(1)*M1~2+2*CDI(2)*M1*M2+CDI(4)*M2~2)
270 F=TSQ*((N-2)/(2*(N-1)))
280 DFN=2:DFD=N-2
285 DIM L$(10)
290 PRINT "T SQUARED = ";TSQ
300 PRINT :PRINT "F = ";F
310 PRINT :PRINT "DFN = ";DFN;"   DFD = ";DFD
320 END
```

P-24 MANOVA for two independent groups.   Section 5.3.

```
20 PRINT "MANOVA - TWO INDEPENDENT GROUPS"
30 PRINT :PRINT "HOW MANY SUBJECTS IN GROUP 1?"
40 INPUT N1
50 PRINT :PRINT "HOW MANY SUBJECTS IN GROUP 2?"
60 INPUT N2
65 PRINT
70 PRINT "ENTER DATA FOR GROUP 1; BOTH NUMBERS   ON ONE LINE; SEPARATED BY A COM
MA"
75 DIM X1(N1),X2(N1),Y1(N2),Y2(N2),L$(10)
80 A1=0:A2=0:S1=0:S2=0
90 FOR I=1 TO N1
100 PRINT "SUBJECT ";I
130 INPUT X1,X2
140 X2(I)=X2:X1(I)=X1
150 NEXT I
160 PRINT "ENTER DATA FOR GROUP 2; BOTH NUMBERS ON ONE LINE"
170 FOR I=1 TO N2
180 PRINT "SUBJECT ";I
200 INPUT Y1,Y2
210 Y2(I)=Y2:Y1(I)=Y1
220 NEXT I
225 B1=0:B2=0:C1=0:C2=0:CX=0:CY=0
230 FOR I=1 TO N1:A1=A1+X1(I):A2=A2+X2(I):S1=S1+X1(I)~2:S2=S2+X2(I)~2:CX=CX+X1(I
)*X2(I):NEXT I
240 FOR I=1 TO N2:B1=B1+Y1(I):B2=B2+Y2(I):C2=C2+Y2(I)~2:C1=C1+Y1(I)~2:CY=CY+Y1(I
)*Y2(I):NEXT I
250 MX1=A1/N1:MX2=A2/N1:MY1=B1/N2:MY2=B2/N2
260 SSX1=S1-(A1~2/N1)
```

```
270 SSX2=S2-(A2~2/N1)
280 SCPX=CX-((A1*A2)/N1)
290 SSY1=C1-(B1~2/N2)
300 SSY2=C2-(B2~2/N2)
310 SCPY=CY-((B1*B2)/N2)
320 SSW1=(SSX1+SSY1)/(N1+N2-2)
330 SSW2=(SSX2+SSY2)/(N1+N2-2)
340 SCPW=(SCPX+SCPY)/(N1+N2-2)
350 DIM CW(4),CWI(4)
360 CW(1)=SSW1:CW(2)=SCPW:CW(3)=SCPW:CW(4)=SSW2
370 CC=1/(CW(1)*CW(4)-CW(2)*CW(3))
380 CWI(1)=CC*CW(4):CWI(2)=CC*(-CW(2)):CWI(3)=CC*(-CW(3)):CWI(4)=CC*CW(1)
390 D1=MX1-MY1:D2=MX2-MY2
400 TSQ=N1*N2/(N1+N2)*(CWI(1)*D1~2+2*CWI(2)*D1*D2+CWI(4)*D2~2)
410 F=TSQ*((N1+N2-3)/(2*(N1+N2-2)))
420 DFN=2:DFD=N1+N2-3
425 PRINT
430 PRINT "T SQUARED = ";TSQ
440 PRINT :PRINT "F = ";F
450 PRINT :PRINT "DFN = ";DFN;"   DFD = ";DFD
460 END
```

P-25   MANOVA for two groups—change scores.   Section 5.4.

```
15 T1=0:T2=0:T3=0:T4=0
20 PRINT "MANOVA FOR DIFFERENCE IN CHANGE SCORES OF TWO GROUPS"
30 PRINT :PRINT "ENTER N FOR GROUP 1"
35 INPUT N1
40 PRINT :PRINT "ENTER N FOR GROUP 2"
42 INPUT N2
45 DIM P1(N1),P2(N1),PO1(N1),PO2(N1)
46 DIM D1(N1),D2(N1),CD(4),CDI(4)
47 DIM P3(N2),P4(N2),PO3(N2),PO4(N2)
48 DIM D3(N2),D4(N2)
50 PRINT :PRINT "ENTER THE DATA FOR GROUP 1"
60 FOR I=1 TO N1
100 PRINT :PRINT "SUBJECT ",I
130 INPUT P1:P1(I)=P1
140 INPUT PO1:PO1(I)=PO1
150 INPUT P2:P2(I)=P2
160 INPUT PO2:PO2(I)=PO2
170 NEXT I
177 PRINT :PRINT "ENTER DATA FOR GROUP 2"
179 FOR I=1 TO N2
181 PRINT :PRINT "SUBJECT ",I:PRINT
183 INPUT P3:P3(I)=P3
184 INPUT PO3:PO3(I)=PO3
185 INPUT P4:P4(I)=P4
186 INPUT PO4:PO4(I)=PO4
188 NEXT I
190 FOR I=1 TO N1:D1(I)=P1(I)-PO1(I):D2(I)=P2(I)-PO2(I):NEXT I
191 FOR I=1 TO N2:D3(I)=P3(I)-PO3(I):D4(I)=P4(I)-PO4(I):NEXT I
193 T3S=0:T4S=0:CP1=0:CP2=0
195 T1S=0:T2S=0
200 FOR I=1 TO N1:T1=T1+D1(I):T2=T2+D2(I):T1S=T1S+D1(I)~2:T2S=T2S+D2(I)~2:CP1=CP
1+D1(I)*D2(I):NEXT I
202 FOR I=1 TO N2:T3=T3+D3(I):T4=T4+D4(I):T3S=T3S+D3(I)~2:T4S=T4S+D4(I)~2:CP2=CP
2+D3(I)*D4(I):NEXT I
210 M1=T1/N1:M2=T2/N1:M3=T3/N2:M4=T4/N2
220 SSX1=T1S-(T1~2/N1):SSX2=T2S-(T2~2/N1):SSCP=CP1-(T1*T2/N1)
225 SSX3=T3S-(T3~2/N2):SSX4=T4S-(T4~2/N2):SSCP2=CP2-(T3*T4/N2)
230 CD(1)=1/(N1+N2-2)*(SSX1+SSX3):CD(2)=1/(N1+N2-2)*(SSCP+SSCP2):CD(3)=CD(2):CD(
4)=1/(N1+N2-2)*(SSX2+SSX4)
240 CC=1/(CD(1)*CD(4)-CD(2)*CD(3))
250 CDI(1)=CC*CD(4):CDI(2)=CC*(-CD(2)):CDI(3)=CC*(-CD(3)):CDI(4)=CC*CD(1)
```

```
255 DD1=M1-M3:DD2=M2-M4
260 TSQ=(N1*N2/(N1+N2))*(CDI(1)*DD1^2+2*CDI(2)*DD1*DD2+CDI(4)*DD2^2)
270 F=TSQ*((N1+N2-2-1)/(2*(N1+N2-2)))
280 DFN=2:DFD=N1+N2-3
285 DIM L$(10)
290 PRINT :PRINT "T SQUARED = ";TSQ
300 PRINT :PRINT "F = ";F
310 PRINT :PRINT "DFN = ";DFN;"  DFD = ";DFD
320 END
```

P-26   MANOVA—more than two groups.   Section 5.5.

```
10 N7=0
20 PRINT "MANOVA - MORE THAN TWO GROUPS"
30 PRINT :PRINT "HOW MANY GROUPS"
40 INPUT G
45 DIM N(G),L$(10)
50 PRINT :PRINT "ENTER THE GROUP N VALUES"
52 FOR I=1 TO G
60 PRINT "GROUP     ";I
70 INPUT N
90 N(I)=N
105 N7=N7+N
110 NEXT I
112 N6=0
113 DIM T1(G,N7),T2(G,N7)
115 FOR I=1 TO G
120 PRINT "ENTER THE DATA FOR GROUP ";I
125 PRINT
130 FOR J=1 TO N(I)
132 PRINT "SUBJECT ";J
135 N6=N6+1
160 INPUT T1,T2
170 T2(I,J)=T2:T1(I,J)=T1
180 NEXT J:NEXT I
190 DIM S1(G),S2(G)
194 DIM S3(G),S4(G),C1(G)
200 FOR I=1 TO G:S1(I)=0:S2(I)=0:NEXT I
204 FOR I=1 TO G:S3(I)=0:S4(I)=0:NEXT I
210 FOR I=1 TO G:FOR J=1 TO N(I):S1(I)=S1(I)+T1(I,J):S2(I)=S2(I)+T2(I,J):NEXT J:
NEXT I
220 FOR I=1 TO G:FOR J=1 TO N(I):S3(I)=S3(I)+T1(I,J)^2:S4(I)=S4(I)+T2(I,J)^2:NEX
T J:NEXT I
240 FOR I=1 TO G:C1(I)=0:NEXT I
250 FOR I=1 TO G:FOR J=1 TO N(I):C1(I)=C1(I)+T1(I,J)*T2(I,J):NEXT J:NEXT I
255 DIM W(4),Q1(G),Q2(G),Q3(G)
260 FOR I=1 TO 4:W(I)=0:NEXT I
263 FOR I=1 TO G:Q1(I)=S3(I)-(S1(I)^2/N(I)):NEXT I
266 FOR I=1 TO G:W(1)=W(1)+Q1(I):NEXT I
270 FOR I=1 TO G:Q2(I)=S4(I)-(S2(I)^2/N(I)):NEXT I
275 FOR I=1 TO G:W(4)=W(4)+Q2(I):NEXT I
280 FOR I=1 TO G:Q3(I)=C1(I)-(S1(I)*S2(I)/N(I)):NEXT I
285 FOR I=1 TO G:W(2)=W(2)+Q3(I):NEXT I
300 W(3)=W(2)
310 D1=0:D2=0:D3=0:D4=0:D5=0
320 FOR I=1 TO G:D1=D1+S1(I):D2=D2+S2(I):D3=D3+S3(I):D4=D4+S4(I):D5=D5+C1(I):NEX
T I
325 DIM T5(4)
330 T5(1)=D3-(D1^2/N7)
340 T5(4)=D4-(D2^2/N7)
350 T5(2)=D5-(D1*D2/N7)
360 T5(3)=T5(2)
370 T9=T5(1)*T5(4)-T5(2)*T5(3)
380 W9=W(1)*W(4)-W(2)*W(3)
390 L=W9/T9
400 F=((1-SQR(L))/SQR(L))*((N7-G-1)/(G-1))
```

```
410 D8=2*(G-1):D9=2*(N7-G-1)
430 PRINT "SUMMARY TABLE INFORMATION"
440 PRINT
450 PRINT "THE W MATRIX"
460 PRINT
470 FOR I=1 TO 2:PRINT W(I),:NEXT I:? :FOR I=3 TO 4:? W(I),:NEXT I
480 PRINT :PRINT "LAMBDA = ";W9;"/";T9;" = ";L
490 PRINT :PRINT "THE T MATRIX"
500 PRINT :FOR I=1 TO 2:PRINT T5(I),:NEXT I:PRINT :FOR I=3 TO 4:PRINT T5(I),:NEX
T I
510 PRINT :PRINT "  F = ";F
520 PRINT :PRINT " DFN = ";D8;"   DFD = ";D9
530 END
```

TEXTBOOK REFERENCE CHART FOR ANOVA AND MANOVA

The following chart is provided as a guide for using *Computational Handbook of Statistics* in conjunction with ten of the most widely used statistics and design texts. The names and section numbers in this book are presented across the top row, with the corresponding pages and names used by the various other textbooks listed directly below them. Texts referred to in the chart include:

Bruning, J. L., and Kintz, B. L. *Computational handbook of statistics*, 3rd ed. Scott, Foresman, 1987.

Edwards, A. L. *Experimental design in psychological research*, 4th ed. Holt, Rinehart & Winston, 1972.

Hays, W. L. *Statistics for the social sciences*, 3rd ed. Holt, Rinehart & Winston, 1981.

Keppel, G. *Design and analysis: A researcher's handbook*, 2nd ed. Prentice-Hall, 1982.

Kirk, R. *Experimental design: Procedures for the behavioral sciences.* Wadsworth, 1968.

Lindeman, R. H., Merenda, P. F., and Gold, R. Z. *Introduction to bivariate and multivariate analysis.* Scott, Foresman, 1980.

Lindquist, E. F. *Design and analysis of experiments in psychology and education.* Houghton Mifflin, 1953.

McNemar, Q. *Psychological statistics*, 4th ed. John Wiley & Sons, Inc., 1969.

Myers, J. L. *Fundamentals of experimental design*, 3rd ed. Allyn & Bacon, 1979.

Tatsuoka, M. M. *Multivariate Analysis.* John Wiley & Sons, Inc., 1971.

Winer, B. J. *Statistical principles in experimental design*, 2nd ed. McGraw-Hill, 1971.

ANOVA Cross References:

Bruning and Kintz	Completely Randomized Design	Factorial Design: Two Factors
Section	2.1	2.2
Edwards	Randomized Group Design	The Factorial Experiment
pages	112–125	154–179
Hays	Model I: Fixed Effects	The Two-Way Analysis of Variance
pages	333–344	350–358
Keppel	Single-Factor Analysis of Variance	Factorial Experiments with Two Factors
pages	44–64	170–183
Kirk	Completely Randomized Design, or CR–k	Completely Randomized Factorial Design, or CRF–pq
pages	99–130	171–217
Lindquist	Simple Randomized Design	Factorial Design, or a Two-Factor A × B Design
pages	47–66	207–219
McNemar	Analysis of Variance: Simple	Analysis of Variance: Complex
pages	288–306	325–338
Myers	Completely Randomized One-Factor Design	Completely Randomized Two-Factor Design
pages	59–92	94–119
Winer	Single-Factor Experiments	Factorial Experiments: Two Factors
pages	149–257	431–449

Factorial Design: Three Factors 2.3	Treatment-by-Levels 2.4
Factorial with More Than Two Levels 184–198	Randomized Block Designs 231–254
More Than Two Experimental Factors 368–369	Randomized Blocks 401–404
The Three-Factor Case 276–298	Nested Factors 265–270
Completely Randomized Factorial Design, or CRF–pqr 217–229	Completely Randomized Factorial Design, or CRF–pq 171–217
Factorial Design, or a Three-Factor A × B × C Design 243	Treatment-by-Levels Design. Also, Groups-Within-Treatments Design 121–155, 172–189
Three-Way Classification 359–363	Not listed
Completely Randomized Three-Factor Design 119–135	Randomized Blocks 146–156
Factorial Experiments: Three Factors 452–464	Treatment by Blocks 431–449

Bruning and Kintz	Treatments-by-Subjects, or Repeated-Measures, Design	Treatments-by-Treatments-by-Subjects Design
Section	2.5	2.6
Edwards	Randomized-Block Design: Repeated Measures	Factorial Experiment with Repeated Measures
pages	259–267, 332–333	272–273
Hays	Repeated Measures Designs	Not Listed
pages	401–404	
Keppel	Single-Factor Design with Repeated Measures	(A × B × S) Design
pages	382–406	463–467
Kirk	Randomized-Block Factorial Design, or RB-k	Randomized-Block Factorial Design, or RBF-pq
pages	131.-150	99–130
Lindquist	Treatments-by-Subjects Design	Treatments-by-Subjects Design
pages	121–155, 172–189	237–238
McNemar	Rows for Persons and Columns for Experimental Conditions	Special Case Where Rows Stand for Persons or Matched Individuals
pages	338–342	364–375
Myers	Repeated Measures Designs	Two-Factor Repeated Measures Design
pages	163–186	187–195
Winer	Single-Factor Design with Repeated Measures	Not Listed
pages	261–305	

Two-Factor Mixed Design: Repeated Measures on One Factor 2.7	Three-Factor Mixed Design: Repeated Measures on One Factor 2.8
Trend Analysis: Trial Means 330–340	Trial Means: A Treatment Factor and an Organismic Factor 347–353
Not Listed	Not Listed
A × (B × S) Design 409–418	Higher Order Factorials (A × B × (C × S) 457–460
Split-Plot Factorial Design, or SPF-p.q. 171–217	Split-Plot Factorial Design, or SPF-pr.q. 217–229
Type I Design 266–273	Type III Design 281–284
Trends and Differences in Trends 389–395	Not Listed
One Between and One Within Subjects Variable 201–210	Two Between and One Within Subjects Variable 210–217
Two-Factor Experiment with Repeated Measures on One Factor 518–539	Three-Factor Experiment with Repeated Measures on One Factor 559–571

Bruning and Kintz	Three-Factor Mixed Design: Repeated Measures on Two Factors	Latin Square
Section	2.9	2.10, 2.11
Edwards	Not Listed	Latin Square
pages		285–325
Hays	Not Listed	Not Listed
pages		Comparisons Among Means
Keppel	Higher Order Factorials: A × (B × C × S) Design	Latin Square
pages	457–460	420–448
Kirk	Split-Plot Factorial Design, or SPF-p. qr.	Simple and Graeco-Latin Square Design, of LS-k and GLS-k
pages	298–307	151–170
Lindquist	Type VI Design	Simple and Graeco-Latin Square: Type II and Type IV Designs
pages		258–265, 273–281, 285–288
McNemar	Not Listed	Latin Squre
pages		383–388
Myers	One Between and Two Within Subjects Variable	Latin Square
pages	217–226	259–287
Winer	Three-Factor Experiment with Repeated Measures on Two Factors	Latin Square
pages	539–559	685–711

MANOVA Cross-References

Bruning and Kintz	Manova Sample-Pop	Manova Matched Pairs	Manova Two Groups
Section	5.1	5.2	5.3
Lindeman, *et al.* pages			
Myers pages	463–469		470–475
Tatsuoka pages	76ff 194ff	79ff	81ff

Multiple Comparisons (Post Hoc Analyses)	Covariance
3.4–3.10	4.9, 4.10
Multiple Comparisons Chapter 8	369–391
413–439	Covariance 523–530
Multiple Comparisons	Covariance
104–125	482–514
Multiple Comparisons	
69–97	455–488
Testing Individual Pairs of Means	
90–96, 146, 164–166	317–339
Selected Contrasts 323–324	413–429
Multiple Comparisons	
356–367	407–431
A Posteriori Test	
185–204	752–787

Manova Change Scores 5.4	Manova More than Two Groups 5.5	Manova Two Factors 5.6
	220–229	229–244
		475–478
81ff	84ff	194ff

Index

A-test, Sandlers, 16–17
Analysis of variance
  designs for, 18–109
    completely randomized, 18, 24–27
      computer program for, 339–340
    dichotomous data, 107–108
    discussion of, 18–23
    factorial
      three factors, 19–20, 32–39
        computer program for, 344–346
      two factors, 19, 27–32
        computer program for, 340–342
    Latin square
      complex, 23, 93–107
      simple, 23, 85–93
    mixed
      three factors (repeated measures on
        one factor), 22, 62–73
        computer program for, 348–351
      three factors (repeated measures on
        two factors), 22–23, 73–85
        computer program for, 351–353
      two factors (repeated measures on
        one factor), 21, 55–62
        computer program for, 346–348
    rank-order data, 108–109
    repeated-measures, 20, 44–48
      computer program for, 346–348
    repeated-measures (two variables), 21,
      48–55
    treatments-by-levels, 20, 39–44
    treatments-by-subjects, 20, 44–48
      computer program for, 346–348
    treatments-by-treatments-by-
      subjects, 21, 48–55
      computer program for, 342–343
  supplemental computations for, 110–
    173
    differences among several
      independent variances, 110,
      115–116
    differences between variances of two
      independent samples, 110, 112–
      113
    differences between variances of two
      related samples, 110, 113–115
  discussion of, 110–111
  Duncan's multiple-range test, 111,
    119–121, 133
  Dunnett's test, 111, 130–132, 133
  F-maximum test for homogeneity of
    variances, 110, 115–116
  F-tests for simple effects, 111, 132–
    145
  homogeneity of independent
    variances, 110, 112–113
  homogeneity of related variances,
    110, 113–115
  Newman-Keuls' multiple-range test,
    111, 122–124, 133
  orthogonal components in tests for
    trend, 111–112, 145–173
  Scheffé's test, 111, 127–129, 133
  summary table for tests of simple
    effects, 133
  t-test for differences among several
    means, 111, 116–119, 133
  Tukey test, 111, 124–127, 133
    table of significant studentized
      ranges for, 320

Blocking and tabling of data, 2–3

Chi-square
  complex, 270, 283–288
  simple, 270, 280–283
    table of centile values for, 298
    computer program for, 332–333
Computer programs
  for chi-square, goodness of fit, 332
  for chi-square, test of independence,
    332–333
  for completely randomized design,
    339–340
  for correlation coefficient, 337
  for covariance analysis, 354–356
  for factorial analysis of covariance (two
    variables), 354–356
  for factorial design (two factors), 340–
    341

368

for factorial design (three factors), 344–346
for frequency distribution, 333
for grouped frequency data, 333–334
for intercorrelations—many variables, 337–338
for Mann-Whitney U Test, 335–336
for MANOVA, matched pairs, 358
for MANOVA, more than two groups, 360–361
for MANOVA, sample and population, 357–358
for MANOVA, two groups, 358–359
for MANOVA, two groups, change scores, 359–360
for mean and standard deviation, 332
for multiple regression—two predictors, 356–357
for Pearson product-moment correlation, 337
for percentile ranks and z scores, 338–340
for rank-ordered distribution, 334–335
for reliability of measurement for two or more variables, 337
for simple ANOVA, 339–340
for simple analysis of covariance (one variable), 354–356
for sums and sums of squares, 331
for t-test for a difference between independent means, 336
for t-test for related measures, 336–337
for three-factor design, 346–348
for three-factor mixed or Type 3 design (repeated measures on one factor), 348–351
for three-factor mixed or Type 6 design (repeated measures on two factors), 351–353
for tables of means, 332
for two-factor mixed, Type 1 or treatments-by-subjects (repeated-measures) design, 346–348
for treatments-by-treatments-by-subjects design, 342–343
Contingency coefficient (C), 270, 283–288
Correlation, 174–229. *See also* the end of each section presenting a particular correlation
correlation ratio (eta), 174–175, 191–192
covariance. *See* Covariance, analysis of difference
dependent correlations, 228–229
independent correlations, 226–228
discussion of, 175–176
Fisher's z transformation of r, 227
table of, 306

Kendall rank-order correlation coefficient (tau), 174–175, 183–187, 194–195
multiple correlation (three variables), 175, 195–197
partial correlation (three variables), 175, 192–193
using Kendall's tau, 194–195
Pearson product-moment correlation coefficient (r), 174, 176–180
computer program for, 337
table of critical values of, 308
point-biserial correlation, 174–175, 187–190
rank-order correlation
Kendall coefficient (tau), 174–175, 183–187
partial, using Kendall's tau, 175, 194–195
Spearman coefficient (rho), 174–175, 178, 180–182
reliability-of-measurement tests
test as individual items (Kuder-Richardson and Hoyt), 176, 223–226
test as a whole (test-retest, parallel forms, split halves), 176, 222–223
computer program for, 337
significance tests
between dependent correlations, 176, 228–229
between independent correlations, 176, 226–228
Covariance, analysis of
factorial, 175, 210–222
computer program for, 354–356
simple, 175, 204–210
computer program for, 354

Data
blocking and tabling of, 2–3
organization of, discussion, 1–2
Duncan's multiple-range test, 111, 119–121, 133
table of significant ranges for, 315
Dunnett's test, 111, 130–132, 133
table for comparison treatment means with a control, 324

F-distribution table, 300
F-maximum test for homogeneity of variances, 110, 115–116
table of critical values for, 315
F-tests for simple effects, 132–145
Factorial analysis of covariance, 175, 210–222
computer program for, 354–356

Factorial designs for analysis of variance
three factors, 19–20, 32–39
two factors, 19, 27–32
computer program for, 340–342
Fisher's z transformation of r, 227
table of, 306

Hoyt's test for reliability of
measurement, 176, 223–226

Kendall rank-order correlation coefficient
(*tau*), 174–175, 183–187
in partial correlation, 175, 194–195
Kuder-Richardson test for reliability of
measurement, 176, 223–226

Latin-square design for analysis of
variance
complex, 23, 93–107
simple, 23, 85–93
Levels design for analysis of variance, 20,
39–44

Mann-Whitney U-test, 270, 275–278
table of critical U values, 310
Mean, standard error of the, 6–7
Multiple correlation (three variables),
175, 195–197
Multivariate analyses of variance
Sample and Population, 230, 231, 235
Matched Pairs, 230, 236–239
Two Groups, 231, 239–244
Two Groups–Change Scores, 231,
245–250
More than Two Groups, 231, 250–256
Two Factors, 231, 256–268

Newman-Keuls' multiple-range test, 111,
122–124, 133
table of significant studentized ranges
for, 320
Nonparametric tests
chi-square
complex, and the contingency
coefficient (*C*), 270, 283–288
simple, and the *phi* coefficient, 270,
280–283
for differences between independent
samples (Mann-Whitney U-test),
270, 275–278
for differences between related samples
(Wilcoxon sign test), 270, 278–280
for randomness, runs, trends, 270,
288–292
for significance of difference between
two proportions, 269, 272–275

for significance of a proportion, 269,
271–272
Normal-curve areas, table of, 294

Organizing data
blocking and tabling, 2–3
range and standard deviation, 4–6
standard error of the mean, 6–7
Orthogonal tests for trend, 111–112,
145–173
table of orthogonal coefficients,
327

Parallel-forms reliability measure, 176,
222–223
Partial correlation
using Kendall's *tau*, 175, 194–195
with three variables, 175, 192–193
Pearson product-moment correlation
coefficient (*r*), 174, 176–180
computer program for, 337
Fisher's z transformation of r,
227
table of, 306
table of critical values of, 308
Phi coefficient, 270, 282–283
Point-biserial correlation, 174–175, 187–
190
Proportions
significance of differences between
two, 269, 272–275
significance of one, 269, 271–272

Randomness, 288–292
Range, 4–6
Rank-order correlations
Kendall's coefficient (*tau*), 174–175,
183–187
partial
using Kendall's *tau*, 175, 194–195
with three variables, 175, 192–193
Spearman's coefficient (*rho*), 174–175,
180–182
Regression Analysis
complex, 175, 200–204
simple, 175, 197–200
Reliability-of-measurement tests
test as individual items (Kuder-
Richardson and Hoyt), 176, 223–
226
test as a whole (test-related parallel
forms, split halves), 176, 228–
229
Runs, 270, 288–292

Sample size
power of hypothesis tests, 6–7

Sandler's *A* test, 16–17
  table of critical values for, 296
Scheffé's test, 111, 127–129
Sign test (Wilcoxon), 270, 278–280
  table of critical values of, 326
Significance tests. *See* each section
  covering a particular topic
Single-case analyses, 270, 288–292
Spearman rank-order correlation (*rho*),
  174–175, 180–182
Split-halves reliability measure, 176,
  222–223
Standard deviation, 4–6
Standard error of the mean, 6–7
Student's *t*. *See* *t*-tests

t table (critical values), 294
t-tests
  for differences between sample mean
    and population mean, 7–9
  for differences among several means,
    108, 113–116
  for differences between two
    independent means, 9–12
    computer program for, 338
  for differences between two related
    measures, 12–16
    computer program for, 338–339
Tabling and blocking of data, 2–4
Test for difference between variances of
  two independent samples, 110, 112–
  113
Test for difference between variances of
  two related samples, 110, 113–115
Test for differences among several
  independent variances, 110, 115–116
Test for significance of difference
  between two proportions, 272–274
Test for significance of a proportion,
  271–272

Test-retest reliability measure, 176, 222–
  223
Tests for equality of variances, 110, 115–
  116
Tests for simple effects, 110, 116–145
Tests for trend, 111–112, 145–173, 270,
  288–292
Textbook Reference Chart, 367–373
Treatments-by-levels design for analysis
  of variance, 20, 39–44
Treatments-by-subjects design for
  analysis of variance, 20, 44–48
  computer program for, 278–281
Treatments-by-treatments-by-subjects
  design for analysis of variance, 20–
  21, 48–55
  computer program for, 342–343
Trend analyses
  orthogonal components, 111–112,
    145–173
  single case analysis, 270, 288–292
Tukey test, 111, 124–127, 133
  table of significant studentized ranges
    for, 320

U-test of Mann-Whitney, 270, 275–278
  table of critical values of, 310

Variance, analysis of. *See* Analysis of
  variance

Wilcoxon sign test, 270, 278–280
  table of critical values of, 328

z (standard deviates), table of, 292
z transformation (Fisher's) of *r*, 227
  table of, 306

The programs from the *Computational Handbook* are also available on diskette for IBM PC, all IBM Compatible Systems, Apple II, and Atari 800 (XL).

You can order this diskette from the author using the order form below. Be sure to specify the computer system you use.

Order Form

Computer System _____

Shipping Information—Please Print

Name _____

Street _____

City _____ State _____ Zip _____

Date _____

_____ diskette(s) to accompany *Computational Handbook*
at: $30.00 each for 1 to 2
      $21.00 each for 3 to 10
      $12.00 each for 11 or more

Check method of payment:

_____ check   _____ money order   _____ cash
(Credit cards and C.O.D. payments are not accepted)

Amount enclosed $_____

Mail to: Compudisk
           c/o B. L. Kintz
           100 Bayside Road
           Bellingham, WA 98225

Price includes postage and handling and applicable sales tax. Price and availability subject to change without notice.
